Jin Chen, Quan Wang (Eds.)
Wind Turbine Airfoils and Blades
GREEN Alternative Energy Resources

GREEN Alternative Energy Resources

Volume 3

Wind Turbine Airfoils and Blades

Optimization Design Theory

Edited by
Jin Chen, Quan Wang

DE GRUYTER

Science Press
Beijing

Editors
Jin Chen
Chongqing University
School of Mechanical Engineering
No. 174 Zheng Street Shapingba District
400044 Chongqing, China
chenjin413@cqu.edu.cn

Quan Wang
Hubei University of Technology
School of Mechanical Engineering
28 Nanli Road
430068 Wuhan, Hubei
People's Republic of China
quan_wang2003@163.com

ISBN 978-3-11-034421-9
e-ISBN (PDF) 978-3-11-034438-7
e-ISBN (EPUB) 978-3-11-038374-4
Set-ISBN 978-3-11-034439-4

Library of Congress Cataloging-in-Publication Data
A CIP catalog record for this book has been applied for at the Library of Congress.

Bibliographic information published by the Deutsche Nationalbibliothek
The Deutsche Nationalbibliothek lists this publication in the Deutsche Nationalbibliografie; detailed bibliographic data are available on the Internet at http://dnb.dnb.de.

Cover image: vencavolrab/iStock
Typesetting: PTP-Berlin, Protago-TEX-Production GmbH, Berlin
Printing and binding: CPI books GmbH, Leck
♾ Printed on acid-free paper
Printed in Germany

www.degruyter.com

Preface

Wind energy, a renewable energy with great potential, has been receiving more and more attention worldwide. One of the most important parts of a wind turbine is the blade, whose value amounts to 20 % of the whole turbine. Good design, reliable quality and excellent performance are key factors for the stable operation of wind turbines. However, the core technology for wind turbine blades (such as the development of wind turbine airfoils and designing an aerodynamic blade shape, etc.) mainly originates from outside China. Therefore, research on optimization theories for wind turbine airfoils and blades plays an important role in improving wind turbine technologies in China.

Books specifically on the design of wind turbine airfoils and blades are rare. Most of the previous books on wind turbines focus on wind turbine aerodynamics, including computational fluid dynamics, blade element momentum theory and wind tunnel experiments, etc. This book, however, comprehensively introduces modern design methods for wind turbine airfoils and blades. In addition, relevant topics related to aerodynamic airfoil performance, aerodynamic shape characteristics of blades and aeroelastic structural characteristics, etc. will also be introduced.

This book is divided into eleven chapters. Chapter 1 introduces relevant research on wind turbine airfoils and blades; Chapter 2 introduces the foundations of aerodynamic theory for wind turbine airfoils and the aerodynamic performance prediction method; Chapter 3 introduces integrated expression theory for wind turbine airfoils in detail; Chapter 4 illustrates the theory of parametric optimization for wind turbine airfoils; Chapter 5 introduces the design and manufacture of an airfoil model, experiments on the wind turbine airfoil and data analysis; Chapter 6 illustrates the aerodynamics of wind turbine rotors and tip-loss corrections; Chapter 7 introduces integrated representations for wind turbine blade shapes; Chapter 8 introduces shape optimization of wind turbine blades; Chapter 9 illustrates the structural optimization of composite wind turbine blades; Chapter 10 introduces the analysis of aeroelastic coupling of the wind turbine blades; Chapter 11 introduces the aeroelastic stability analysis of two-dimensional airfoil sections for wind turbine blades.

Based on the aerodynamic theory of wind turbine airfoils and blades, this book creatively introduces a conformal transformation and proposes parametric optimization design methods for different airfoil series with the maximum thickness of 12–40 %. The aerodynamic performance of the newly-designed airfoils was validated by wind tunnel experiments. The mathematical model with the optimization objectives of high COE of the wind turbine was established, based on modified BEM theory. Furthermore, the evaluation of optimization result was illustrated. The newly-designed wind turbine airfoil series was adopted for the whole blade span. The optimization model for the wind turbine blade based on performance and light weight was established. Then the performance of the initial and the optimized blade was compared. The initial layup

DOI 10.1515/9783110344387-001

design for the 2 MW wind turbine blade was based on elastics mechanics theory, while the inner structure layer design under the aerodynamic load was constructed with the use of a particle swarm algorithm and finite element method. Finally, the coupling mechanism between the vibration of the wind turbine blade and its aerodynamics were discussed, which includes the influence of blade vibration speed on blade load, the redistribution of loads on the 2D airfoil section, the flutter analysis of the 2D airfoil section and dynamic stall aeroelastic analysis of the wind turbine blade.

This book has been accomplished collectively. Chapters 1 and 2 were written by Jin Chen and Quan Wang; Chapter 3 was written by Xudong Wang and Shiqiang Zhang; Chapter 4 was written by Xudong Wang, Shiqiang Zhang, Jiangtao Cheng and Quan Wang; Chapter 5 was written by Jin Chen and Shiqiang Zhang; Chapter 6 and 7 were written by Xudong Wang and Shiqiang Zhang; Chapter 8 was written by Xudong Wang and Quan Wan; Chapters 9 to 11 were written by Quan Wan and Xudong Wang. All chapters were compiled by Jin Chen. The book was initially written in Chinese. Chapters 2 and 3 were translated by Quan Wang and Haoyu Wang. The translation of the Preface, Chapter 1 and Chapters 4 to 11 was undertaken by Zhenye Sun who also corrected several mistakes.

The authors acknowledge the support of the National Natural Science Foundation (51175526, 50775227), 863 Plan (2012AA051301), Key Project of Chongqing Natural Science Foundation, etc. Without the support of these projects, our research on wind turbine airfoil and blade optimization design could not have been carried out. Additionally, we are deeply grateful to Professor Wen Zhong Shen of the Technical University of Denmark, researcher Peter Eecen from Netherlands' Nation energy Research Center and Professor Ning Qin of the University of Sheffield. Thanks to the research platforms and data resources provided, our research could be successfully completed (Dr. Xudong Wang, Shiqiang Zhang, Jiangtao Cheng, and others spent one-year research visits working on wind turbines at the above institutes). What is more, we would like to thank Zhihong Zou, Mingfeng Liang and Le Tong for their help on process diagrams.

In spite of our best efforts, there might still be mistakes and flaws in this book, so please do not hesitate to contact and correct us.

Jin Chen
30th September, 2016
Chongqing University

Contents

1 Introduction

1.1 Introduction

As a kind of clean and renewable energy, wind energy is inexhaustible and is attracting increasing attention from all over the world. Of the global wind energy reserves, 2×10^7 MW are available from the whole 2.74×10^9 MW, which is 10 times more than can be extracted from exploitable water resources. Wind energy is abundant in China, with the energy on land amounting to 2.53×10^5 MW.

With innovations in wind turbine theories, the application of new materials and the development of manufacturing technologies, wind power technology is continuously improving, with unit capacity increasing from the 10-kilowatt scale to the multi-megawatt (MW) scale. The average unit capacity of wind turbines has increased to twenty times that of twenty years ago. Before 2008, large-scale wind turbine manufacturing produced only few products. German Repower manufactured the first 5 MW prototype in 2004. German Enercon has recently developed the second generation of 6 MW direct-drive wind turbines whose diameter is increased to 127 m from 112 m (4.5 MW). In recent years, more and more companies have started to design and develop large-scale wind turbines. Danish Vestas is developing the 4 MW-Micon offshore wind turbine. Spanish Gamesa has developed a 4.5–5 MW model. German BARD has developed a 5 MW wind turbine series and three have been installed on land and offshore. In 2009, BARD announced that it would start developing 6.5 MW wind turbines. Siemens has completed testing a 3.6 MW direct-drive conception wind turbine. Dutch Darwind is developing a 5 MW direct-drive wind turbine. American Clipper is going to cooperate with Britain to develop 7.5 MW and 10 MW units. AMSC, which is cooperating with the US Department of Energy, plans to manufacture a 10 MW unit with superconducting generators. So the large-scale wind turbine is becoming the mainstream of future market development and applications. Meanwhile, prompted by the large-scale developments in international wind industries, Chinese wind energy companies have also become involved in the fierce competition among large-scale wind turbines. In 2009, China made a breakthrough in research and manufacturing MW-scale wind turbines. For example, a 3 MW wind power unit developed by Shenyang Industrial University has been successfully manufactured. 2.5 MW and 3 MW wind turbine units developed by Goldwind have been put into trial operation in a wind farm [2, 3]. The 6 MW offshore wind turbine unit developed by Sinovel in 2011 has been successfully manufactured. And the 6 MW offshore wind turbine developed by National Power has also finished development and tests. In addition, Goldwind, DEC, Haizhuang, XEMC and so others are developing or have developed wind turbines whose capacity exceeds 5 MW, indicating that research and manufacturing of large-scale wind turbines in China has stepped into a new era.

DOI 10.1515/9783110344387-002

Although the development of the global wind industry has, to some extent, been affected by the global economic crisis in recent years, newly installed capacity is still on the increase. According to statistics from the Global Wind Energy Council (GWEC), after the slower rate of increase of wind power in 2010, in 2011 the newly installed capacity globally amounted to 40 564 MW, which represented annual growth of more than 20 % [2]. Compared with capacity installed in 2010 of 18.94 GW, the annual capacity installed in China decreased in 2011 by 6.9 % to 17.63 GW. After rapid growth, the Chinese wind power market is entering a period of steady development. By the end of 2011, the Chinese wind market, with 45 894 wind turbines installed annually and capacity of 62.36 GW, continues to be the largest worldwide. The capacity installed annually in China since 2001 is shown in Fig. 1.1.

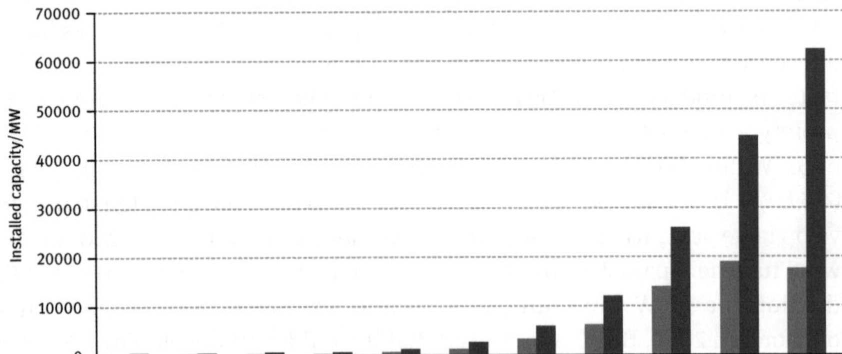

	2001	2002	2003	2004	2005	2006	2007	2008	2009	2010	2011
■ Newly installed capacity	41.7	66.3	98.3	196.9	506.9	1287.6	3311.3	6153.7	13803.2	18928	17630.9
■ Total installed capacity	381.2	447.5	545.8	742.7	1249.6	2537.2	5848.5	12002.2	25805.4	44733.4	62364.3

Fig. 1.1: Development of wind power in China.

During the years 2005–2010, many companies from abroad with great strength sought potential partners in China. For example, American GE and Danish Vestas have set up factories or cooperated with Chinese enterprises. Although Chinese enterprises have achieved initial successes. improved technologies and produced many wind power products, the core technologies of wind power are almost all controlled by large international companies. The blade is one of the most crucial parts of a wind turbine and its value amounts to 20 % of that of the whole wind turbine. Good design, reliable quality and superior performance are the decisive factors to ensure normal and stable operation of wind turbines.

1.2 The research in China and worldwide

The design theories for wind turbine airfoils and aerodynamic blade shapes are the decisive factors in wind turbine power performance and aerodynamic load characteristics. Structural design of composite blades is the crucial factor affecting stiffness characteristics of blades. Blade aeroelastic complex multidisciplinary coupling and collaborative method are used to investigate the aerodynamic performance and structural stiffness characteristics. At present, the analysis, design and optimization of the blade's shape and structure are mainly based on specific geometric contours.

1.2.1 Research on wind turbine airfoils

The development of wind turbine airfoils started with the application of low-speed aviation airfoils, such as glider airfoils, FX-77 airfoil and NASA LS airfoil, etc. In order to adapt to the requirements of wind turbine operating conditions, companies abroad started to develop wind turbine airfoils from the mid-1980s and by now have developed several series of airfoils, such as American NREL-S series, Danish Risø series, Dutch DU series and Swedish FFA-W series. Tangler and Somers [3] from the National Renewable Energy Laboratory (NREL) designed 35 kinds of NREL-S airfoils which have great lift-to-drag ratio under different conditions, based on Eppler theory and the inverse design method. Delft University, Energy research Center of the Netherlands (ECN) and National Aerospace Laboratory of the Netherlands (NLR) have designed the DU airfoil series with relative thickness of 15–40 % based on the hybrid design method. The geometric compatibility and deep stall characteristics were considered during the design of the airfoils. By limiting the thickness of the upper surface and having the S-type trailing edge [4, 5], the lift coefficient of the airfoil has been increased. Compared with traditional aviation airfoils, these airfoils have much better aerodynamic characteristics. Danish Risø National Laboratory, using the direct design method and coupling XFOIL software, has developed wind turbine airfoils [6, 7] for various operating conditions and control modes. It is hard to determine the proper target pressure and velocity distribution in inverse design. Different to the traditional inverse design method, Risø serial airfoils adopted a B-spline curve to control the airfoil profile. Based on CFD technology and optimization methods, the airfoils were optimized directly with the maximum lift-to-drag ratio as the objective, which remedies the shortcomings of inverse design methods and makes multidisciplinary design possible. In addition, Sobieczky [8] and Hájek [9] had proposed PARSEC and improved PARSEC design methods by which the airfoil shape had been researched by controlling the airfoil geometric parameters. What is more, Bjork [10], Soemarwoto [11], Habali [12], Nielsen [13], Baker [14] and Bermudez [15] have also contributed to the design of wind turbine airfoils. In China, research on airfoil design started quite late. He Dexin [16] from China Aerodynamics Research and Development Center applied the improved ideal Stratford pressure distribution on the recovery area of the suction surface and designed a new kind of airfoil series for low Reynolds number using the theory of Weber. The experiments finished at a Reynolds number of 5.6×10^5. Qiao Zhide, Song Wenping, et al. [17] from Northwestern Polytechnical University developed the NPU-WA airfoil series under a high Reynolds number, with high lift aerodynamic performance, for MW-scale wind turbines. Wind tunnel experiments were conducted and analyzed to illustrate the excellent aerodynamic performance of this airfoil series. Additionally, many scholars in China have also reported valuable research which is worthy of attention [18–22].

1.2.2 Research on aerodynamic shape and performance of wind turbine blades

For the optimization of aerodynamic blade shape, local improvements and modifications of the blade's shape are commonly made based on initial blade shape parameters in order to achieve high performance. Fuglsang [23] from Danish Risø laboratory optimized blades for a 1.5 MW wind turbine, taking cost of energy (COE) as the objective function and considering extreme loads and fatigue strength. It has been proved that the COE is decreased by 3.5 %. The Turkish scholar Varola [24] proposed installing a steering plate on the blade to increase the rotor speed and thus increase the output power. The Tunisian scholar Badreddine [25] conducted an optimization of the blade with output power as the objective function. Rajakumar et al. [26] proposed an iterative method to optimize angle of attack, twist and chord in order to improve the power coefficient of the rotor. Liu et al. [27] proved a new method to optimize the change rate of chord and twist, which linearizes the chord and twist of the blade. It not only improves the performance of the rotor, but also improves blade manufacture. Maki et al. [28] reported in 2012 on systematically designing and optimization of rotors utilizing a multi-objective method and analyzed the results, including consideration of COE and power production. In China, Liu Xiong from Shantou University [29] designed a 1.3 MW wind turbine blade considering the actual probability distribution of wind in a wind farm, with the objective function of annual average power output and the use of a genetic algorithm. Han Zhonghe [30] from North China Electric Power University has proposed a method of setting the initial value of induction factor through optimization. Then the optimization for a blade of a 1.5 MW wind turbine was performed, with the objective of maximum power output and annual power production. Wang Tongguang [31] from Nanjing University of Aeronautics and Astronautics executed a multi-objective optimization for the blade of a 5 MW wind turbine utilizing an improved genetic algorithm and gave the Pareto optimal solution set.

1.2.3 Research on structural design of composite wind turbine blades

The main research objectives for structural design and optimization of composite wind turbine blades consist of internal structural parameters (skin thickness, web thickness, width, location, etc.), aerodynamic characteristics, stability, strength, stiffness and fatigue. Fuglsang [32, 33] from Danish Risø National Laboratory adopted XFOIL software and numerical analysis in the optimized design of the Risø-A1, Risø-P and Risø-B1 airfoil series. For the Risø-A1 airfoils for application with stall wind turbines, the weight of the airfoil was reduced by 4 % through increasing the thickness-to-chord ratio. The fatigue loads of the blade were decreased and the annual power generation was increased by 15 %. For Risø-P airfoils, which can be applied to variable pitch wind turbines, the optimization should be started from the leading edge of the blade. Additionally, experiments on the performance of those airfoils were conducted and the effectiveness of the theoretical optimization was validated. Jureczko [34] from Poland researched the variation of blade natural frequency with respect to structural parameters and material properties. An optimization and analysis of the blade's dynamic performance were made. Structural analyses for the blade of a 750 kW wind turbine under various load and fatigue conditions were conducted by Kong et al. [35] and compared with experimental data. Israeli scholar Shokrieh [36] studied and simulated the fatigue failure of a wind turbine blade of 23 m, and predicted the order of blade laminate damage. British scholars Maherii et al. [37, 38] studied the distribution of shell along chord and span and optimized the blade by analysis of loads. Jensen et al. [39] carried out structure tests and numerical simulations on a composite material blade of 34 m and got the accurate location of failure. Paluch et al. [40] adopted the finite element method and genetic algorithms to optimize composite material structure and applied the method to a simplified blade model to reduce weight. Johansen et al. [41] optimized the failure performance of geometric linear/nonlinear composite material structure. They carried out finite element analysis and compared the results with experimental data, which validated the nonlinear characteristics of wind turbine blades. Barroso et al. [42] studied the failure mechanism of blades of a 300 kW wind turbine, including initial failure position, separation of laminates and the final failure condition. Ping et al. [43] analyzed the performance of blades under extreme conditions of ice-load. Lund [44] from Aalborg University performed topology optimization for shell structures made of composite materials and carried out topology optimization for a simplified wind turbine blade model. Overgaard et al. [45] performed static load tests and simulations for the single ply model and the laminate model respectively. The failure results were analyzed and assessed. Egyptian scholars Maalawia et al. [46] studied the modes and loading responses for different blades under different load conditions. Then optimization of the blade was performed from a dynamics point of view. In China, Li Deyuan [47] from Shantou University analyzed the vibration modes of the blade considering blade rotation. Chen Yuyue et al. [48] from Shanghai FRP Research Institute discussed several problems in the design of large-scale wind turbine blades, including the aerodynamic design, the blade root and section design, load conditions, strength and fatigue analysis, bulking analysis, evaluation of the dynamic characteristics of the blades. Liao et al. [49] from the Chinese Academy of Sciences had optimized the layout of the blade by a combination of FAST software and improved PSO. Then the effectiveness of this method was validated with FOCUS5 software. Additionally, many scholars had contributed to the research [50–70].

1.2.4 Research on aeroelastic performance of wind turbine blades

Aircraft wing aeroelastic theories provide references for aeroelastic research on wind turbine blades. In 1999, Greek scholar Chaviaropoulos [71] studied the section flapwise/edgewise coupled aeroelastic stability and applied the eigenvalue method to analyze the stability of a two-dimensional (2D) airfoil section of a wind turbine blade. And in 2003, Chaviaropoulos et al. [72] studied the nonlinear stability of a 2D airfoil section and applied NS equations to stability analysis of classical flutter and dynamic stall flutter. In 2004, Hansen [73] from Denmark Risø Laboratory combined the finite element method and blade element theory to develop the stability analysis software of HAWCSTAB and carried out eigenvalue analysis on a blade of a 600 kW wind turbine. The predicted aerodynamic damping matched the experimental result well. In 2006, Hansen et al. [74] summarized research on wind turbine aerodynamics and aeroelastics and presented the future direction for aeroelastic research. Hoogedoorn et al. [75] analyzed the aeroelastic mechanism at wind change condition and studied the influences of the airfoil thickness, camber and pitch angle and other parameters on the elastic stiffness. Baxevanou et al. [76] proposed a new numerical model solving the aeroelastic characteristics of a wind turbine blade under typical flutter conditions. Jacques et al. [77] proposed an efficient calculation model of blade aerodynamic performance and structure load based on the vortex method, which adopted implicit structural and aerodynamic coupling equations. Aeroelastic properties of a flexible blade can be accurately calculated with yaw angle below 30°. Alireza et al. [78] presented an aerodynamic structural coupling analysis method based on the finite element method. Calculations for a flexible prebending and re-twisted blade were performed and the calculation time, compared with the traditional finite element analysis method, was shortened, and the error was restricted below 1 %. Sarkar et al. [79] studied nonlinear aeroelastic behavior of a 2D symmetrical blade at dynamic stall condition and found the existence of superharmonic vibration and quasi-harmonic vibration of the blade under flapwise and edgewise vibration conditions. Additionally, many scholars abroad have conducted much research on aerodynamic and aeroelastic performance [80–87]. In China, Cao Renjing et al. [88] built a mathematical model for aeroelastic stability and sensitivity analysis for blades and synthetically considered the impacts of aerodynamic and structural parameters on aeroelastic stability. Jin Yan et al. [89] developed a fluid-solid coupling numerical method and studied the self-excited oscillation aeroelastic problems of airfoils. In addition, they analyzed the vibration reduction technology on the back of airfoil with injection. Ren Yongsheng et al. [90, 91] studied the dynamic stall-inducted flutter of a blade using a time-domain numerical integration method and researched nonlinear aeroelastic stability of a composite material blade. The research above illustrates the aeroelastic analysis of wind turbine blades and airfoils from different perspectives. However, a breakthrough has not been achieved since the coupling effects of airfoil profile design, aerodynamic shape design and inner structural design were not considered.

In conclusion, the designs of airfoil, blade shape and blade structure were normally independent of each other in most of the designs. First of all, the design and optimization of the airfoil profile were completed. Then the blade's aerodynamic shape was designed to have high performance. Finally, the design and optimization of blade structures were carried out to obtain a wind turbine blade of high performance. However, most of the studies have not implemented a complete integrated design system for wind turbine airfoils, and the aerodynamic shape and structure of the blade.

Therefore, studies on design and optimization methods for airfoils and blades from a more integrated perspective are needed. Studies have been conducted on the innovative integrated expression of airfoils. Using series theory in the integrated expression of airfoils, airfoils of arbitrary shape can theoretically be represented by this theory. And in the optimization process, the performance of the airfoils is not restricted by the initial airfoil. The multidisciplinary coupling effects between aerodynamic characteristics, power characteristics of the wind turbine, noise characteristics, etc. were discussed. Based on the new airfoil design method using integrated expression, the new airfoil series with multidisciplinary advantages was obtained after optimization. The integrated representation of the 3D blade surface was established and described. And aerodynamic shape optimization of wind turbine blades was performed adopting the new airfoils obtained from optimization using integrated expression. Based on the integrated representation of blade surface and ANSYS, a parametric model for composite blade structure optimization was established. Optimization design has been carried out on parameters like ply thickness, order, and direction of composite material blade. A preliminary analysis of blade aeroelastic characteristics has been made. Problems such as aeroelastic integrated optimization of wind turbine aerodynamic shape and composite structure have become the final frontierof blade design which need to be urgently solved. And further research is needed. Researching these problems is of vital importance to form a complete theory system for wind turbine airfoil, aerodynamic shape and structural design and then to design wind turbine airfoils and blades of good aerodynamic performance, structural strength, and aeroelastic stability. It is also of great importance to industriythat airfoils and blades with high aerodynamic performance, high structural strength and good aeroelastic stability be designed.

2 Aerodynamic characteristics of wind turbine airfoils

2.1 Introduction

The current urgent issue is how to improve the efficiency of wind turbines and reduce the cost of wind power. The efficiency of a wind turbine is mainly determined by the blade, and the aerodynamic performance of the airfoil directly influences the aerodynamic performance of the wind turbine blade.

Compared with traditional aviation airfoils, wind turbine airfoils have different operating conditions and performance requirements. Here are some examples: wind turbine blades usually run at a relatively high Reynolds number (usually in the order of 10^6–10^{10}), which can cause the characteristics of the airfoil boundary layer to change; the deep stall characteristics of airfoils are very important because wind turbine blades run at a large inflow angle; when the yawing movements of an operating wind turbine make the inflow angle of each cross section of the blade change cyclically, the dynamic stall characteristics of the airfoils must be considered; if the wind turbine operates in an atmosphere close to ground, then dust, gravel, rain and oil will increase the surface roughness, affecting the aerodynamic performance of the wind turbine airfoil; considering the structural strength and rigidity, the relative thickness of wind turbine airfoils is greater than that of traditional aviation airfoils and can reach about 30 % near the blade root. Therefore, it is important to understand changes in the flow field around the wind turbine blade and the aerodynamic parameters of wind turbines in the process of airfoil design [92].

2.2 Basic theory of wind turbine airfoils

The geometric parameters of wind turbine airfoils are firstly introduced in this section. The concepts of Reynolds number, Mach number and the boundary layer which are associated with wind turbine airfoils are discussed. The boundary layer flow equations are given. The parameterized solved model for the aerodynamics of airfoils is further derived using numerical methods.

2.2.1 Geometric parameters of airfoils

Fig. 2.1 shows the geometric shape and parameters of a wind turbine airfoil.
Mean line: the connection of the inscribed circle center around the airfoil.
Leading edge: the forward point of the mean line for the airfoil.
Leading edge radius: the radius of inscribed circle for the airfoil leading edge.

DOI 10.1515/9783110344387-003

Trailing edge: the last point of the mean line for the airfoil.
Trailing edge angle: the angle between the upper surface and lower surface of the trailing edge.
Trailing edge thickness: the thickness for the airfoil trailing edge.
Chord: the connection between the leading edge and the trailing edge.
Thickness: the diameter of the inscribed circle for an airfoil.
Camber: the maximum vertical distance between the mean line and the chord line.

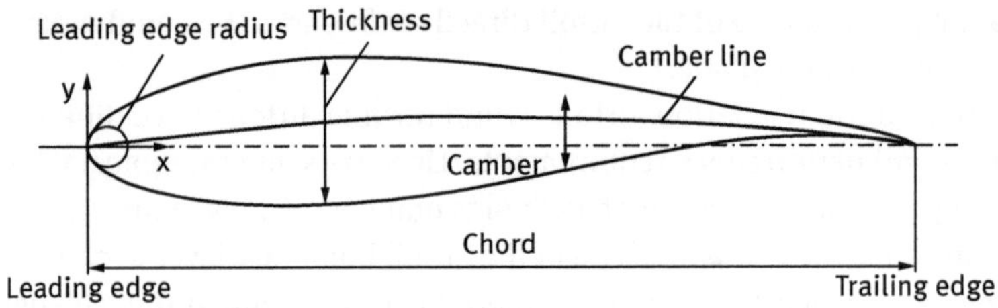

Fig. 2.1: Geometric parameters of an airfoil.

2.2.2 Reynolds number

The most important fluid factor for low-speed airfoil aerodynamic performance is the fluid viscosity. The viscosity generates lift force indirectly, generates drag force directly, and causes fluid separation. This effect is expressed by the Reynolds number. When a fluid flows, the ratio of inertial forces F_g to viscous forces (internal friction) F_m is called the Reynolds number [93], symbolized by *Re*. *Re* is a dimensionless quantity:

$$Re = \frac{F_g}{F_m} = \frac{\rho V_0 l}{\mu}. \tag{2.1}$$

In most cases, viscosity is expressed as $\nu = \mu/\rho$ and called kinematic viscosity.

$$Re = \frac{V_0 l}{\nu}, \tag{2.2}$$

where
V_0: average velocity of the flow
l: characteristic length of the flow
ν: kinematic viscosity
μ: dynamic viscosity
ρ: density of the fluid

From equation (2.2), the Reynolds number *Re* depends on three parameters, namely the fluid velocity, the characteristic length and the viscosity. When *Re* is large, the viscous effects are low. Otherwise, when *Re* is small, the viscous effects are high. So Reynolds number *Re* is a basic parameter to describe the state of fluid motion, which is divided into two major categories: laminar flow and turbulent flow. Laminar flow occurs at low Reynolds number, and the fluid flows in parallel layers, with no disruption between the layers, so the velocity field and the stress field experience a gradual and continuous change with time and space. Turbulent flow occurs at high Reynolds number, and is a flow regime characterized by chaotic property changes including low momentum diffusion, high momentum convection, and rapid variation of flow field in space and time [94].

Because the power can range from 0.7×10^6 up to 10×10^6, wind turbine airfoils work over a large range of Reynolds numbers. Over this range, the performance of airfoils will be changed by variations of jet turbulence, the vibration of the airfoil or the roughness of the airfoil's surface. The Reynolds number has an important influence on airfoil lift-to-drag force characteristics. For instance, as the Reynolds number increases, the slope of the lift curve increases, the maximum of the lift coefficient increases, the stall angle of attack (AOA) increases, and the minimum drag coefficient decreases [95]. But some airfoils have the opposite properties, i.e., as the Reynolds number increases, the slope of the lift curve decreases, the maximum lift coefficient decreases, and the minimum drag coefficient increases.

2.2.3 Mach number

The ratio of flow velocity (v) to the local speed of sound (v_c) is called the Mach number [94–96], Ma:

$$Ma = \frac{v}{v_c}. \tag{2.3}$$

Because wind turbines work in the air, their Mach number depends on the ratio of wind velocity to the speed of sound in air. Therefore, when ambient temperature and the speed of sound are constant, the greater the wind velocity, the greater the Mach number; when the wind velocity is constant, the ambient temperature and the speed of sound change, and the Mach number changes. So the Mach number is a local value and reflects a gaseous state like the speed of sound. But it reflects dynamic flow, and has more practical value than the speed of sound which reflects the size of static gas compressibility. In operating conditions, the Mach number around the airfoil is generally less than 0.3, and the air flow can be treated as an incompressible flow within this range [97].

2.2.4 Boundary layer

The reason for separated flow at higher Reynolds numbers is the existence of a thin boundary layer of slow moving fluid, close to the body surface, within which viscous forces predominate. Outside this layer the flow behaves almost inviscidly. The drag on the body caused directly by viscosity is quite small, but the effect on the flow pattern is profound.

2.2.4.1 Boundary layer separation

Referring to Fig. 2.2 (a), the inviscid flow pressure distribution around a cylinder has the following characteristics: fore and aft the cylinder, the pressure is high; above and below the cylinder, the pressure is low. The fluid on the downstream side is slowing down against an adverse pressure gradient and, at the wall boundary, it slows down exactly to a standstill at the rear stagnation point. In real flow the boundary layer, which has already been slowed down by viscosity, comes to a halt well before the stagnation point is reached and the flow begins to reverse under the action of the adverse pressure. At this point, where the pressure is still low, the boundary layer separates from the body surface forming a wake of stagnant, low-pressure fluid (Fig. 2.3), and the resulting pressure distribution is thereby dramatically altered as shown in Fig. 2.2 (b). The high pressure acting at and around the forward stagnation point is no longer balanced by the high pressure at the rear and so a drag-wise pressure force is exerted.

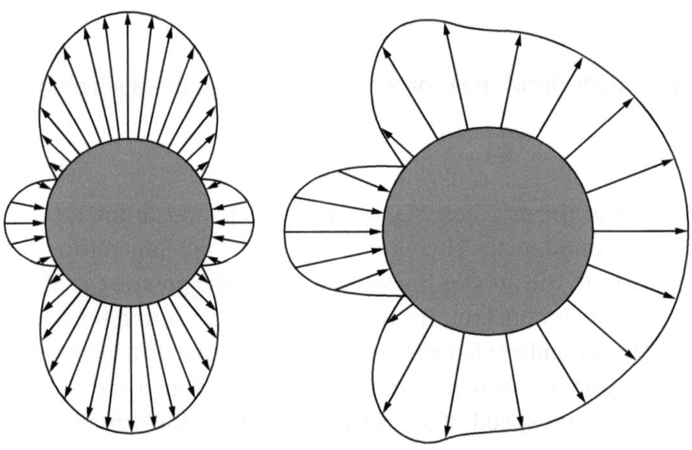

(a) Inviscid (b) Viscid

Fig. 2.2: Pressure distribution around a cylinder.

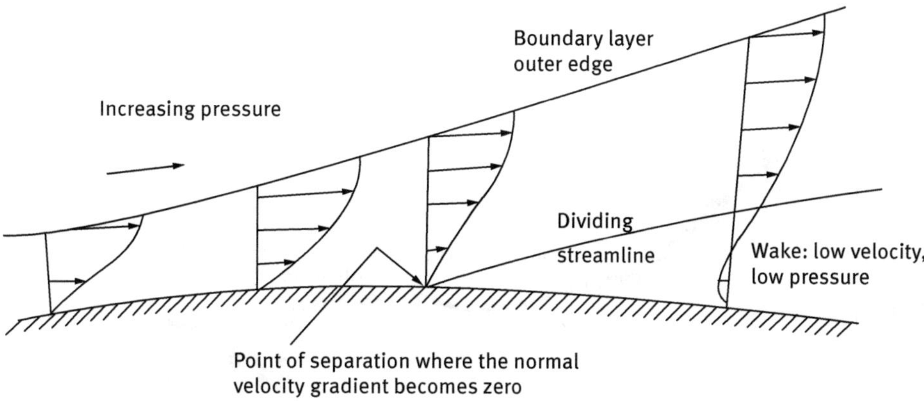

Fig. 2.3: Separation of a boundary layer.

2.2.4.2 Laminar and turbulent boundary layers

A boundary layer grows in thickness from the forward stagnation point, or leading edge. Initially, the flow in the layer is ordered and smooth but, at a critical distance l from the stagnation point, characterized by the critical Reynolds number $Re_{crit} = \rho ul/\mu$, the flow begins to become turbulent (Fig. 2.4). This turbulence causes mixing of the boundary layer with faster moving fluid outside, resulting in enhanced energy and delaying of the point of separation. The result is to reduce the pressure drag, because the low pressure stagnant rear area is reduced; to increase the viscous drag, because the velocity gradient at the surface is increased; and increase the boundary layer thickness.

The drag coefficient, therefore, varies with Re in a complex fashion (Fig. 2.5). The critical Re is never reached for small bodies at low speed and so separation takes place early. For large bodies, or high speed, turbulence develops quickly and separation is delayed.

Turbulence can be artificially triggered by roughening the body surface or simply by using a "trip wire". General flow turbulence tends to produce turbulent boundary layers at Reynolds numbers ostensibly below the critical value and this certainly seems to happen in the case of wind turbine blades.

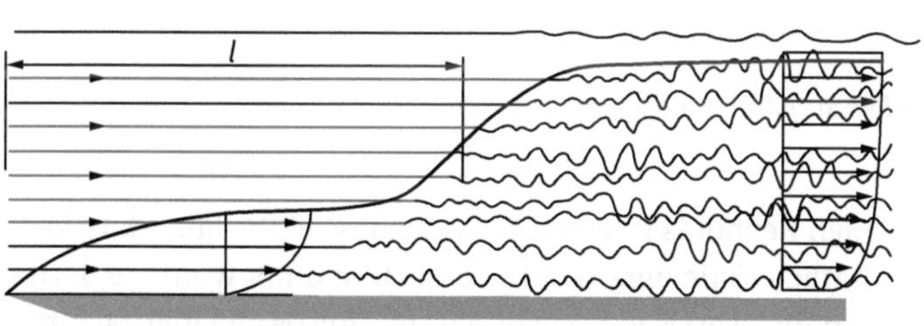

Fig. 2.4: Laminar and turbulent boundary layers.

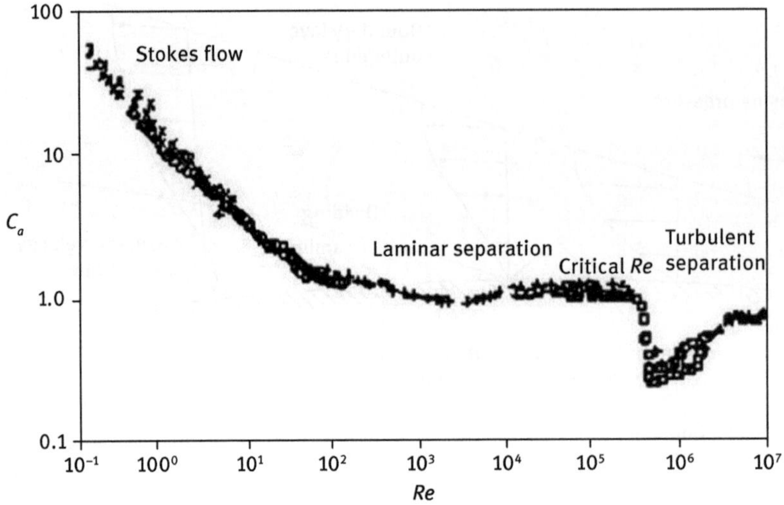

Fig. 2.5: The drag coefficient as a function of Reynolds number.

2.2.4.3 Boundary layer flow equation

In physics, the NS (Navier–Stokes) equations describe the motion of viscous fluid substances. Because the boundary layer's thickness (δ) is far smaller than the characteristic length of the airfoil, and the velocity component in the x-direction changes along with the normal direction are greater than that with the tangential direction, the NS equations can be significantly simplified to give the Prandtl boundary layer equations, which are the basic equations for the processing of boundary layer flow. The schematic diagram of an airfoil's boundary layer is shown in Fig. 2.6.

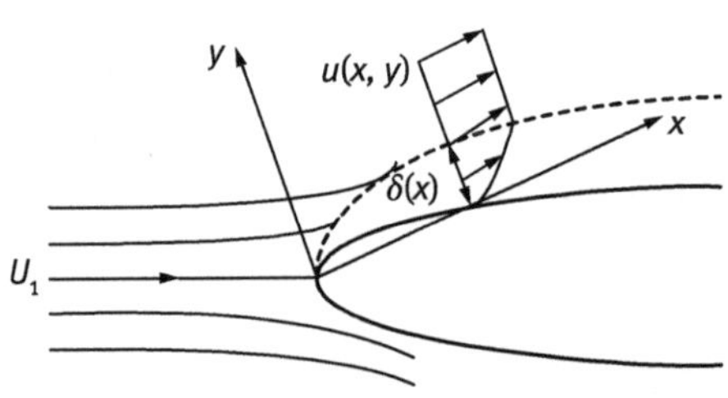

Fig. 2.6: Schematic diagram of an airfoil's boundary layer.

For boundary layer flow at high Reynolds number, the thickness of the boundary layer is much smaller than the characteristic length of the object. In other words, δ/L (the relative thickness of the boundary layer) is very small. Viscous forces and inertial forces are of the same order within the boundary layer.

For a wind turbine airfoil, the Prandtl boundary layer equations can be obtained by analyzing magnitude orders [95]:

$$\frac{\partial u}{\partial x} + \frac{\partial v}{\partial y} = 0, \tag{2.4}$$

$$\frac{\partial u}{\partial t} + u\frac{\partial u}{\partial x} + v\frac{\partial u}{\partial y} = \frac{\partial U_1}{\partial t} + U_1\frac{\partial U_1}{\partial x} + \frac{1}{\rho}\frac{\partial \tau}{\partial x}. \tag{2.5}$$

By formally integrating equation (2.5) across the boundary layer ($u = v = 0$, $\tau = \tau_w$ at $y = 0$ and $u = U_1$, $y = \tau = 0$, at $y = \infty$), the well-known von Karman momentum integral equation results [96]:

$$\frac{x}{\theta}\frac{\mathrm{d}\theta}{\mathrm{d}x} = \frac{x}{\theta}\frac{C_\mathrm{f}}{2} - (H + 2 - Ma^2)\frac{x}{U_1}\frac{\mathrm{d}U_1}{\mathrm{d}x}, \tag{2.6}$$

where u is the flow's tangential velocity within the boundary layer, v is the flow's normal velocity within the boundary layer, U_1 is inflow velocity, τ is skin friction, t is time, C_f is skin friction coefficient, $H = \delta^*/\theta$ is a shape parameter of the boundary layer, δ^* is the displacement thickness of the boundary layer, θ is the momentum thickness of the boundary layer.

If equation (2.6) is first multiplied through by local velocity u and then integrated, the momentum and kinetic energy equations can be written as

$$\frac{x}{H^*}\frac{\mathrm{d}H^*}{\mathrm{d}x} = \frac{x}{\theta}\frac{2C_\mathrm{D}}{H^*} - \frac{x}{\theta}\frac{C_\mathrm{f}}{2} - \left(\frac{2H^{**}}{H^*} + 1 - H\right)\frac{x}{U_1}\frac{\mathrm{d}U_1}{\mathrm{d}x}, \tag{2.7}$$

where $H^* = \theta^*/\theta$, $H^{**} = \delta^{**}/\theta$ are also shape parameters, θ^* is the kinetic energy thickness, δ^{**} is the density thickness, C_D is dissipation coefficient.

The momentum and shape parameter equations (2.6) and (2.7) are valid for both laminar and turbulent boundary layers, as well as for free wakes. The fundamental difficulty with them is the value of the three shape parameters (H, H^*, H^{**}), dissipation coefficient C_D, and skin friction coefficient C_f. Here, due to lack of space, we will not derive the boundary layer parameters in detail.

2.2.5 Potential flow solving method for an arbitrary airfoil

Wind turbine airfoils work at low Mach numbers ($Ma < 0.3$), which means the flow is without separation in a certain angle of attack (AOA) range, so the viscosity has little influence on the pressure distribution, lift and moment, and a reasonable solution can be obtained approximately by inviscid flow theory. But when thickness or AOA is large, the theory is no longer applicable. In this section, a viscous flow numerical solution method – the panel method – will be introduced and is applicable to both thick and thin airfoils [98, 99].

The airfoil's surface is divided into infinitesimal pieces, called panels; each panel is a surface vortex or point source with unknown intensity; then a point (often referred to as control point) will be selected from each panel, at which the panel cannot be penetrated; then linear equations with the unknown strength are obtained. These equations determine intensity, and calculate the pressure, and the lift and torque characteristics.

The airfoil contour and wake trajectory are discretized into flat panels, originating from the lower wing surface in anticlockwise direction, with N panel nodes on the airfoil. Each airfoil panel has a linear vortices distribution defined by the node values γ_j $(j = 1, \ldots, n)$. Each airfoil and wake panel also has control points $P_i(x_i, y_i)$, $i = 1, \ldots, n$, as shown in Fig. 2.7.

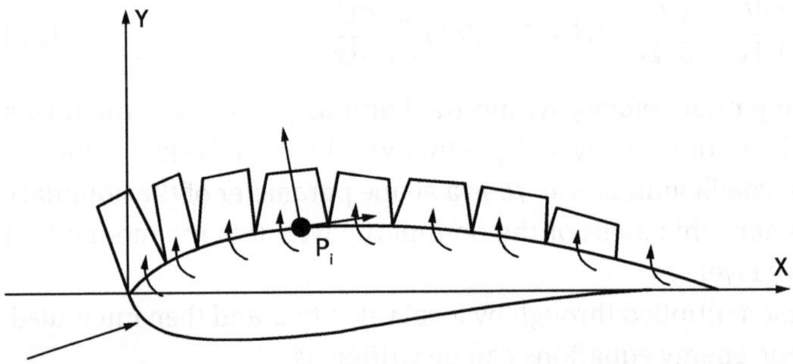

Fig. 2.7: The numerical method of obtaining airfoil surface vortices' values.

The stream function at the point i on the panel j is specified in the form

$$d\varphi_{ij} = -\frac{\gamma_j}{2\pi} \int_{s_j} \theta_{ij}\, ds_j, \tag{2.8}$$

where

$$\theta_{ij} = \arctan \frac{x_i - y_j}{x_i - x_j},$$

so the stream function is also written as

$$\varphi_i = -\sum_{i=1}^{n} \frac{\gamma_j}{2\pi} \int_{s_j} \theta_{ij}\, ds_j. \tag{2.9}$$

The flow normal velocity components have to be

$$(v_n')_i = \frac{\partial \varphi_i}{\partial n_i} = -\sum_{j=1}^{n} \frac{\gamma_j}{2\pi} \int_{s_j} \frac{\partial \theta_{ij}}{\partial n_i}\, ds_{ij}. \tag{2.10}$$

Then, the boundary conditions at the control point i

$$v_\infty \cos \beta_i - \sum_{j=1}^{n} \frac{\partial \theta_{ij}}{\partial n_i}\, ds_j = 0, \tag{2.11}$$

where β_i is the angle between the stream and the outer normal of panel i.

To meet the conditions of the trailing edge, the first control point on the lower surface and the last n-th control point on the upper surface should be close to the trailing edge, in the other words, the distance between these two surfaces vortices is very short. Trailing edge conditions can be approximately written as

$$\gamma_1 = -\gamma_n. \tag{2.12}$$

By equations (2.11) and (2.12), the value γ_j can be obtained. The tangential velocity and the pressure coefficient of each control point are then determined, respectively,

$$\upsilon_{s_i} = \upsilon_\infty \sin\beta_i + \frac{\gamma_i}{2} - \sum_{\substack{j=1 \\ j\neq i}} \frac{\gamma_i}{2\pi} \int_{s_j} \frac{\partial \theta_{ij}}{\partial n_i} \, \mathrm{d}s_j, \tag{2.13}$$

$$C_{p_i} = 1 - \left(\frac{\upsilon_{s_i}}{\upsilon_\infty}\right)^2. \tag{2.14}$$

This numerical method includes the combined effects of AOA, camber and thickness. As long as the value of n is large enough, the numerical solutions are in good agreement with the experimental results. For instance, as Fig. 2.8 shows, for the wind turbine airfoil S809 [100], the numerical solutions of the pressure coefficient distribution is in good agreement with the exact potential flow theoretical results.

For the above calculation, the flow is treated as incompressible flow. In the case of high flow velocity, by using the Karman–Tsien model, the compressible pressure and the velocity coefficient on the airfoil surface can be approximated by the value of

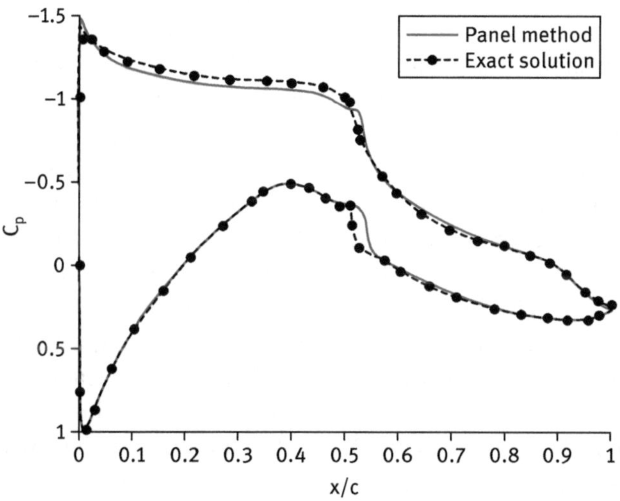

Fig. 2.8: Comparison between exact solution and panel method.

incompressible flow [99]

$$C_p = \frac{C_{p,inc}}{\sqrt{1 - Ma} + \frac{Ma^2}{1+\sqrt{1-Ma^2}} \cdot \frac{C_{p,inc}}{2}}, \tag{2.15}$$

where $C_{p,inc}$ is the pressure coefficient of incompressible flow, Ma is the Mach number.

2.3 Aerodynamic characteristic of airfoils

2.3.1 Pressure coefficient of the airfoil

The reacting force is generated by the wind interacting with the upper and lower airfoil. There are two kinds of surface force, one is the normal force, i.e., pressure; the other is the tangential force, i.e., friction. Usually, the component of the force perpendicular to the direction of motion is called lift. The component parallel to the direction of motion is called drag.

Changes in pressure along the airfoil surface are often expressed as the pressure coefficient C_p for the chord curve (Fig. 2.8). The pressure coefficient is defined as

$$C_p = \frac{p - p_0}{\frac{1}{2}\rho V_0^2}, \tag{2.16}$$

where p is the pressure to be evaluated, p_0 is the pressure in the freestream (i.e., remote from any disturbance).

When the flow hits the airfoil surface, there is a stagnation point at the leading edge. The stagnation point is a point where the local velocity of the fluid is zero, and where the pressure is at its maximum value, in the other words, the pressure coefficient $C_p = +1.0$. The flow over the stagnation line, starting from the stagnation point, first travels a short distance, then bypasses the leading edge, and goes along the upper surface to the trailing edge. The flow will reach a high local velocity at the leading edge. The larger the AOA is, the more rearward the stagnation point is, and the higher the local velocity that the flow reaches at the leading edge. The minimum pressure corresponding to the peak velocity could reach a high negative value. Meanwhile, for the same airfoil, the greater the AOA is, the more forward the lowest pressure point is. Therefore, the flow velocity accelerates around the leading edge, beginning with the stagnation point, and completes acceleration in a very short distance, thus leading to the deceleration segment, adverse pressure gradient segment, being lengthened. In the deceleration segment, in the first part of about 10 % of the chord length, there will be a very fast deceleration phase, and the negative value of C_p will reduce by half; in the second part of about 90 % of chord length, C_p will reach a small positive value at the trailing edge. So, if the AOA is too large, and peak velocity is very high, then the first part of the airfoil will have a serious adverse pressure gradient, which will destroy the laminar boundary layer, resulting in flow separation, and decreasing the lift.

2.3.2 Lift coefficient

Airfoil lift characteristics usually mean the curve of lift coefficient variation vs AOA. The lift coefficient C_l is defined as

$$C_\mathrm{l} = \frac{L}{\frac{1}{2}\rho V_0^2 c}, \tag{2.17}$$

where L is the lift, ρ is the density of the flow, V_0 is wind speed, and c is the length of the airfoil chord, often simply called the chord.

Lift is composed of the pressure force from local elements of the upper and lower surfaces of the airfoil, in the direction perpendicular to the flow. Thus, the lift coefficient can also be expressed as

$$C_\mathrm{l} = \int_0^1 (C_{\mathrm{p,l}} - C_{\mathrm{p,u}}) \cos\alpha \, \mathrm{d}\bar{x}, \tag{2.18}$$

where $C_{\mathrm{p,u}}$ is the pressure distribution coefficient for the upper surface of the airfoil, $C_{\mathrm{p,l}}$ is the pressure distribution coefficient for the lower surface of the airfoil, and $\bar{x} = x\cos\alpha + y\sin\alpha$. In the same chord location, the distance between the upper and lower surface of the pressure distribution represents its effective pressure difference which generates lift at this point.

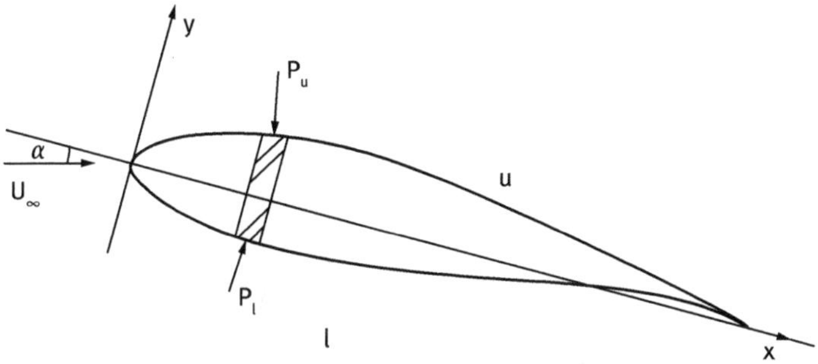

Fig. 2.9: Schematic drawing for airfoil aerodynamic calculation.

In Fig. 2.9, u is the upper surface, l is the lower surface.

Lift characteristic related to the flow around the airfoil can generally be divided into the attached flow area, stall area and deep stall area according to the AOA. In the attached flow area, the range of the AOA is about from –10° to 10°; in the stall area the range is about from 10° to 30°; in the deep stall area the range is about from 30° to 90°. When the AOA increases to about 10°, flow separation is initiated, because with increasing AOA the lift coefficient first slowly increases and then gradually decreases [97]. Fig. 2.10 shows the lift characteristic of the NACA 64418 airfoil in the smooth working conditions (boundary layer: $Re = 6.0 \times 10^6$, $Ma = 0.15$, $\alpha \in [-5°, 20°]$).

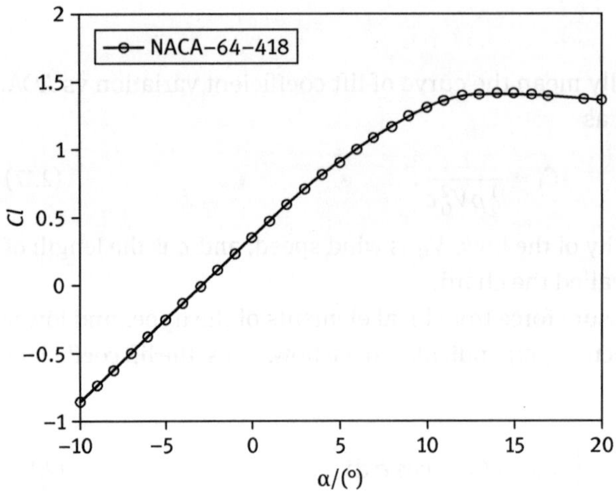

Fig. 2.10: Lift coefficient for the NACA 64418 airfoil.

2.3.3 Drag coefficient

Airfoil drag characteristics usually mean the curve of drag coefficient C_d variation vs AOA, but can also refer to the curve of drag coefficient vs lift coefficient. Drag coefficient is defined as

$$C_d = \frac{D}{\frac{1}{2}\rho V_0^2 c}, \tag{2.19}$$

where D is the drag, ρ is the density of the flow, V_0 is wind speed, and c is the length of the airfoil.

Airfoil drag comprises two fluid dynamic drags: skin friction and pressure drag. In the attached flow area, the drag is mainly skin friction, and the drag coefficient increases slowly with increasing AOA; in the airflow separation area, the main resistance is pressure drag, and the drag coefficient will increase rapidly with increasing AOA.

Obtaining the pressure distribution of the airfoil surface by a numerical method, we use the Squire–Young equation to calculate the drag coefficient C_d, which is available for the wake (non-trailing edge):

$$C_d = 2\theta_i = 2\theta\left(\frac{u}{U_1}\right)^{(H+2)/2}, \tag{2.20}$$

where θ is the momentum thickness, θ_i is the momentum thickness at infinity, u is flow velocity at the end of the trail edge, H is shape parameter, and U_1 is flow velocity.

The drag calculated from the Squire–Young equation (2.20) is the total viscous drag of the airfoil, including skin friction $C_{d,f}$ and pressure drag $C_{d,p}$, which are expressed as

$$C_{d,f} = \int C_f \, d\bar{x}, \tag{2.21}$$

$$C_{d,p} = C_d - C_{d,f}, \tag{2.22}$$

where $C_{d,f}$ is skin friction coefficient under free dynamic pressure, $C_{d,p}$ is the pressure drag coefficient. $C_{d,p}$ is deduced from the drag coefficient C_d and the skin friction coefficient $C_{d,f}$, rather than by calculating the integral of the surface pressure, because pressure drag magnitude is very small in the main work area of the AOA, and it is susceptible to variable transmission deviation in the calculating process.

Fig. 2.11 shows the drag characteristics for the NACA 64418 airfoil in smooth working conditions (boundary conditions: $Re = 6.0 \times 10^6$, $Ma = 0.15$, $\alpha \in [-5°, 20°]$).

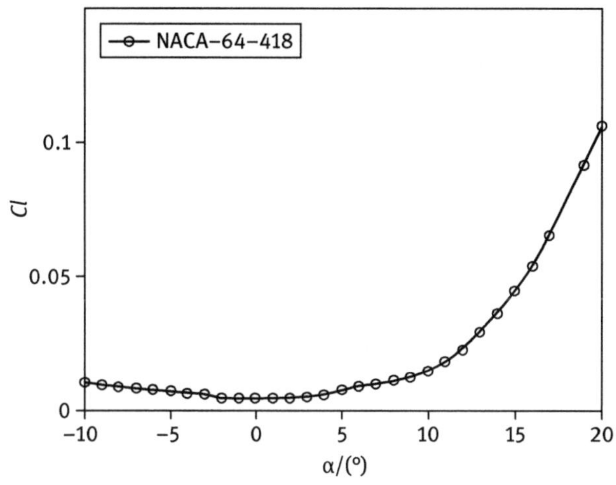

Fig. 2.11: The drag characteristics for the NACA 64418 airfoil.

2.3.4 Pitching moment coefficient

In theoretical mechanics, a plane force system is a synthesis of a force and a torque at a specified point. Pressure distributing on the airfoil surface also can be a synthesis of a force and a torque at a specified point. This torque is called the pitching moment. This point is called the acting point of force in aerodynamic studies, as the airfoil aerodynamic center is incremental at the point of lift, i.e., the torque at this point does not change with AOA. Similar to the lift and drag coefficients, the pitching moment coefficient is defined as follows

$$C_m = \frac{M}{\frac{1}{2}\rho V_0^2 c^2}. \tag{2.23}$$

Airfoil pitching moment characteristics usually means the curve of pitching moment coefficient C_m vs AOA. In the attached flow area, this curve varies linearly, and the pitching moment coefficient remains unchanged around the aerodynamic center.

In practice, the reference center of the pitching moment is generally at $\frac{1}{4}$ chord away from the leading edge. The pitching moment coefficient-AOA curve is nearly a straight line before a stall. After a stall, flow separation occurs, and the curve is no longer straight. The pitching moment characteristics curve can be calculated after pressure distribution has been obtained:

$$C_m = \int -C_p[(x - x_{ref})\,dx + (y - y_{ref})\,dy], \tag{2.24}$$

where (x_{ref}, y_{ref}) are the pitching moment reference center coordinates.

Fig. 2.12 shows the pitching moment coefficient for the NACA 64418 airfoil in smooth working conditions (boundary conditions: $Re = 6.0 \times 10^6$, $Ma = 0.15$, $\alpha \in [-5°, 20°]$).

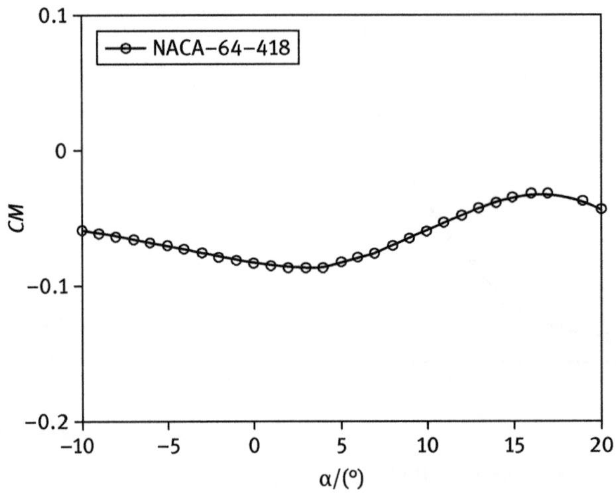

Fig. 2.12: The pitching moment coefficient for the NACA 64418 airfoil.

2.4 Stall on airfoils

Stall properties of airfoils refer to their performance around $C_{l,max}$. Due to the fluid viscoitys, the boundary layer near the edge exists even though the AOA is small. But the separation has little impact on the aerodynamic performance of the airfoils. Not only is the pressure distribution on the entire airfoil basically the same as that value in inviscid theory, but also the pressure distribution near the trailing edge shows no obvious discrepancy. But the airflow over the upper surface separates considerably around $C_{l,max}$. The separation zone spreads from the trailing edge to the leading edge, and can extend up to the front half of the airfoil surface. Originally, the lift coefficient gradually increases with increasing AOA, and circulation around the airfoil will also increase according to Kuta–Joukowsky theory. With the growth of the circulation, the front stagnation point on the lower surface will move backwards. If the front stagnation point is far away from the leading edge, the flow velocity of the airflow from the leading edge to the upper surface is greater. In other words, with increasing AOA, not only does the suction of whole airfoil strengthen, but also the suction peak near the leading edge rises more quickly than elsewhere. Passing by the peak point, the flow slows down, which is against the adverse pressure gradient. The greater the peak is, the greater the adverse pressure gradient is, and the more difficult it is for the flow to decelerate. It not only leads to turbulent flow in the boundary layer, but also the flow is separated because of the declining speed which is unable to withstand the averse pressure when the AOA increases to a certain extent. Then, the flow is divided into two parts: the interior and outside of the flow separation zone. At the separation boundary, the static pressure will be the same everywhere. The main stream of the flow is no longer slowed down and experiences pressurization after separation. In the separation region, because of the main stream on the free border constantly taking away mass by viscous action, it will cause the airflow from the back to fill the central portion, and form the reflux of the central portion. So a separation zone is basically a dead zone. However, it is not absolutely without flow, and it still has some reflux. As we know, the main stream on the upper surface is decelerated with the pressure increased from the lowest pressure point towards the trailing edge. The pressure in the flow separation zone is virtually equal to the main stream pressure at the point of separation. The more powerful the separation is, the further forward the separation point is, and the lower the pressure in the separation zone is. Since the pressure on the rear section of the upper airfoil surface is extremely low, and the AOA is large, there will be a big pressure drag. In other words, with larger AOA, the drag coefficient increases very fast around the stall AOA. In addition, for low velocity flow, the separation affects not only half of the airfoil but the entire flow field. In this case, the main stream is equal to a noncollapsible trailing edge airfoil. Circulation is smaller than before separation. Both the suction in front and the peak value are reduced. Therefore, the lift coefficient will be reduced after separation.

Once airfoil boundary layer separation causes airfoil stall, even if the stall AOA immediately goes back to the former state, the airfoil boundary layer will not immediately be restored to its state before the separation. This phenomenon is known as flow hysteresis. In Fig. 2.13, if the AOA α exceeds the critical angle, C_l is suddenly down. Later, with increasing AOA, C_l is continuously changing. When the AOA is reduced down from the stall AOA, C_l corresponds to a significant increase, shown in Fig. 2.13 as line BC. But the lift coefficient is still different to the original. If after reaching point C the AOA α continues to reduce, the curve C_l will jump to the original curve. The flow has changed from separation to attached bubbles at this time. If the AOA increases at this time, the curve will follow the line DA.

2.5 Roughness properties of airfoils

Due to erosion by dust, oil and raindrops, the surface of a wind turbine blade, especially the leading edge, becomes rough. Airfoil surface roughness has an important impact

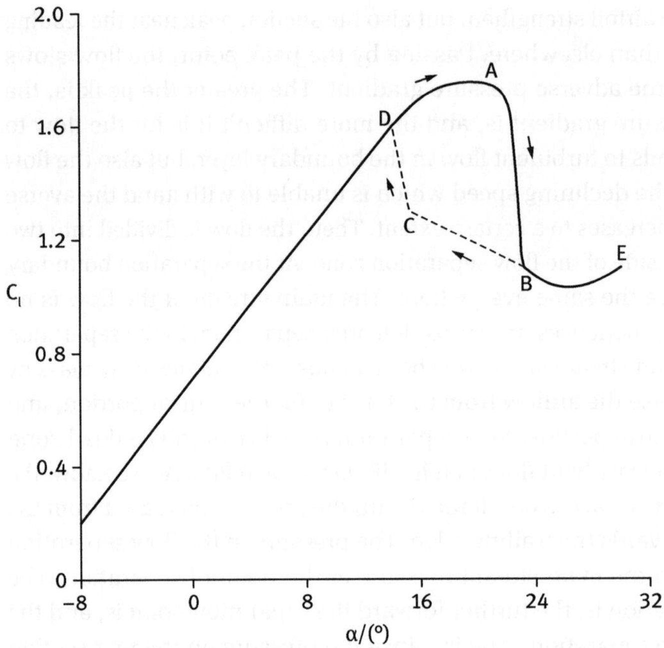

Fig. 2.13: Curve of flow hysteresis.

on the airfoil's aerodynamic characteristics. Surface roughness moves the boundary layer transition point forward, increases the boundary layer thickness, and reduces the airfoil camber, therefore reducing the maximum lift coefficient. In addition, surface roughness can change the laminar boundary layer into a turbulent boundary layer. As a result, the friction will be increased. In theoretical calculations and tests, in order to simulate airfoil surface roughness, generally a fixed position on the upper and lower surfaces of the airfoil will be selected to make the laminar flow transition. Normally, the position is at 5 % chord from the leading edge on the upper surface, 10 % from the leading edge on the lower surface [101].

Pollution and destruction are inevitable, and have an important influence on wind turbine power loss. Pollution at the leading edge causes multiple stalls of wind turbines. As a result, it reduces the rated output power and the maximum output power. The value of actual power loss depends on the design parameters of the rotor blades, the airfoil spanwise distribution and optimum tip speed ratio of the rotor and so on. According to tests, if the leading edge of the middle section $(0.42 < r/R < 0.57)$ and the proximal section $(0.2 < r/R < 0.42)$ for wind turbine blades is pullulated, the power loss will reach up to 8 % when the tip speed ratio is up to 8.

2.6 Influence of geometric parameters on aerodynamic characteristics

Airfoil geometry has a direct impact on the airfoil's aerodynamic characteristics. For wind turbine airfoils, the main geometric parameters affecting airfoil aerodynamic characteristics are leading edge radius, relative thickness and the string position of the maximum thickness, the trailing edge, and so on.

2.6.1 Influence of the leading edge radius of an airfoil

The leading edge radius (i.e., the leading edge bluntness) has an important influence on the maximum lift coefficient for the airfoil. Usually, the difference between the y-coordinate value at 6 % and the y-coordinate value at 0.15 % of the chord on the upper surface represents the blunt part of the airfoil's leading edge. When the other geometric parameters of the airfoil remain unchanged, the maximum lift coefficient increases due to the leading edge radius increasing. When the leading edge radius is small, the airfoil pressure gradient along the suction side is large. With increasing AOA, the flow pressure will quickly reach a negative peak near the leading edge. The transition from laminar flow to turbulent flow will occur near the leading edge under the adverse pressure gradient effect. At the same time, it is easy to produce flow separation, which will reduce the maximum lift coefficient [92].

2.6.2 Influence of the maximum relative thickness and its position

The maximum relative thickness of an airfoil and its chordwise position have a great influences on the aerodynamic performance of the airfoil. If the maximum relative thickness is too large or small, the maximum lift coefficient will be limited. For instance, with the NASA LS series airfoil, when the maximum relative thickness is at around 15 % of the chord, the airfoil has a maximum lift coefficient; for the NACA series airfoil, with a maximum relative the thickness at around 13 %, the airfoil has a maximum lift coefficient. On one side, if the position of the maximum relative thickness is close to the leading edge, the maximum lift coefficient will be large; on the other side, it would cause flow transition, and the drag coefficient would increase. In the same airfoil series, when the relative thickness increases, the minimum drag coefficient increases. In addition, when the position of maximum relative thickness is further back, the minimum drag coefficient will be reduced. The relative thickness has little effect on the pitching moment coefficient [92].

2.6.3 Influence of the maximum camber and its position

Under normal circumstances, if the camber is increased, the maximum lift coefficient will be increased, particularly if the leading edge radius is small or the airfoil is thin. However, with increasing camber, the increase in the lift coefficient becomes slow, and the drag coefficient increases at the same time. And the increments of drag coefficients are different with different airfoils. Further, when the position of maximum camber is further forward, the maximum lift coefficient becomes large [92].

2.7 Influence of Reynolds number on aerodynamic characteristics

Changes in the Reynolds number reflect changes in the state of the boundary layer and affect flow separation, which will change the aerodynamic characteristics especially for the maximum lift coefficient. When the Reynolds number is small due to the existence, the development and collapse of the leading edge separation bubble, the maximum lift coefficient is very sensitive to the Reynolds number. When the Reynolds number is large, the stall AOA of the airfoil increases with the Reynolds number. Therefore, the maximum lift coefficient of the airfoil will also increase. When the Reynolds number is greater than 6.0×10^6, the variation of airfoil stall AOA and maximum lift coefficient with Reynolds number will be flatter. As well as being related to Reynolds number, the maximum lift coefficient also affects the minimum drag coefficient. When the AOA is small, airfoil drag depends on friction, and its value is associated with the location of the turning point. When the Reynolds number increases, it can postpone the laminar flow separating, and reduce friction. In addition, when the AOA is small, the pressure drag is reduced with increasing Reynolds number. Thus, the minimum drag coefficient decreases correspondingly [92]. It should be noted that the influence of the Reynolds number on the airfoil aerodynamic characteristics is related to its geometric properties, surface roughness, flow turbulence and so on.

2.8 Method of predicting aerodynamic performance of airfoils

The airfoil's aerodynamic characteristics are directly related to the performance of the blades. Currently, the methods of obtaining airfoil data are mainly the viscous-inviscid iterative method, which considers the potential flow and viscous boundary layer equations, CFD methods, wind tunnel experiments and so on.

2.8.1 Introduction to XFOIL and RFOIL

The XFOIL program developed by the American scholar Dr. Drela is used to calculate the airfoil's aerodynamic performance. For a given airfoil, if the AOA and Reynolds number are given, XFOIL can calculate the pressure distribution coefficient C_p, lift coefficient C_l and drag coefficient C_d. In addition, XFOIL can also simulate the transition and separation of the boundary layer [99].

RFOIL has been co-developed by ECN (Energy research Centre of the Netherlands), DUT (Delft University of Technology) and NLR (National Aerospace Laboratory of the Netherlands) and is commercial software dedicated to the design of wind turbine airfoils and the analysis of their aerodynamic characteristics. Compared with XFOIL, the main advantage of RFOIL is improved computing stability and accuracy of airfoil aerodynamic performance in the stall region; its calculation process is similar to XFOIL, and the software interface is shown in Fig. 2.14. It prompts the user to enter the boundary conditions to calculate the aerodynamic characteristics at the software interface.

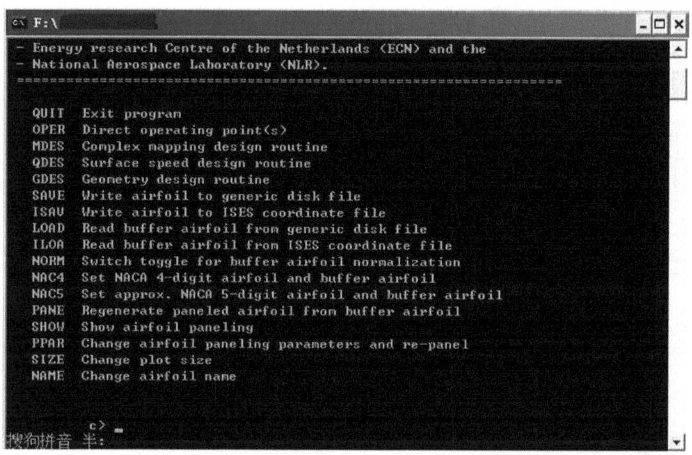

Fig. 2.14: Interface of RFOIL software.

2.8.2 Airfoil aerodynamic performance calculation cases

Compared with other fluid calculation software, these two software programs have some advantages, such as simple interface, easy to operate, fast convergence and short runtime in the calculation of viscous and non-viscous fluids. In particular, when the AOA is large, it is easier to achieve convergence than with other procedures. Therefore, XFOIL and RFOIL are widely used in airfoil design and analysis.

Using XFOIL and RFOIL, the aerodynamic performance of the famous Dutch DU 91-W2-250 airfoil was calculated and analyzed, and was compared with experimental data. The results show that the maximum relative thickness of airfoil is 0.25, i.e., the ratio of absolute thickness to chord length of the airfoil is 25 %.

In the free transition working conditions, $Re = 1.0 \times 10^6$, the aerodynamic characteristics for the DU 91-W2-250 airfoil (the airfoil profile was shown in Fig. 2.15) compared with test results is shown in Fig. 2.16–2.18. In Fig. 2.16 and 2.17, compared with experimental values, the results calculated by XFOIL and RFOIL show lift and drag coefficients in the linear phase are very consistent. The deviation of the maximum lift coefficient corresponding to AOA is relatively small. Compared with XFOIL, the results also show that RFOIL can better simulate the airfoil stall condition. When the DU 91-W2-250 airfoil is stalled ($\alpha > 9.5°$), flow instability and separation have a great impact on airfoil aerodynamic characteristics. Currently, it is difficult to accurately and completely simulate the stall process with numerical methods, especially for three-dimensional unsteady flow after stall. So the calculated values deviate from the test, but converge to the trend after stall.

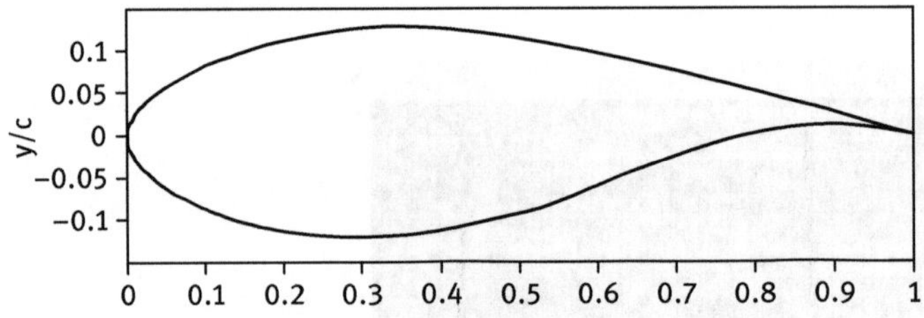

Fig. 2.15: Profile of the DU 91-W2-250 airfoil.

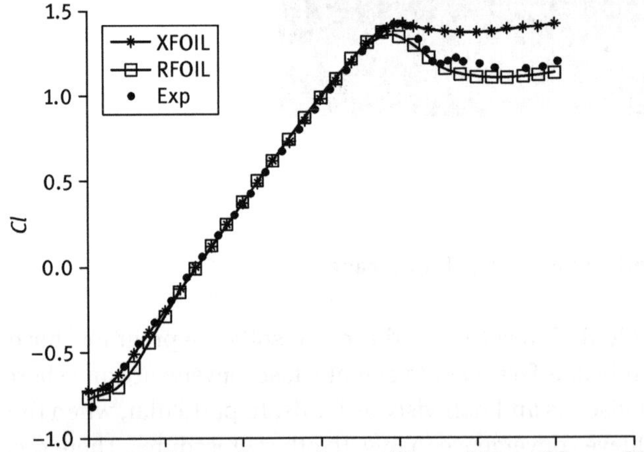

Fig. 2.16: Comparison of lift coefficients obtained from theoretical calculations and test results.

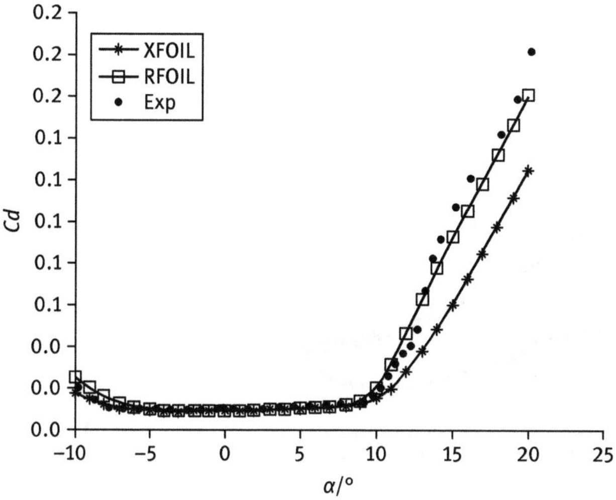

Fig. 2.17: Comparison of drag coefficients obtained from theoretical calculations and test results.

As shown in Fig. 2.18, the reason for a larger difference around the maximum lift-to-drag ratio is that the value of the denominator (drag coefficient) is small, so a small difference in the denominator can lead to a large change in the lift-to-drag ratio. In other words, the lift-to-drag ratio is more sensitive to the drag coefficient.

Fig. 2.19 shows the pressure distribution obtained from test results, XFOIL calculated data and RFOIL results for the DU 91-W2-250 airfoil in freedom transition condition

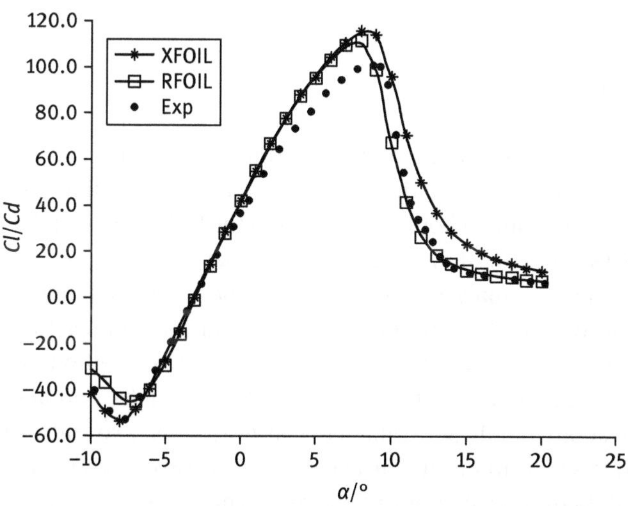

Fig. 2.18: Comparison of lift-to-drag ratio obtained from theoretical calculations and test results.

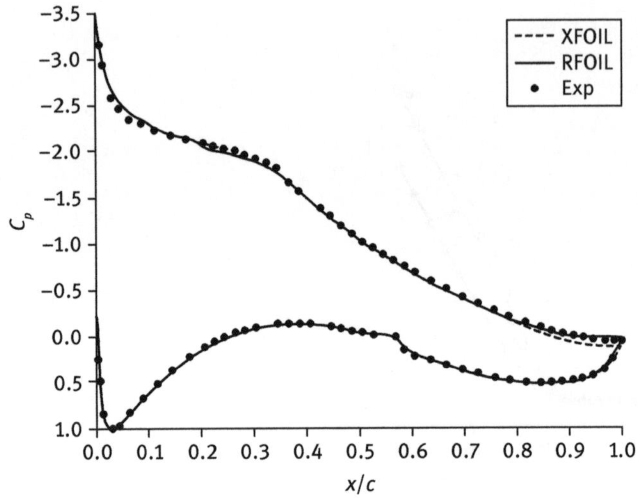

Fig. 2.19: Comparison of pressure distribution at maximum lift-to-drag ratio ($\alpha = 9.74°$).

of the main AOA at Reynolds number $Re = 1.0 \times 10^6$. Comparing the pressure distribution characteristics at the maximum lift-to-drag ratio conditions when AOA = 9.74°, the theoretical result is in good agreement with the experimental test which validates the XFOIL and RFOIL theoretical models. The above calculation and analysis compared with the experimental data provide a theoretical calculation method and the feasibility of further development for aerodynamic performance analysis and optimization of new airfoils.

2.9 Chapter conclusions

This chapter discussed the basic concepts of Reynolds number and Mach number for wind turbine airfoils. The flow characteristics of boundary layer separation were analyzed. The integral equation for laminar boundary layer and the turbulent boundary layer for a wind turbine airfoil surface was derived. The variation pattern of the flow field around the wind turbine airfoils was analyzed. The airfoil's aerodynamic parameters, such as pressure distribution coefficient, lift coefficient, drag coefficient, and pitching moment coefficient were calculated theoretically based on the panel method and potential flow theory. This chapter explores the effect of airfoil geometry and aerodynamic parameters on the aerodynamic performance of airfoils. Then two software programs (XFOIL and RFOIL) were introduced. The accuracy of the two programs was verified with experimental data through computing the aerodynamic performance of an airfoil. The two programs are suitable for direct design methods for airfoils and provide aerodynamic tools for the design of wind turbine airfoils.

3 Integrated expressions of wind turbine airfoils

3.1 Introduction

The section shape of a wind turbine blade is called the wind turbine airfoil and it has great influence on the performance of the wind turbine. There are two main kinds of wind turbine airfoils. One is aircraft airfoils such as NACA 230, NACA 44, NACA 632, NACA LS and FX airfoil series, to name a few. Since then, some airfoils specially tailored for wind turbines have been introduced, such as the NRELS airfoil series designed by the Americans, the DU airfoil series designed by the Dutch, the Risø airfoil series designed by the Danish and the FFA-W airfoil series designed by the Swedes.

The important key element for the power characteristics and the load features of a wind turbine is the design theory for the wind turbine airfoils, which is why international scholars are studying it. The expressions for current airfoil are implemented by discrete points, not by a specific function. The design of new airfoils is also based on the original airfoil coordinates. The local airfoil profiles are adjusted to obtain a new airfoil with better aerodynamic performance. Therefore, some subjects for the design of wind turbine airfoils must be studied, such as the functional characteristics of the airfoil profile, the mathematical expression of the function, the aerodynamic characteristics of the geometric curve for the mathematical expression, the shape optimization of the geometric curve and so on.

3.2 Transformation theory of airfoils

In order to study the aerodynamic performance of airfoils, it is essential to understand the flow around the airfoil surface. Because the shape expression of the airfoil is complex, if the flow is solved using the analytical method, it is difficult to get an accurate analytical answer. Conformal mapping can transform a complex boundary curve into a simple boundary. The simple boundary flow can be easily obtained, then according to the transform relationship, the original flow field with complex boundary can be calculated by reversing.

3.2.1 Conformal transformation [93]

It is assumed that there is a ζ-plane outside of the z-plane, which is also a complex variable:

$$\zeta = \xi + \mathrm{i}\eta. \tag{3.1}$$

There is a certain relationship between them:

$$\zeta = f(z). \tag{3.2}$$

DOI 10.1515/9783110344387-004

Thus, a corresponding relationship is established for the points on the two planes. When there is a shape on the z-plane, there also must be a shape on the ζ-plane which is transformed. In general, the angle between two segments of the corresponding shapes equals the angle in the original shape, so it is called conformal mapping [94].

Suppose there is a triangle PQR on the z-plane, it can be transformed into another triangle $P'Q'R'$ on the ζ-plane by the transform relation, as shown in Fig. 3.1. The corresponding points of P, Q and R are P', Q' and R'.

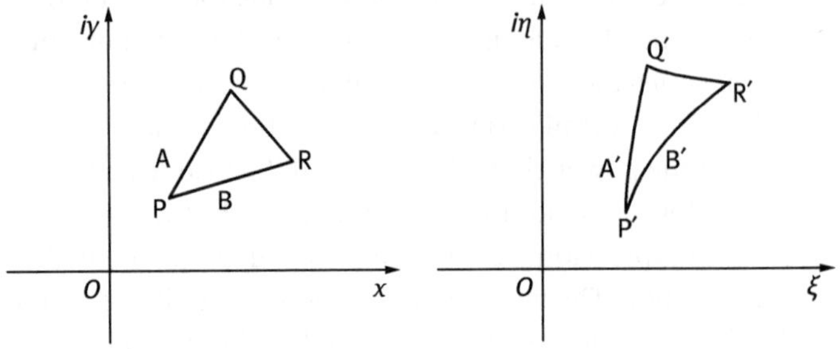

Fig. 3.1: Conformal mapping.

Starting from point P on the z-plane, a microsegment along the direction of PQ and PR is made respectively. Correspondingly, a microsegment along the direction of $P'Q'$ and $P'R'$ are made respectively from point P' on the ζ-plane. The relation between them is:

$$\frac{\mathrm{d}\zeta}{\mathrm{d}z} = f'(z). \tag{3.3}$$

This derivative is also a complex function, which can be expressed by $a\mathrm{e}^{\mathrm{i}\alpha}$, that is,

$$\mathrm{d}\zeta = a\mathrm{e}^{\mathrm{i}\alpha}\,\mathrm{d}z. \tag{3.4}$$

It is can be found from equation (3.4) that the module of $\mathrm{d}\zeta$ and $\mathrm{d}z$ differs by a multiple a, and the argument differs by an angle α:

$$\left.\begin{aligned} |\mathrm{d}\zeta| &= a|\mathrm{d}z|, \\ \arg \mathrm{d}\zeta &= \arg \mathrm{d}z + \alpha. \end{aligned}\right\} \tag{3.5}$$

Derivative $f'(z)$ is only a function of points. After taking a given point P, for all the corresponding segments that start from point P and are transformed into plane ζ, the module will be a times larger, and the argument will increase by α. So the angle between PA and PB always equals the angle between $P'A'$ and $P'B'$ on the ζ-plane. Since there is a straight angle at the straight line in the original graphic, the curve after transformation is also bound to be a continuous curve.

3.2.2 Joukowsky transformation of airfoils [93]

The complex function flowing around the circle on the z-plane can be expressed as

$$w(z) = v_0\left(z + \frac{a^2}{z}\right). \tag{3.6}$$

Here, taking the Joukowsky transformation,

$$\zeta = z + \frac{a^2}{z}. \tag{3.7}$$

After transforming the flow field in equation (3.6) of the z-plane by equation (3.7), on the ζ-plane it becomes

$$w(\zeta) = v_0\zeta. \tag{3.8}$$

This is a straight uniform flow field which is parallel to the horizontal axis and has a flow velocity of v_0.

If the radius of the circle is enlarged a little bit, and only treated with the transformation equation once, the center of the circle will still be located on the x-axis, as shown in Fig. 3.2. Then a symmetrical airfoil with thickness on the ζ-plane is obtained, namely the so-called Joukowsky symmetrical airfoil.

As shown in Fig. 3.2, assuming the radius of the original circle is d, where $d = a(1+\varepsilon)$, and the circle's center is located at $(-\varepsilon a, 0)$, there are relationships between the coordinates of any point P on the circle $z = re^{i\theta}$ and the radius of circle, as given below:

$$d^2 = r^2 + a^2\varepsilon^2 + 2a\varepsilon r\cos\theta. \tag{3.9}$$

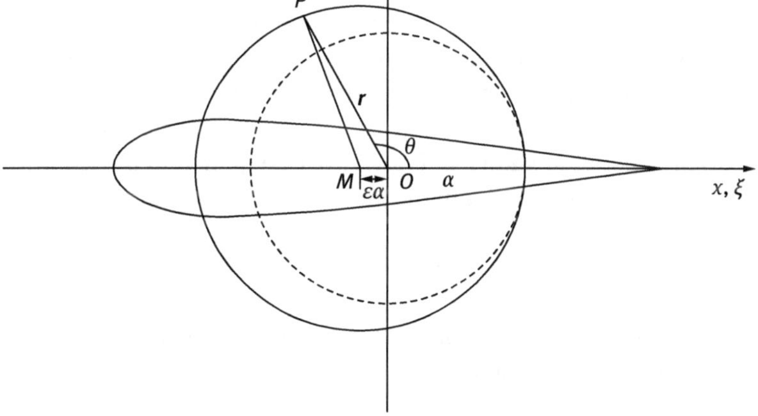

Fig. 3.2: Joukowsky transformation of an airfoil.

Crossing out ε^2 and solving r, we obtain

$$r = a[1 + \varepsilon(1 - \cos\theta)]. \tag{3.10}$$

Taking equation (3.10) into equation (3.7) yields:

$$\left.\begin{aligned} \xi &= \left(r + \frac{a^2}{r}\right)\cos\theta = 2a\cos\theta, \\ \eta &= \left(r - \frac{a^2}{r}\right)\sin\theta = 2a\varepsilon(1 - \cos\theta)\sin\theta. \end{aligned}\right\} \tag{3.11}$$

The chord length of this airfoil is $4a$, maximum thickness locates at $\theta = \arccos(-1/2)$, namely the location of $\xi = -a$, and the maximum thickness is

$$2\eta_{\max} = 3\sqrt{3}a\varepsilon. \tag{3.12}$$

When the attack angle $\alpha = 0°$, the fluid field that passes this airfoil on the z-plane can be expressed as

$$w(z) = v_0\left[(z + \varepsilon a) + \frac{a^2(1+\varepsilon)^2}{(z + \varepsilon a)}\right]. \tag{3.13}$$

In addition, moving up the center of original circle a little bit, the transformed graphic will be a generic Joukowsky airfoil with thickness and camber.

3.2.3 Theodorsen method [93]

Although the Joukowsky airfoil looks almost like a real airfoil by appearance, strictly speaking its coordinate relations are incorrect, so it is not applied in practice. In the 1930s, Theodorsen proposed a method which can calculate real airfoils. Then Fourier series can be used to transform it into a perfect circle. The number of terms in the series can be set to any number, a higher accuracy can be achieved with more terms.

Put a given airfoil on the ζ-plane, connect a trailing edge point and a point near the leading edge (it can be the midpoint of the leading edge curvature center and leading edge point) as a line, and put this line on the ζ-axis, then take the midpoint of the connection as the original point, as shown in Fig. 3.3 (a). One-quarter of the length of the connection line is set as the value of a in the Joukowsky transformation. Use the transformation equation

$$\zeta = z + \frac{a^2}{z}. \tag{3.14}$$

Transforming the given airfoil onto the z'-plane, a near circle is obtained, shown in Fig. 3.3 (b). This graphic can be expressed by

$$z' = a\mathrm{e}^{\varphi + \mathrm{i}\theta}, \tag{3.15}$$

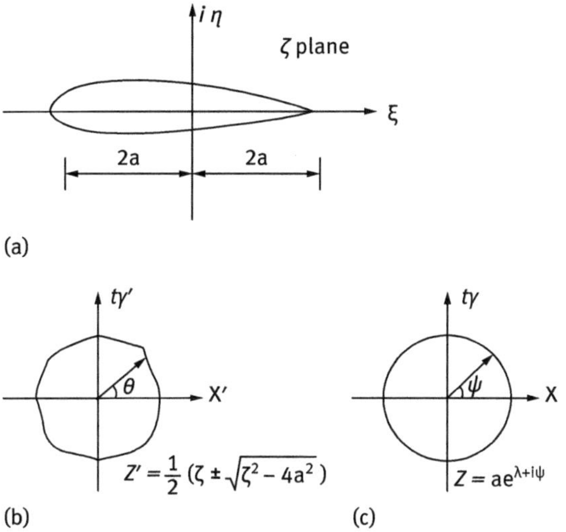

Fig. 3.3: Schematic drawing of airfoil transformation.

where θ is amplitude, ae^{φ} is the radius vector length of this near circle, it is close to a constant and varies in the vicinity of this constant. So the index φ of e is a function of θ: $\varphi = \varphi(\theta)$, φ and θ can be seen as coordinates on the z'-plane. The relationship between φ, θ and the original airfoil is

$$\zeta = ae^{\varphi+i\theta} + ae^{-\varphi-i\theta} = a(e^{\varphi} + e^{-\varphi})\cos\theta + ia(e^{\varphi} - e^{-\varphi})\sin\theta = \xi + i\eta. \tag{3.16}$$

Then the near circle on the z'-plane is transformed into a circle on the z-plane, as shown in Fig. 3.3 (c). Assuming the radius of this circle is ae^{λ}, the z-coordinate can be expressed as

$$z = ae^{\lambda+i\psi}, \tag{3.17}$$

where ψ is the argument on the z-plane. Defining the transformation relation between the z'-plane and the z-plane as

$$z' = z\exp\left(\sum_{0}^{\infty}\frac{C_n}{z^n}\right), \tag{3.18}$$

where C_n is plurality, $C_n = A_n + iB_n$, then,

$$\frac{z'}{z} = \exp\left(\sum_{0}^{\infty}\frac{C_n}{z^n}\right). \tag{3.19}$$

It is also known that

$$\frac{z'}{z} = \frac{e^{\varphi+i\theta}}{e^{\lambda+i\psi}} = e^{\varphi-\lambda+i(\theta-\psi)}. \tag{3.20}$$

Based on equations (3.19) and (3.20), the relationship on the z-plane can be derived:

$$\varphi - \lambda + \mathrm{i}(\theta - \psi) = \sum_{1}^{\infty}(A_n + \mathrm{i}B_n)\frac{1}{r^n}(\cos n\psi - \mathrm{i}\sin n\psi), \tag{3.21}$$

where $z = re^{\mathrm{i}\psi}$, $r = ae^{\lambda}$. These are two Fourier series:

$$\left.\begin{aligned} \varphi - \lambda &= \sum_{1}^{\infty}\left(\frac{A_n}{r^n}\cos n\psi + \frac{B_n}{r^n}\sin n\psi\right), \\ \theta - \psi &= \sum_{1}^{\infty}\left(\frac{B_n}{r^n}\cos n\psi - \frac{A_n}{r^n}\sin n\psi\right). \end{aligned}\right\} \tag{3.22}$$

These two functions are conjugate. Then each coefficient can be solved by the method of Fourier series determinate coefficients.

3.3 Integrated expression of airfoil profiles

In order to facilitate the calculation of the airfoil's aerodynamic performance, the Theodorsen method first used two conformal mappings to transform an arbitrary airfoil into a near circle. And then it uses Fourier series to obtain a perfect circle. However, in the second transformation step, the solving process of Fourier series coefficients is too long, which costs too much computation time. Therefore, a new kind of integrated expression for airfoil profile is proposed to omit the second transformation step in the Theodorsen method. The trigonometric series or Taylor series is used to express near circularity directly. This makes the airfoil expression more concise, and reduces the computation time significantly, which also provides the possibility for global optimization of the airfoil.

From the previous section it is known that a circle on the z-plane can be transformed into an airfoil on the ζ-plane by changing the position of the circle's center and using the Joukowsky airfoil transformation:

$$\zeta = f(z) = z + a^2/z. \tag{3.23}$$

Meanwhile, the circle needs to pass the point $x = a$ in order to ensure the existence of a sharp airfoil trailing edge, where a is $\frac{1}{4}$ of the airfoil's chord length. This actually transforms the airfoil with a complex shape into a simple typical flow pattern.

Airfoil coordinates on the ζ-plane can be expressed as

$$\begin{cases} x = (r + a^2/r)\cos\theta, \\ y = (r - a^2/r)\sin\theta. \end{cases} \tag{3.24}$$

where r is the radius vector of the airfoil, θ is the argument.

3.3.1 The trigonometric series representation of airfoil shape function

From the airfoil geometric theory presented above it can be shown that the airfoil can be expressed by conformal mapping. But a real airfoil cannot be obtained by the simple transformation. Here, the shape is expressed by

$$z = a\exp(\varphi(\theta) + \mathrm{i}\theta). \tag{3.25}$$

If $\varphi(\theta)$ is defined as a constant, the expression above is a circle, the transformed result is a symmetrical airfoil without camber and thickness. If $\varphi(\theta)$ is a changeable unknown function, infinitely many airfoils with camber and thickness can be obtained by taking different $\varphi(\theta)$. $r = a\exp(\varphi(\theta))$ is the radius vector of the curve.

Trigonometry is used in this expression:

$$\begin{aligned}\varphi(\theta) &= a_1(1-\cos\theta) + b_1\sin\theta + a_2(1-\cos\theta)^2 + b_2\sin^2\theta + \cdots \\ &\quad + a_k(1-\cos\theta)^k + b_k\sin^k\theta + \cdots, \qquad k = 1, 2, \ldots, n.\end{aligned} \tag{3.26}$$

This expression satisfies the condition of $\varphi(0) = 0$, which ensures the sharp trailing edge property of the airfoil. The z-curve expression can be obtained by combining expressions (3.26) and (3.25).

3.3.2 The Taylor series representation of airfoil shape function

Although the Joukowsky airfoil and the Carmen–Terry Fuzi airfoil, which is based on the Joukowsky airfoil, are close to a practical airfoil, their coordinate relationship still cannot be expressed. Theodorsen solved this problem [102]. The idea is that, since using the Joukowsky transformation, the airfoil that was transformed from a circle is almost like a low speed airfoil; conversely, using the same transformation equation to transform the existing practical airfoil back, the resulting graphic cannot be a perfect circle, but it will not be too far off, shown in Fig. 3.4.

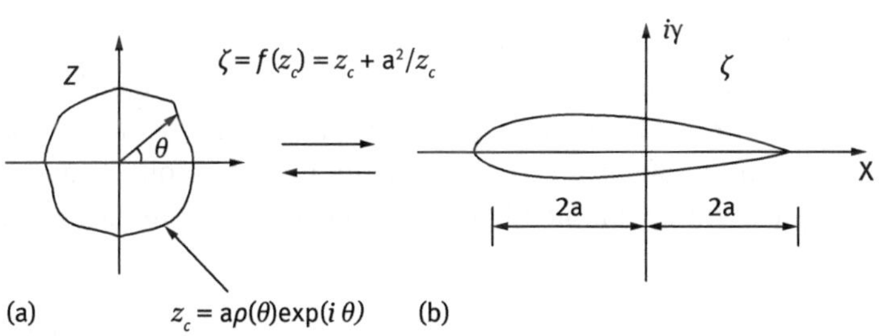

Fig. 3.4: Transformation of generic airfoil.

Based on Theodorsen method, a simple and universal expression is proposed here:

$$z_c = a\rho(\theta)\exp(i\theta), \tag{3.27}$$

where θ is the argument, $\rho(\theta)$ is the radius vector of a near circle, which is close to a constant and varies in the vicinity of this constant. By choosing different $\rho(\theta)$, infinitely many kinds of airfoil with different thickness, chamber, leading edge radius and trailing edge angle can be achieved by transformation.

Combining equations (3.27) and (3.24),

$$\begin{cases} x = a\left(\rho + \dfrac{1}{\rho}\right)\cos(\theta), \\ y = a\left(\rho - \dfrac{1}{\rho}\right)\sin(\theta). \end{cases} \tag{3.28}$$

Under the condition of knowing the shape and coordinates of an airfoil, solving the relationship between shape function ρ and argument θ:

$$\rho + \frac{1}{\rho} = \frac{x}{a\cos(\theta)}, \tag{3.29}$$

$$\rho - \frac{1}{\rho} = \frac{y}{a\sin(\theta)}. \tag{3.30}$$

Solving equations (3.29) and (3.30), we get:

$$\sin^2\theta = \frac{1}{2}\left(h + \sqrt{h^2 + \frac{y^2}{a^2}}\right), \tag{3.31}$$

$$\rho = \frac{1}{2a}\left(\frac{x}{\cos\theta} + \frac{y}{\sin\theta}\right), \tag{3.32}$$

where

$$h = \frac{-x^2 - y^2 + 4a^2}{4a^2}.$$

Then the corresponding airfoil coordinates of shape function ρ and argument θ can be derived by equations (3.31) and (3.32).

According to Taylor series and other thoughts [103], an arbitrary mathematical expression of a function curve can be expanded into a series, whose geometric and analytic properties can be controlled by adjusting and optimizing the coefficients of the series. Through the integration of research on a large number of airfoils it can be found that $\rho(\theta)$ can be expressed by a simple high order polynomial. Here the airfoil shape function is named

$$\rho(\theta) = C_0 + C_1\theta + C_2\theta^2 + C_3\theta^3 + \cdots + C_k\theta^k + \cdots, \tag{3.33}$$

where C_k is the coefficient of the polynomial.

3.4 Airfoil profile analysis using integrated expressions

For the airfoil shape function expressed by trigonometry (equation (3.26)), the normalized value of airfoil chord length is taken to be 1. Different coefficients are used to verify that the expression has airfoil properties.

3.4.1 Type I airfoil profile

Taking $a_1 = 0.1$, $b_1 = 0.05$, $k = 1$ and with other coefficients defined as 0, then we obtain that

$$\varphi(\theta) = 0.1 \times (1 - \cos\theta) + 0.05 \times \sin\theta. \tag{3.34}$$

By equations (3.34), (3.24) and (3.25) we can get:

$$\begin{cases} x = a \times (\exp(0.1 \times (1 - \cos\theta) + 0.05 \times \sin\theta) \\ \qquad + \exp(-0.1 \times (1 - \cos\theta) - 0.05 \times \sin\theta)) \times \cos\theta, \\ y = a \times (\exp(0.1 \times (1 - \cos\theta) + 0.05 \times \sin\theta) \\ \qquad - \exp(-0.1 \times (1 - \cos\theta) - 0.05 \times \sin\theta)) \times \sin\theta. \end{cases} \tag{3.35}$$

As we can see in Fig. 3.5, this curve has airfoil properties.

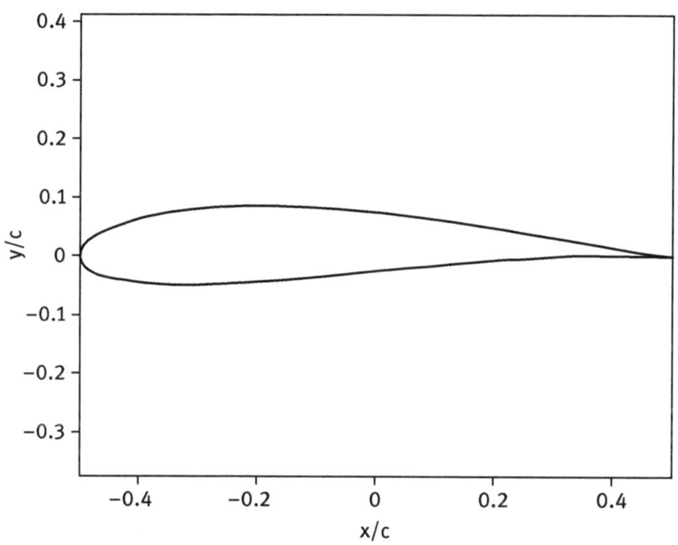

Fig. 3.5: Type I airfoil.

3.4.2 Type II airfoil profile

Taking $a_2 = 0.05,\ b_2 = 0.05,\ k = 2$ and with other coefficients defined as 0, then

$$\varphi(\theta) = 0.05 \times (1 - \cos\theta)^2 + 0.05 \times \sin^2\theta. \tag{3.36}$$

By equations (3.36), (3.24) and (3.25) we can get:

$$\begin{cases} x = a \times (\exp(0.05 \times (1 - \cos\theta)^2 + 0.05 \times \sin^2\theta) \\ \qquad + \exp(-0.05 \times (1 - \cos\theta)^2 - 0.05 \times \sin^2\theta)) \times \cos\theta, \\ y = a \times (\exp(0.05 \times (1 - \cos\theta)^2 + 0.05 \times \sin^2\theta) \\ \qquad - \exp(-0.05 \times (1 - \cos\theta)^2 - 0.05 \times \sin^2\theta)) \times \sin\theta. \end{cases} \tag{3.37}$$

As we can see in Fig. 3.6, this curve also has airfoil properties.

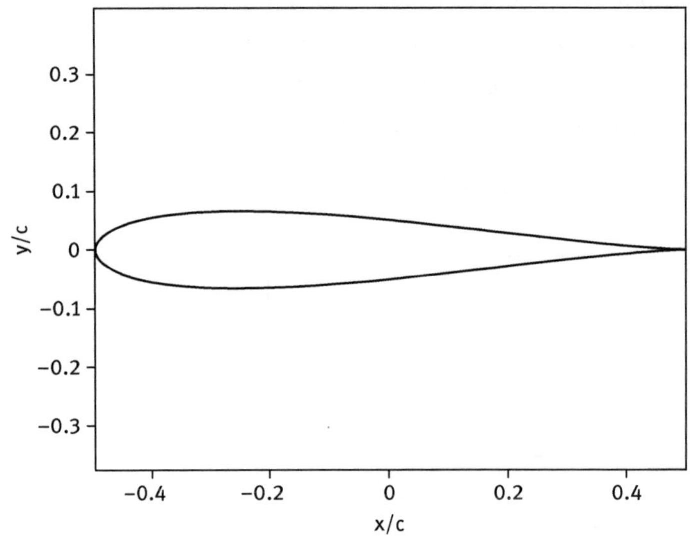

Fig. 3.6: Type II airfoil.

3.4.3 Type III airfoil profile

Taking $a_3 = 0.03,\ b_3 = 0.05,\ k = 3$ and with other coefficients defined as 0, then:

$$\varphi(\theta) = 0.03 \times (1 - \cos\theta)^3 + 0.05 \times \sin^3\theta. \tag{3.38}$$

By equations (3.38), (3.24) and (3.25) we can get:

$$\begin{cases} x = a \times (\exp(0.03 \times (1 - \cos\theta)^3 + 0.05 \times \sin^3\theta) \\ \qquad + \exp(-0.03 \times (1 - \cos\theta)^3 - 0.05 \times \sin^3\theta)) \times \cos\theta, \\ y = a \times (\exp(0.03 \times (1 - \cos\theta)^3 + 0.05 \times \sin^3\theta) \\ \qquad - \exp(-0.03 \times (1 - \cos\theta)^3 - 0.05 \times \sin^3\theta)) \times \sin\theta. \end{cases} \tag{3.39}$$

As shown in Fig. 3.7, this curve also has airfoil properties.

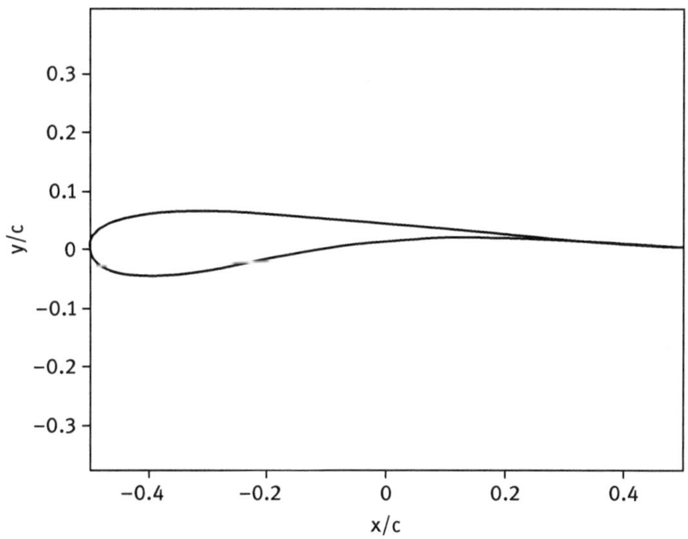

Fig. 3.7: Type III airfoil.

3.5 Versatility properties for integrated expression of airfoils

For the airfoil shape function expressed by trigonometry (equation (3.26)), the expression of existing airfoils is achieved by discrete points, i.e. no specific form of expression function is proposed. The airfoil datas in the airfoil plane can be expressed by [104, 105]

$$\zeta = x + yi, \tag{3.40}$$

where x is the horizontal axis, y is the vertical axis.

It is known from equations (3.25) and (3.40):

$$\cosh\varphi = \frac{x}{2a\cos\theta},$$
$$\sinh\varphi = \frac{y}{2a\sin\theta},$$
$$2\sin^2\theta = p + \sqrt{p^2 + \left(\frac{y}{a}\right)^2}, \tag{3.41}$$

where

$$p = 1 - \left(\frac{x}{2a}\right)^2 - \left(\frac{y}{2a}\right)^2.$$

From equation (3.41) it can be seen that once the airfoil coordinate values are known, based on the number of terms in the series and some selected discrete points, integrated equation coefficients can be achieved by taking the coordinate points into equations (3.41), (3.24) and (3.25), thus implementing the integrated expression of an airfoil. Therefore, this integrated equation is suitable for the theoretical expression of any airfoil.

For a better demonstration, two kinds of airfoils (NACA 64418 and S809) which are commonly used in wind turbines were selected. The coefficients of the integrated equation were determined by reverse solving, which validated the versatility of the integrated equation and also provided references for optimizing the airfoils.

3.5.1 First-order fitting

Firstly, the two airfoils are expressed by a first-order term. The obtained coefficients are shown in Tab. 3.1.

Tab. 3.1: First-order fitting coefficients.

	NACA 64418	S809
a_1	0.11349	0.15554
b_1	0.04482	0.02791

Fig. 3.8 and 3.9 were obtained respectively by putting the coefficients into the integrated equation. The line represents the integrated expression, and the boxes represent the real airfoils. It can be seen from the figure that first-order fitting cannot well express the existing airfoils.

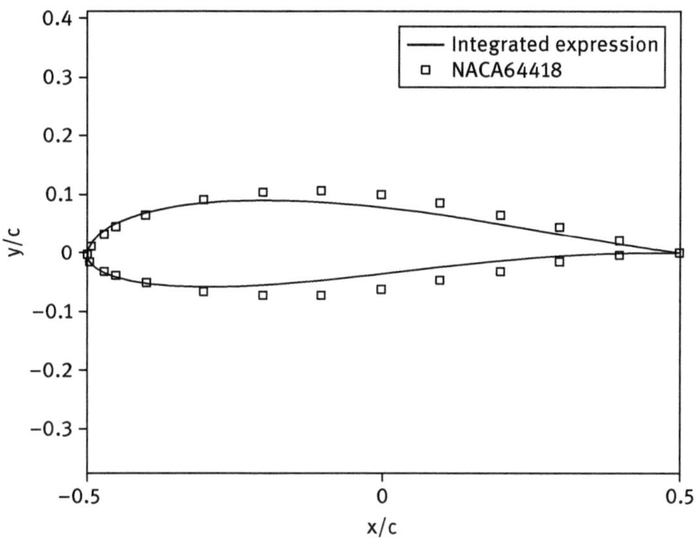

Fig. 3.8: First-order fitting of NACA 64418 airfoil.

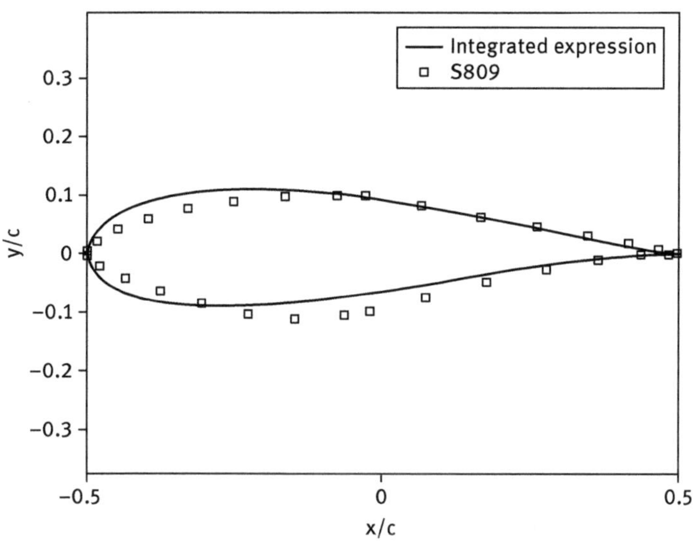

Fig. 3.9: First-order fitting of S809 airfoil.

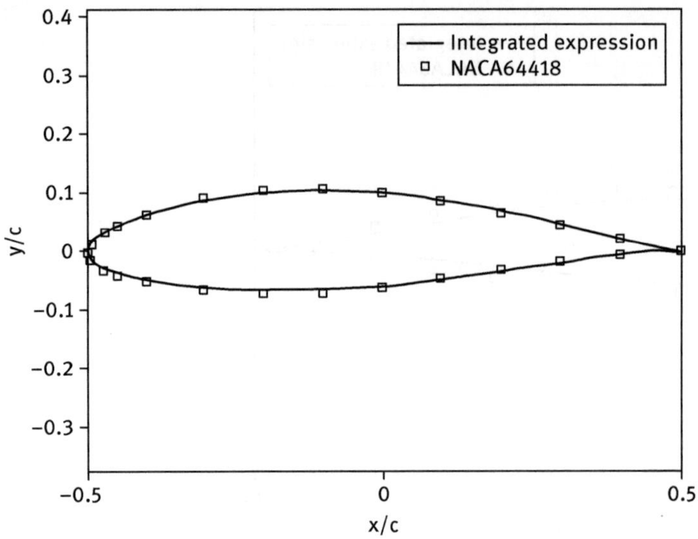

Fig. 3.10: Second-order fitting of NACA 64418 airfoil.

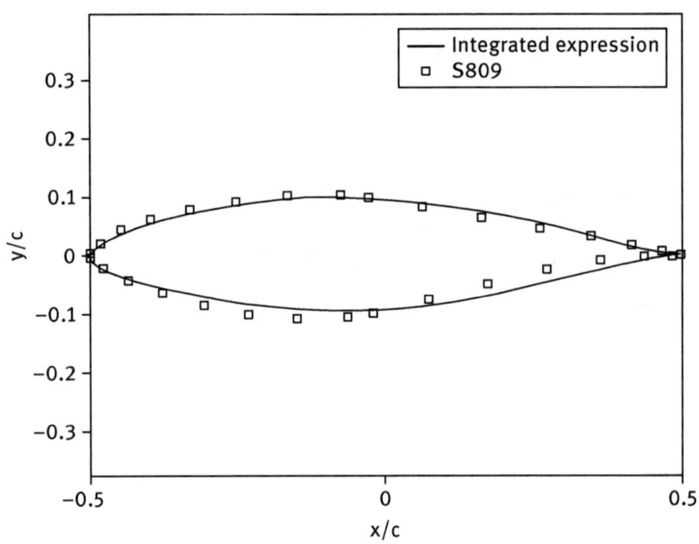

Fig. 3.11: Second-order fitting of S809 airfoil.

3.5.2 Second-order fitting

The second-order term was used. The coefficients obtained are shown in Tab. 3.2.

Tab. 3.2: Second-order fitting coefficients.

	NACA 64418	S809
a_1	0.54913	0.47162
b_1	0.04321	0.00282
a_2	−0.23045	−0.20091
b_2	−0.16144	−0.08267

Fig. 3.10 and 3.11 were obtained respectively by putting the coefficients into the integrated equation. Now it can be seen that the airfoil is approaching the actual airfoil, and the difference is decreasing.

3.5.3 Third-order fitting

And then the third-order term is used. The obtained coefficients are shown in Tab. 3.3.

Tab. 3.3: Third-order fitting coefficients.

	NACA 64418	S809
a_1	0.89482	1.47486
b_1	0.04960	0.04733
a_2	−0.35665	−0.55641
b_2	−0.35445	−0.65773
a_3	−0.02444	−0.07436
b_3	−0.01001	−0.05212

Fig. 3.12 and 3.13 were obtained respectively by putting the coefficients into the integrated equation. Now it can be seen that the airfoil is almost identical with the existing airfoils.

By the integrated expressions shown above, the versatility of the integrated equation has been well demonstrated. It also proves that airfoil profiles can be changed and controlled by controlling the order of the series. This provides references for the optimization of an airfoil.

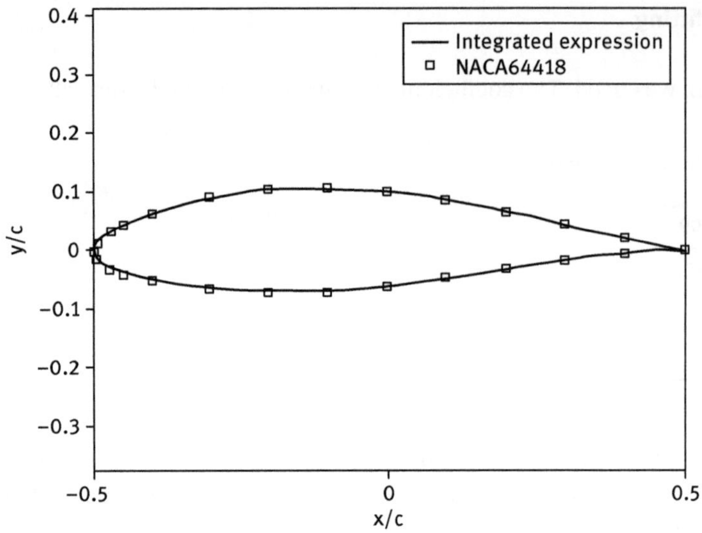

Fig. 3.12: Third-order fitting of NACA 64418 airfoil.

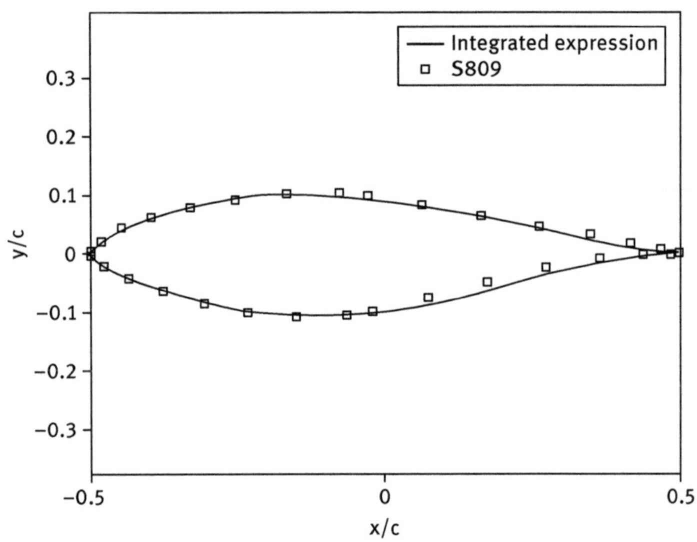

Fig. 3.13: Third-order fitting of S809 airfoil.

3.6 Control equation of shape function

Considering the shape function of an airfoil, which is expressed by a Taylor series, the geometric characteristics of this airfoil expression are studied, including airfoil sharp trailing edge characteristics, horizontal offset characteristics, vertical offset characteristics, and design space, etc.

3.6.1 Characteristics of airfoil sharp trailing edge

Based on the Joukowsky airfoil transformation it is known that the near circle in the z-plane has to pass a transformation singularity $(a, 0)$ to ensure that the airfoil has a sharp trailing edge. This also ensures the closure property of the transformed shape. Taking the corresponding angle θ into equation (3.33), and then taking it into (3.27):

$$\begin{cases} a\rho(0) = a, \\ a\rho(2\pi) - a. \end{cases} \tag{3.42}$$

The second control equation of shape control can be obtained:

$$\begin{cases} C_0 = 1, \\ \sum_{k=1}^{n} 2^k \pi^k C_k = 0. \end{cases} \tag{3.43}$$

3.6.2 Horizontal offset characteristics

The transformed shape is determined not only by the specific transformation equation, but also by the position of the original shape in the z-plane. To make sure that the profile has airfoil characteristics, the original shape must be biased in the direction of the x-axis. Assuming the near circle to offset εa in the negative direction of x, taking the values of θ and horizontal offset into equation (3.33), and then taking it into (3.27):

$$a\rho(\pi) = a\,(1 + \varepsilon). \tag{3.44}$$

The second control equation of shape control can be obtained:

$$\sum_{k=0}^{n} \pi^k C_k = 1 + \varepsilon. \tag{3.45}$$

3.6.3 Vertical offset characteristics

When the original shape is only offset in the x-direction, the transformed shape will still be a symmetric airfoil. If an asymmetric airfoil is wanted (a camber airfoil), the original shape also needs to be offset by a microvalue in the direction of the y-axis. Assuming the near circle to offset Δa in the positive direction of y, the values of θ and vertical offset are taken into equation (3.33), and then into (3.27):

$$ap\left(\frac{\pi}{2}\right) = ap\left(\frac{3\pi}{2}\right). + \Delta a \tag{3.46}$$

The third control equation of shape control can be obtained:

$$\sum_{k=0}^{n}\left(\frac{\pi}{2}\right)^k C_k = \sum_{k=0}^{n}\left(\frac{3\pi}{2}\right)^k C_k + \Delta. \tag{3.47}$$

Combining equations (3.43), (3.45) and (3.47), the control equations of shape function for airfoil design are derived:

$$\begin{cases} C_0 = 1, \\ \sum_{k=1}^{n} 2^k \pi^k C_k = 0, \\ \sum_{k=1}^{n} \pi^k C_k = \varepsilon, \\ \sum_{k=0}^{n} \frac{(3^k - 1)\pi^k}{2^k} C_k = -\Delta. \end{cases} \tag{3.48}$$

3.6.4 Design space

By changing the coefficients of shape function, the offset of x- and y-axis and the shape of the near circle can be changed. Furthermore, the thickness, camber, radius of the leading edge and angle of the trailing edge of the airfoil can also be changed. Hence, airfoil profiles with different shapes and properties can be designed.

Fig. 3.14 shows the changes to airfoils generated by controlling the different offset values of the near circle. In Fig. 3.14 (a), the vertical offset value is fixed as constant $\Delta = 0.0$, while the horizontal offset ε varies from 0.02 to 2.0. Correspondingly, the transformed airfoils are symmetric and change from flat to circle. In Fig. 3.14 (b), the horizontal offset value is fixed as constant, namely $\varepsilon = 0.1$, while the vertical offset Δ varies from 0.0 to 0.6. Then the corresponding airfoils change from a symmetric airfoil with thickness of 0.1 to a big camber airfoil with a camber of 0.13. What can be obtained from the figure is that the airfoil designed by this method can completely cover a nearly circular space. In other words, all the airfoil profiles in this space can be expressed. In

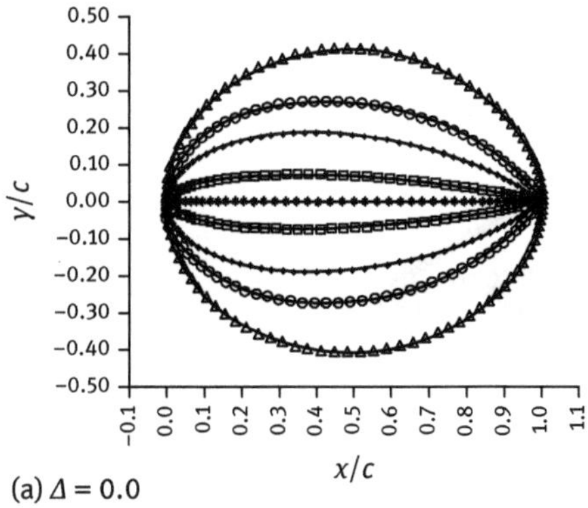

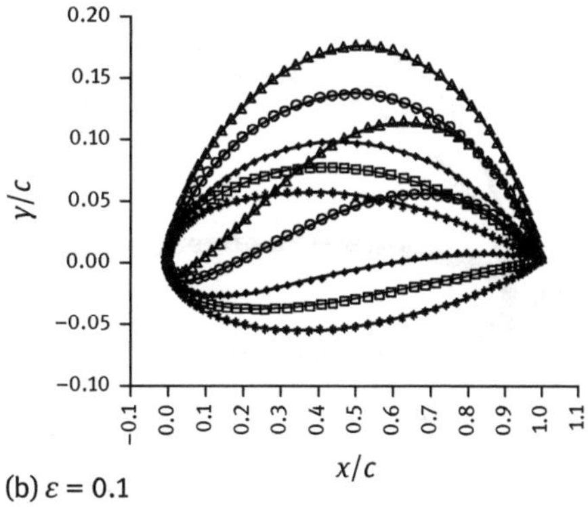

Fig. 3.14: Design space of an airfoil.

Fig. 3.14, c represents the chord length, i.e., the distance between leading edge and trailing edge, x/c represents the ratio of horizontal coordinate and chord length, y/c represents the ratio of vertical coordinate and chord length.

3.7 Convergence analysis of integrated expression of airfoils

Based on the research on the shape function control equation (Section 3.6), 30 kinds of popular airfoil data were used in a parametric integration study. These airfoils include NACA 4-digit airfoil, NACA 63, 64, 65 series airfoil family, FFA-W3 airfoil family, FX airfoil family and DU airfoil family [7, 11, 106].

Fig. 3.15 shows the integrated expression of several typical airfoils which are expressed by the parametric method, where OP (the Order of Polynomial) represents the order of polynomial. It was shown that this method has good integration property for different kinds of airfoils.

The study of integration convergence characteristics of the parameterized airfoil profile includes two aspects: the first one is geometric convergence property, and the second one is aerodynamic property. Here, we take the FX 66-S196-V1 airfoil as an example to study the convergence characteristics of the parameterized integrated expression. Using the least squares method to fit the airfoil shape function, we study the geometric convergence of 13 airfoils and aerodynamic convergence with OP ranging from 3 to 15.

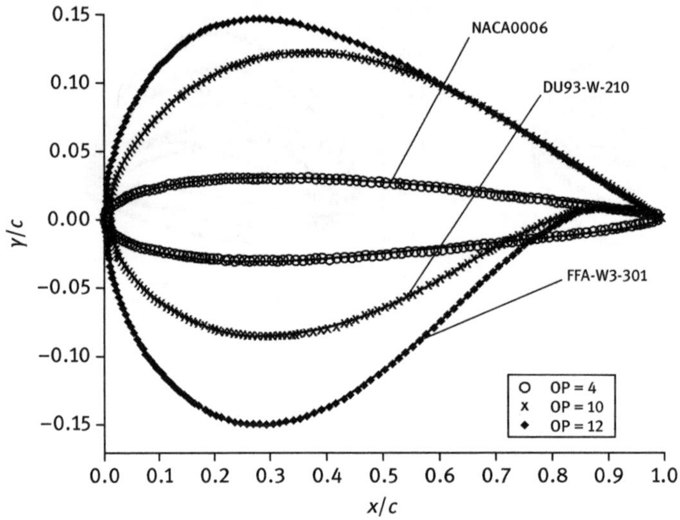

Fig. 3.15: Parametric integration for several kinds of airfoils.

3.7.1 Convergence characteristic of airfoil shape

Fig. 3.16 shows the airfoil profile and shape function curve of FX 66-S196-V1. This airfoil is a laminar flow airfoil with thickness of 19 designed by Althauls and Wortmann. This characteristic of the airfoil is that the maximum velocity positions on both upper and lower airfoil surface are very rearward within the vicinity of designed attack angle so that it has a low drag coefficient and high lift-to-drag ratio (the maximum value reaches 157) in working condition. So it is widely used in wind turbine blades.

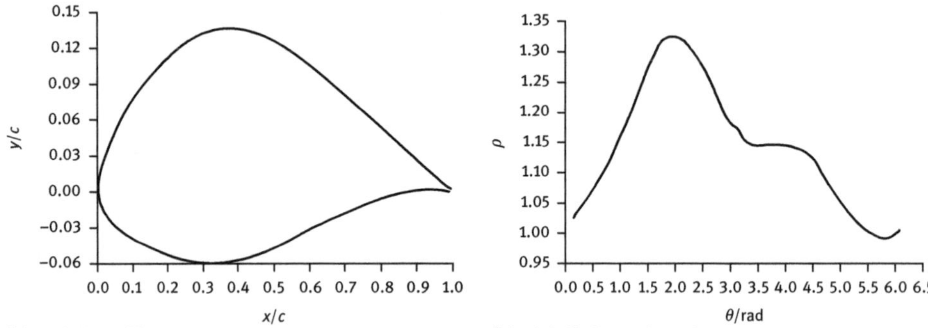

(a) Airfoil profile (b) Airfoil shape function

Fig. 3.16: The airfoil profile and shape function of FX 66-S196-V1.

Geometric convergence analysis includes residual analysis and standard deviation analysis between different orders' fitted shape functions curve and original shape function curve; residual analysis and standard deviation analysis between fitted airfoil profiles with original airfoil profile.

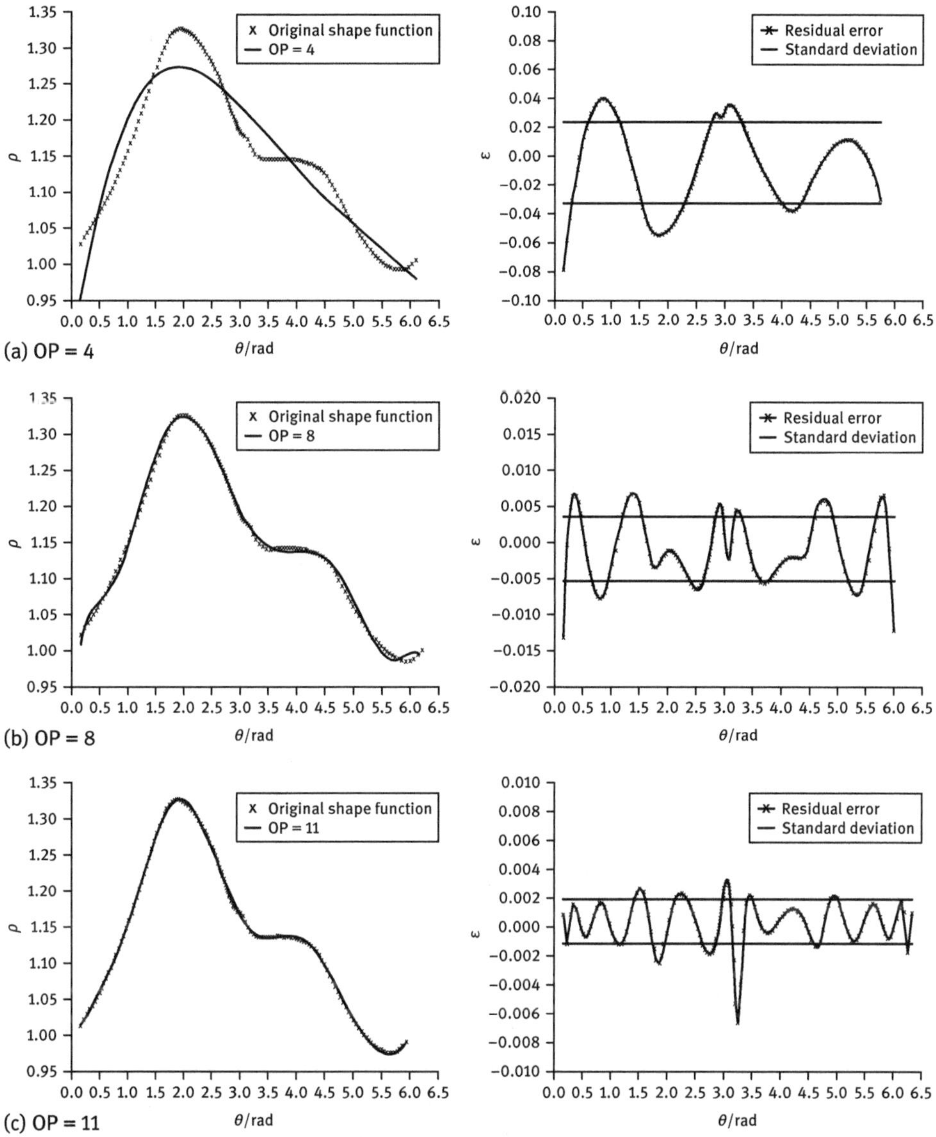

Fig. 3.17: Convergence characteristics of shape function of FX 66-S196-V1.

Fig. 3.17 shows the comparison between fitted airfoils of different orders and the original airfoil. The comparison of residuals and standard deviation are also shown in Fig. 3.17. With the increasing fitting order, the fitting function gradually approaches the original function.

Tab. 3.4 shows the standard deviation of shape functions of several different airfoils which were integrated in different order. The standard deviation reflects the quality of the fitting in different orders (convergence property). The airfoils are ranked according to the convergence property from top to bottom. Among them, the NACA 0006 airfoil belongs to the symmetric airfoils, and has the best convergence property. The S809 airfoil has the worst convergence property, when OP = 15, the corresponding shape function's standard deviation is 1.51×10^{-3}.

Tab. 3.4: Standard deviation of shape functions ($\times 10^{-3}$).

Airfoil	**Order of polynomial (OP)**					
	4	5	6	7	8	9
NACA 0006	1.070	1.070	0.624	0.624	0.313	
DU 93-W-210	15.700	15.400	11.800	11.800	3.810	3.350
FX 66-S196-V1	29.900	28.100	14.100	13.800	4.480	4.470
FFA-W3-301	21.900	18.800	14.900	14.900	8.090	6.630
NACA 63215	12.800	12.400	6.310	5.870	2.510	2.220
NACA 4412	12.800	5.3900	5.360	4.350	3.860	3.820
LS(1)-0413	16.000	12.700	11.900	5.850	5.750	5.740
S809	38.000	37.300	22.300	22.20	10.600	9.010
	10	11	12	13	14	15
NACA 0006	0.305	0.305	0.223	0.223	0.192	0.192
DU 93-W-210	1.350	1.280	1.250	1.230	1.020	1.000
FX 66-S196-V1	1.970	1.610	1.550	1.540	1.36	1.350
FFA-W3-301	3.410	2.690	1.310	1.280	0.829	0.786
NACA 63215	2.080	1.780	1.770	1.600	1.540	1.380
NACA 4412	3.670	2.610	1.770	1.700	1.600	1.470
LS(1)-0413 5.320	2.670	2.570	2.570	2.560	1.330	
S809	5.980	4.500	2.600	2.230	1.830	1.510

The study found that, for the FX 66-S196-V1 airfoil, the fitted airfoil was relatively consistent with the original airfoil when OP = 8, and almost completely coincided with the original airfoil when OP = 11. The standard deviation of the upper airfoil surface is 4.044×10^{-4}, while that for the lower airfoil surface is 2.874×10^{-4}. The comparison of coordinates, residuals, and standard deviation are shown in Fig. 3.18.

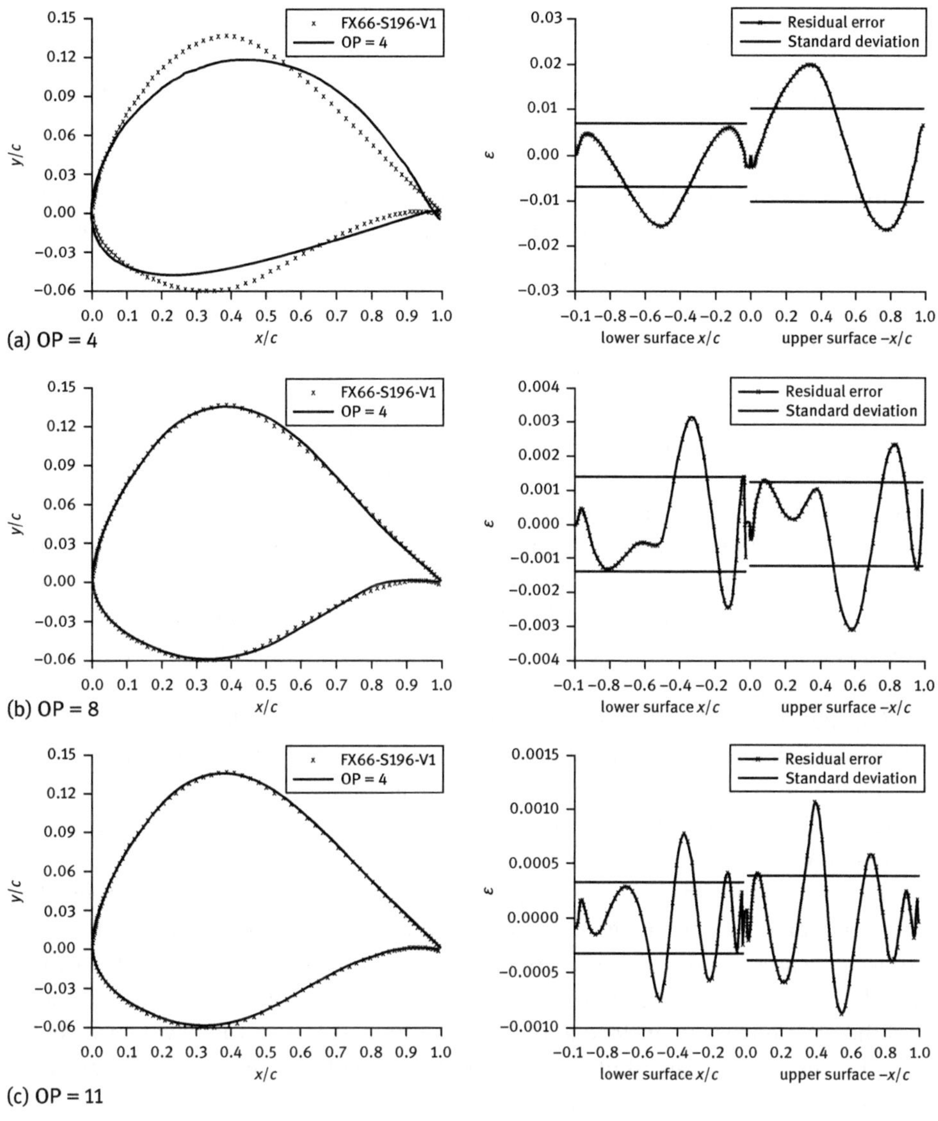

Fig. 3.18: Convergence characteristics of airfoil shape of FX 66-S196-V1.

3.7.2 Convergence characteristic of airfoil aerodynamic performance

To test the aerodynamic properties of the fitted airfoil, the aerodynamic performance of the airfoil is analyzed and studied using RFOIL software in different conditions [4, 5, 109–112].

In this section, RFOIL software is applied to analyze the target airfoil and fitted airfoil with different orders. Fig. 3.19 shows the lift coefficients (C_L) and drag coefficients (C_D) in the attack of angle (α). The calculation results and experimental results [106] are compared in the same conditions. It can be seen from Fig. 3.19 that, for the FX 66-S196-V1 airfoil, the results calculated by RFOIL are identical with test values for lift coefficients and the linear stage of drag coefficients. And the results also show that RFOIL can simulate airfoil stall conditions well, which means that stall conditions have good convergence. Since entering the stall condition, instability and separation conditions of the gas flow will greatly influence the characteristics of airfoil aerodynamics. It is difficult to accurately simulate the process of airfoil stall with the currently used numerical calculation methods. So, after entering the stall area, the calculated value deviates from the test value, but the two trends converge. In Fig. 3.19, with increasing fitting order, the aerodynamic calculation results of the fitted airfoil approach the target airfoil. The lift characteristics and drag characteristics of the 11 order fitted airfoil agree well with the target airfoil calculation results.

For the FX 66-S196-V1 airfoil, the pressure distribution under different angles of attack was calculated with variations in the order of fitted airfoil from 4 to 15. In the calculation process, aerodynamic conditions are maintained completely in accord with the target airfoil. The pressure distributions at the angles of attack –2°, 2°, 5° and 9°, calculated with $OP = 4$, $OP = 8$ and $OP = 11$ are given in Fig. 3.20.

For $OP = 4$, the difference in the upper surface pressure distribution between fitted airfoil and target airfoil is large. However, the difference is relatively small on the lower surface. The pressure distribution variation trend is in accordance with the target airfoil. For $OP = 8$, the difference in pressure distribution on the upper surface is reduced. And the difference is very close to that of the target airfoil on the lower surface. For $OP = 11$, the pressure distribution agrees quite well with the target airfoil at the attack angles of –2°, 2° and 5°. But there are some slight differences at 9° attack angle when entering the stall condition.

The study found that, for most of the commonly used wind turbine airfoils, the 11 order approximation of the airfoil was in good agreement with the target airfoil in terms of aerodynamic performance. And the differences in lift and drag characteristics are very small in the same working conditions. Hence, the target airfoil can be substituted with an 11 order approximation for the design and analysis of wind turbines.

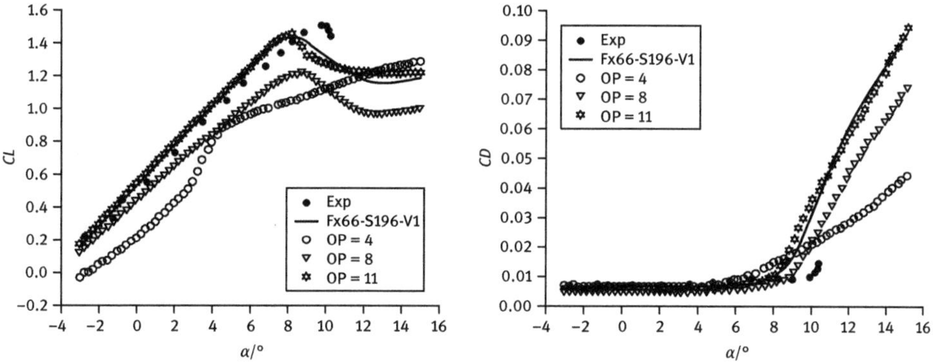

(a) The comparison of lift coefficient

(b) The comparison of drag coefficient

Fig. 3.19: Comparison of aerodynamic performance between calculation and test.

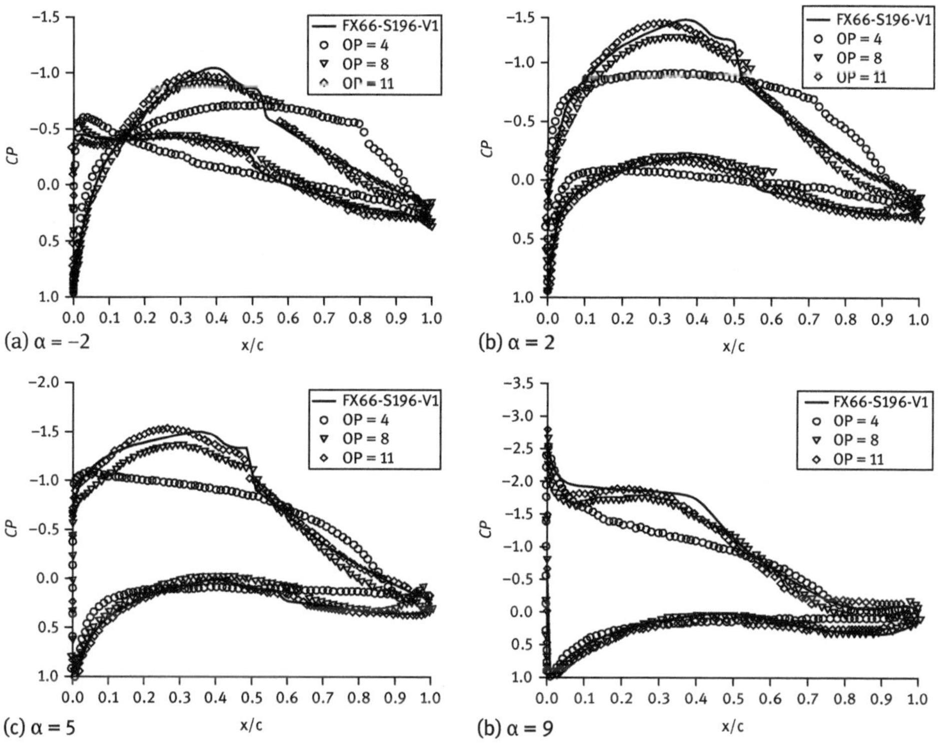

(a) α = −2

(b) α = 2

(c) α = 5

(b) α = 9

Fig. 3.20: Comparison of pressure distribution of integrated airfoil.

3.8 Chapter conclusions

In this chapter, on the basis of research on wind turbine airfoil geometry, an airfoil general integrated expression equation is proposed based on conformal mapping and series theory. Then the airfoils with different shapes can be obtained by varying the coefficients of the control equation. For the airfoils expressed by a trigonometric function, three kinds of typical airfoil profiles were achieved by solving the equation coefficients. Hence the feasibility of this method was verified. Finally, the versatility of the integrated expression was studied theoretically by the expression and comparison of two kinds of commonly used airfoils. According to the Taylor series representation of the airfoil shape function, the coverage property in airfoil design space was discussed. It was shown that the airfoil profile designed by this new method can be completely covered with a circular space. Then, all airfoil profiles in that space can be expressed. Furthermore, the geometric and aerodynamic properties for the wind turbine airfoil parametric integrated expression method were studied, and the results show that the geometry and aerodynamic characteristics of the airfoil will gradually approach the target airfoil with increasing the airfoil fitting order. Therefore, the two kinds of airfoil integrated equations proposed have broken the limit of inherent characteristics of the original airfoil. This is not for the improvement of design and modification of the mathematical model of airfoil profiles, but to start from the root of fundamental element of airfoil profile property, airfoil profile shape function, to study and find an airfoil mathematical model with good properties. This broadens the ideas of wind turbine airfoil design. It is also highly significant theoretically and very important for industrial applications.

4 Theory of parametric optimization for wind turbine airfoils

4.1 Introduction

It is necessary to accurately predict the energy capture of wind turbines and this is influenced by the aerodynamic design of the wind turbine's airfoils and blades. One important key element in the aerodynamic design of the wind turbine blade is the use of specially designed airfoils to improve the power coefficient and to reduce the cost of energy. It is well known that the aerodynamic design of wind turbine blades is mainly concerned with airfoil design. Hence, this chapter focuses on establishing a mathematical model and a design method for optimizing wind turbine airfoils with high aerodynamic performance.

Since the 1980s, many wind turbine airfoils have been designed. In the mid-1980s, Tangler and Somers from the National Renewable Energy Laboratory proposed the NREL airfoil series [113–116]. The DU airfoil series [5] with thickness from 15 to 40 % was designed by the Wind Energy Research Institute of Delft University, in association with the Energy research Centre of the Netherlands (ECN) and the National Aerospace Laboratory of the Netherlands (NLR) using the mixed inverse design method. The structural and geometric compatibility and deep stall features were considered in the design process for the DU airfoil series. Compared with traditional airfoils, the DU airfoil families exhibited better aerodynamic characteristics. From the mid-1990s, four airfoil families (Risø-A1, Risø-P, Risø-B1 and Risø-C1) were designed by Risø National Laboratory in Denmark [33]. These airfoils can be adapted to different power sizes, operating conditions and control modes for wind turbines. Furthermore, the Swedish scholar Bjork proposed the FFA airfoil series [10]. These wind turbine airfoils not only improve the efficiency of wind energy, but can also be adapted to different operating conditions.

Most wind turbine airfoils have been designed by using traditional inverse methods where the airfoil surface flow was prescribed at specified operational conditions, and a shape was achieved which could generate these surface conditions. This design method is indirect and time consuming and it makes it more difficult to obtain an airfoil with optimal performance from a variety of possible airfoils [106, 117, 118]. Therefore, it is necessary to find a more direct and efficient design method to design wind turbine airfoils that is based on the general parametric representation function (Shape Function). The general characteristics of airfoils are studied in this chapter. Finally, by controlling the coefficients of the Shape Function, different airfoils with different characteristics are designed.

DOI 10.1515/9783110344387-005

4.2 Design requirements of wind turbine airfoils

The design of wind turbine airfoils is the primary task of rotor design. Airfoils with high lift coefficient and high lift-to-drag ratio can improve the power coefficient of the rotor and reduce the cost of energy for the wind turbine. However, with extensive application of wind energy, the noise emitted by wind turbines will affect the environment and people nearby. Therefore, the design of high performance wind turbine airfoils with a low noise level has become more and more important.

The design of wind turbine airfoils should meet aerodynamic and structural requirements. However, the two requirements have many contradictions, such as high lift-to-drag ratio and large thickness of airfoil, high maximum lift and low leading edge roughness sensitivity, high lift and low noise etc. [119]. Therefore, the design process of airfoils means finding a good balance between these conflicting requirements. Additional requirements, such as the power range, control and spanwise position on blade, should also be considered in airfoil design. For small wind turbines, generally thinner airfoils with a bigger chord are adopted. For large wind turbines, thicker airfoils are favored to reduce rotor weight and cost. For fixed-pitch stall wind turbines, the maximum lift coefficient of the airfoil at the blade tip needs to be limited to ensure reliable stall control. For modern wind turbines, which usually adopt variable pitch control, the airfoils should have larger lift-to-drag ratio, especially before stall, to let the rotor reach its maximum power coefficient at all wind speeds. The requirements of airfoils at different spanwise locations are summarized in Tab. 4.1 [6]. Airfoils have higher lift-to-drag ratio at the tip and most of the noise is emitted there, which means that the airfoils must be controlled at low noise level. The thick airfoils adopted at the blade root can increase the structural stiffness and reduce the weight of the blade. For airfoils which are to be applied in the main power production area (near the 75 % blade span), high lift-to-drag ratio, low roughness sensitivity, smooth transition to stall and a larger lift at stall are required.

Tab. 4.1: Requirements of airfoils at different spanwise locations.

	Blade root	Middle span	Blade tip
Maximum relative thickness δ	> 27	27–21	21–15
Structural property S	Very important	Important	Less important
Geometrical compatibility G	Important	Important	Important
Roughness sensitivity R	–	–	Very important
Design lift coefficient C_L	–	Less important	Very important
Maximum $C_{L,max}$ and deep stall performance	–	Less important	Very important
Low noise N (dB)	–	–	Very important

4.2.1 Structural and geometric compatibility

The maximum relative thickness of the airfoil is the most important structural parameter. In addition, the chordwise position of the maximum relative thickness and the shape of the airfoil are also important factors. The forward or backward shift of chordwise position of the maximum relative thickness can have a great influence on the structural characteristics of the airfoil. When it moves forward (towards the leading edge), the airfoil shear center and gravity center move towards the aerodynamic center, thus easing aeroelastic instabilities. However, the distance to the trailing edge becomes larger, resulting in a larger maximum stress. In addition, the shape and position of the cap bearing the load can exert *N* important influence on the shape of the airfoil near the maximum relative thickness [120].

The geometric compatibility of airfoils within the same airfoil series is of great importance. a smooth transition of airfoil thickness can ensure the smoothness of the blade profile so as to minimize the influence of the 3D effects. Moreover, good geometric compatibility also requires that the radius of curvature (at the airfoil leading edge area and the trailing edge area of pressure surface) is smooth and continuous.

4.2.2 Insensitivity of the maximum lift coefficient to leading edge roughness

Dust, insect debris, manufacturing error, blade corrosion and other factors can usually lead to increased roughness of the blade's leading edge. This can cause an earlier transition from laminar turbulence in the advancing airfoil boundary layer, an increase in boundary layer thickness and an increase in an adverse pressure gradient near the trailing edge. Then early airflow separation will happen in this region leading to a decline in the maximum lift coefficient. The early transition also leads to increased drag, a decreased lift-to-drag ratio and finally a decreased power coefficient.

In order to make the maximum lift coefficient insensitive to the roughness of the leading edge, a rational leading edge shape should be designed to ensure that the transition point on the suction surface is as close to the leading edge as possible when the critical angle of attack is reached. This will make sure that the flow on the leading edge is either in the transition state or in turbulence state, which means the variation in the lift coefficient is small. But this will also limit the maximum lift coefficient [120].

4.2.3 Design lift coefficient

The choice of the design lift coefficient is closely related to blade solidity, the design point of the wind turbine and the requirements of off-design conditions. Although wind turbines often work in off-design conditions such as stall, yaw, and extreme turbulence, the performance of wind turbines during the off-design conditions is more relevant to the control strategies. In most cases, the design lift coefficient is expected to reach the maximum lift coefficient so as to ensure a large lift-to-drag ratio [120].

4.2.4 The maximum lift coefficient and deep stall characteristics

For the blade root, the airfoil with larger maximum lift coefficient is expected so as to reduce the solidity of the blade. As the power contribution of the blade root is very limited, so the higher lift coefficient cannot be obtained at the expense of reducing the structural strength of the blade root. A vortex generator can be used to improve the lift coefficient, if possible, with the combination of Gurney flaps in applications [120].

For airfoils used at blade tip, the maximum lift coefficient is influenced by the structure and load requirements, power control strategy and deep stall characteristics. Generally speaking, it is not easy to obtain desired values. If a smaller COE of a wind turbine is expected, it has been shown that the airfoil at tip should have a higher maximum lift coefficient as long as the deep stall characteristics changes smoothly [119]. For the traditional stall control mode, the deep stall characteristics may increase the risk of stall induced vibration, which will lead to higher dynamic load. For a conventional pitch regulated wind turbine, running in stall condition is to be strictly avoided and a higher maximum lift coefficient is also reasonable.

4.2.5 Low noise

The noise of wind turbine mainly comes from the wide band noise when the turbulent flow is passing the trailing edge of the airfoil. This type of noise mainly depends on the wind speed and on the thickness and shape of the boundary layer of the trailing edge region. Inflow velocity is the main factor which is determined by the rotor speed. However, the trailing edge boundary layer thickness is closely related to airfoil shape. So airfoil design should reduce the boundary layer thickness as much as possible to reduce the noise [120].

4.3 Single object optimization of wind turbine airfoils

4.3.1 Objective function

The lift-to-drag ratio is the most important performance index for wind turbine airfoils, which can greatly affect the efficiency of the rotor and the performance of the whole wind turbine. For airfoil design, it is generally expected that lift is large and drag is very small. So the lift-to-drag ratio of the airfoil is adopted as the objective function, thus the optimization object is established

$$f(x) = \max(C_l/C_d), \tag{4.1}$$

where C_l is the lift coefficient of the airfoil, and C_d is the drag coefficient of the airfoil.

4.3.2 Design variables

According to the common characteristics of the airfoil profile, the general function $\varphi(\theta)$ based on the trigonometric series was described in Chapter 3:

$$\begin{aligned}\varphi(\theta) = {} & a_1(1-\cos\theta) + b_1\sin\theta + a_2(1-\cos\theta)^2 + b_2\sin^2\theta + \cdots \\ & + a_k(1-\cos\theta)^k + b_k\sin^k\theta + \cdots, \quad k = 1, 2, \ldots, \infty.\end{aligned} \tag{4.2}$$

The first 6 coefficients of the shape function $\varphi(\theta)$ are selected as the design variables.

$$X = (x_1, x_2, x_3, x_4, x_5, x_6). \tag{4.3}$$

The variation ranges of the variables are determined according to the shape of the airfoil.

4.3.3 Design constraints

In the process of optimization, when the value of each variable exceeds a certain range, the representation will no longer have the shape characteristics of the airfoil, so the scope of the variable is bound

$$X_{\min} \le X \le X_{\max}. \tag{4.4}$$

The layout of the wind turbine airfoil along the spanwise direction of the blade is an important part of the aerodynamic design of wind turbine blades. Near the blade root, the high strength and stiffness of the airfoil are the main requirements which lead to the selection of a thicker airfoil, followed by the inferior requirement on aerodynamic performance. The main power production region of the blade is located around 75 % along the blade radius, where airfoils with maximum relative thickness of around 18 % are usually adopted. High lift-to-drag ratio and low sensitivity to roughness are required. Therefore, three kinds of airfoil with maximum relative thicknesses of 15 %, 18 % and 21 % in the first half of the blade were selected in this optimization. In the optimization of the airfoil, the constraints are applied respectively

$$t/c = 15\,\%, 18\,\%, 21\,\%, \tag{4.5}$$

where t is the maximum thickness of airfoil, c is the airfoil chord.

For the requirements concerning the structure and strength characteristics of the airfoil, in addition to the maximum relative thickness, the chordwise lactation of maximum thickness should be considered. Generally, the aerodynamic center locates at about $\frac{1}{4}$ of the chord. Therefore, the maximum camber airfoil position is controlled at

$$0.2 \leq x/c \leq 0.3. \tag{4.6}$$

4.3.4 Optimization method with MATLAB

The optimization objective is selected of high lift-to-drag ratio so as to improve the efficiency of wind turbine. It is a typical nonlinear single objective optimization problem with multiple constraints.

$$\min_{\mathtt{x}} \mathtt{f(x)} \quad \text{such that} \quad \begin{cases} \mathtt{C(x)} \leq 0, \\ \mathtt{Ceq(x)} = 0, \\ \mathtt{A} \cdot \mathtt{x} \leq \mathtt{b}, \\ \mathtt{Aeq} \cdot \mathtt{x} = \mathtt{beq}, \\ \mathtt{lb} \leq \mathtt{x} \leq \mathtt{ub}, \end{cases} \tag{4.7}$$

where $\mathtt{x}$, $\mathtt{b}$, $\mathtt{beq}$, $\mathtt{lb}$, $\mathtt{ub}$ are vectors, $\mathtt{A}$, $\mathtt{Aeq}$ are matrices, $\mathtt{C(x)}$, $\mathtt{Ceq(x)}$ are the functions of the return vector, $\mathtt{f(x)}$ is the objective function. The function `fmincon` in MATLAB can be used

$$\mathtt{x} = \mathtt{fmincon(fun, x0, A, b, Aeq, beq, lb, ub, nonlcon, options,)} \tag{4.8}$$

where `fun` is the objective function, `x0` is the initial value. The role of `nonlcon` is to calculate the nonlinear inequality constraint $\mathtt{C(x)} \leq 0$ and equality constraint $\mathtt{Ceq(x)} = 0$ using the vector $\mathtt{x}$.

Based on the above mentioned optimization model, the shape function equations of the airfoil shape were solved. The aerodynamic characteristics of the airfoil were calculated with the XFOIL software, which is fast and widely used. In order to save time and directly control the lift and drag characteristics and the shape parameters of the airfoil, shape function equations are coupled with XFOIL solver.

4.3.5 Optimized results

Three airfoils were obtained through optimization. They are CQU-DTU-A15, CQU-DTU-A18 and CQU-DTU-A21 with relative thicknesses of 15 %, 18 % and 21 % respectively. The main aerodynamic parameters of the airfoils are listed in Tab. 4.2.

Tab. 4.2: Aerodynamic parameters of the airfoils.

Maximum relative thickness	Chordwise location of maximum camber	Reynolds number $Re \times 10^6$	Maximum lift coefficient C_l	Maximum lift-to-drag ratio (C_l/C_d)
0.15	0.25	1.6	1.86	143.92
0.18	0.25	1.6	1.87	150.09
0.21	0.23	1.6	1.96	130.10

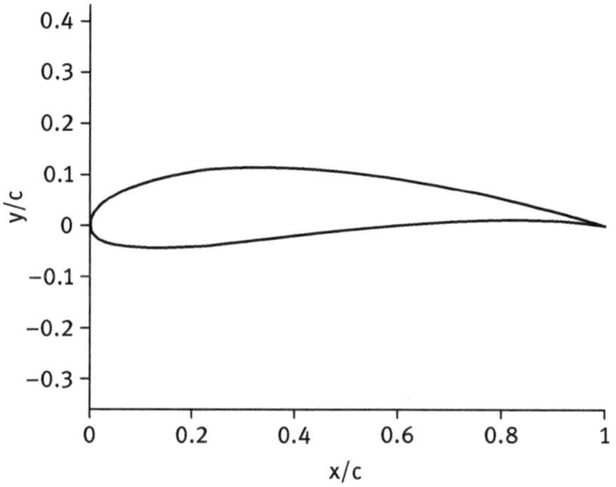

Fig. 4.1: The optimized airfoil with maximum thickness of 15 %.

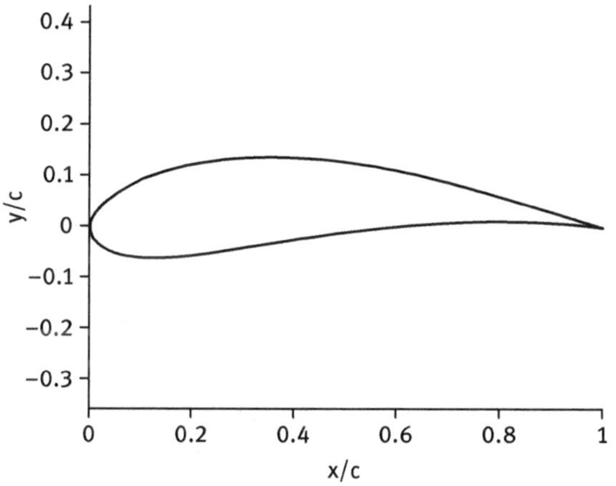

Fig. 4.2: The optimized airfoil with maximum thickness of 18 %.

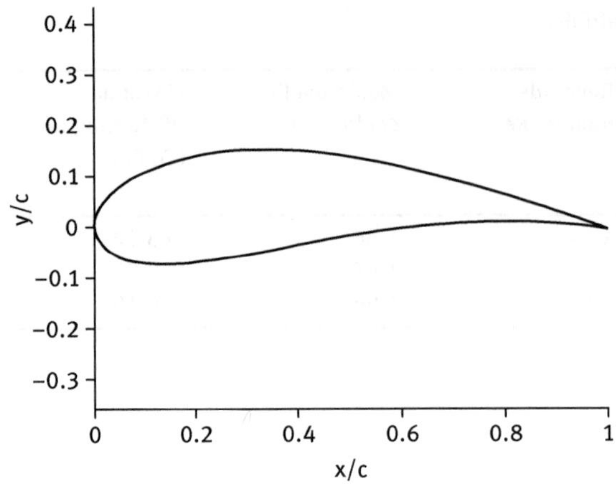

Fig. 4.3: The optimized airfoil with maximum thickness of 21 %.

The shape of CQU-DTU-A15, CQU-DTU-A18 and CQU-DTU-A21 airfoils are respectively shown in Fig. 4.1–4.3. The maximum thicknesses of CQU-DTU-A15 and CQU-DTU-A18 are both at 25 % of chord. When the Reynolds number $Re = 1.6 \times 10^6$, the maximum lift coefficient of 1.86 is achieved at AOA of 18° and the maximum lift-to-drag ratio reaches 143.92 at AOA of 6.5° for the CQU-DTU-A15 airfoil. When the Reynolds number $Re = 1.6 \times 10^6$, the maximum lift coefficient of CQU-DTU-A15 is also achieved at AOA of 18° with a value of 1.87 and the maximum lift-to-drag ratio of 150.09 is reached at AOA of 5.5°. However, the maximum thickness of the CQU-DTU-A21 airfoil is located at 23 % of chord. When the Reynolds number is $Re = 1.6 \times 10^6$, the airfoil has maximum lift coefficient of 1.96 (at AOA of 18°) and maximum lift-to-drag ratio of 130.10 (at AOA of 6°).

4.3.6 Roughness sensitivity of the optimized airfoils

The surface roughness of the airfoil can exert an important influence on the aerodynamic characteristics of the airfoil. The boundary layer transition position will move towards the leading edge under the roughness condition. It will cause the increase of boundary layer thickness, the increase of an adverse pressure gradient near the trailing edge, the decrease of effective camber and the decrease of maximum lift coefficient. In addition, surface roughness can convert the laminar boundary layer into a turbulent boundary layer which leads to an increase in drag. Therefore, for blade design, it is necessary to select the airfoil with low roughness sensitivity to ensure the output power of the rotor.

To simulate the surface roughness of an airfoil in XFOIL with Reynolds number of $Re = 1.6 \times 10^6$, the fixed transition point was set at 5 % chord on the upper surface of the airfoil and the at 10 % chord on the upper surface. The comparison of lift coefficient and drag coefficient under smooth and rough conditions for CQU-DTU-A15 are shown in Fig. 4.4. It can be seen that the lift coefficient of the rough airfoil only decreases slightly compared to the smooth airfoil and the drag coefficient increased a little, which means that the roughness sensitivity of the airfoil is low. The comparison of lift coefficient and drag coefficient for CQU-DTU-A18 is shown in Fig. 4.5. When compared to the smooth surface of the airfoil, the decrease in the lift coefficient and the increase in the drag coefficient are larger in the AOA range from 5° to 15°, which means that the roughness sensitivity of the CQU-DTU-A18 airfoil is higher than that of CQU-DTU-A15. The comparison of lift coefficient and drag coefficient for CQU-DTU-A21 is shown in Fig. 4.6. The decrease in the lift coefficient and the increase in the drag coefficient are obvious at all AOA. Compared with CQU-DTU-A15 and CQU-DTU-A18, the CQU-DTU-A21 airfoil is the most sensitive to roughness, which is because the thickness of the CQU-DTU-A21 airfoil is the largest and is more likely to cause flow separation.

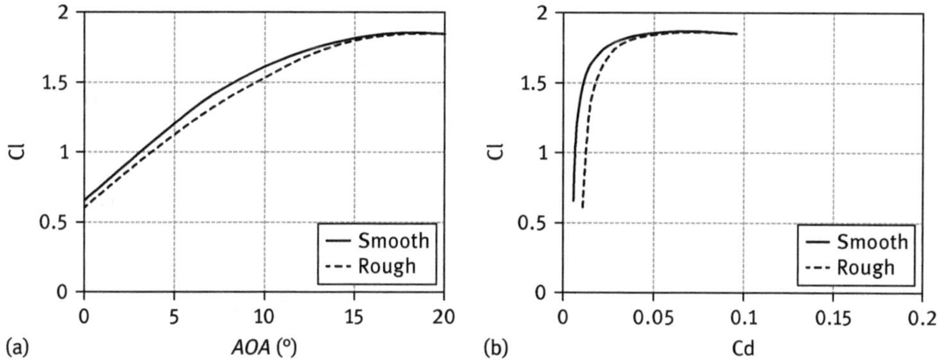

Fig. 4.4: The lift and drag characteristics of the CQU-DTU-A15 airfoil.

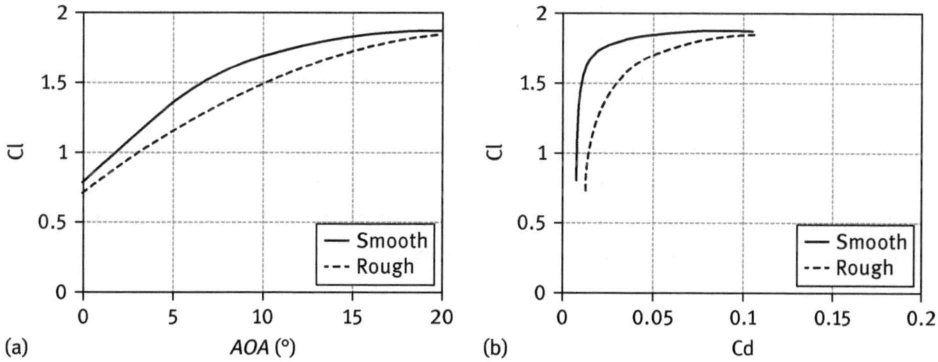

Fig. 4.5: The lift and drag characteristics of the CQU-DTU-A18 airfoil.

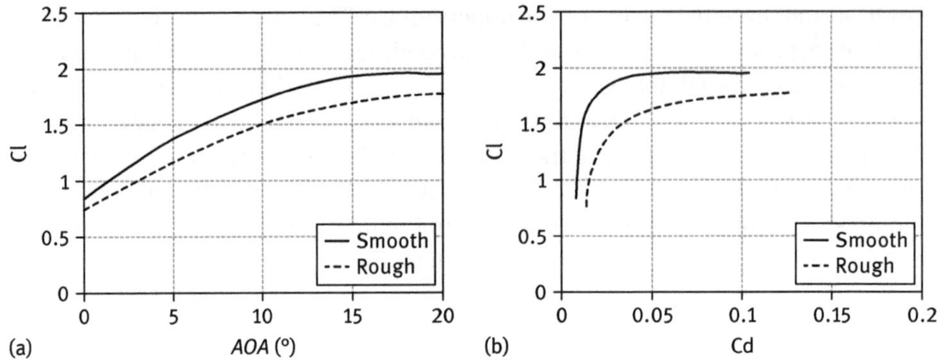

Fig. 4.6: The lift and drag characteristics of the CQU-DTU-A21 airfoil.

The aerodynamic characteristics of the three airfoils with Reynolds number $Re = 6.0 \times 10^6$ are shown in Fig. 4.7–4.9. As can be seen from Fig. 4.7, the stall performance of the smooth CQU-DTU-A15 airfoil is particularly good with increasing lift coefficient until AOA of 20°. Compared with the smooth airfoil, the maximum lift-to-drag ratio of the rough CQU-DTU-A15 airfoil decreased from 161.13 to 119.18. Similar to the CQU-DTU-A15 airfoil, the stall performance of CQU-DTU-A18 and CQU-DTU-A21 airfoils are good too. It can be seen from Fig. 4.8 and Fig. 4.9 that the maximum lift-to-drag ratio of CQU-DTU-A18 and CQU-DTU-A21 are reduced from 181.41 and 101.12 to 165.85 and 93.54 respectively under leading edge roughness.

The aerodynamic performance under turbulent flow is also an important factor to consider. Under Reynolds number $Re = 1.6 \times 10^6$, the comparison of lift coefficient and lift-to-drag ratio under laminar and turbulent conditions for CQU-DTU-A15, CQU-DTU-A18 and CQU-DTU-A21 airfoils are shown in Fig. 4.10–4.12 respectively. The lift coefficients of airfoil CQU-DTU-A15 in laminar and turbulent flow conditions almost remain unchanged, as shown in Fig. 4.10. However, the lift-to-drag ratio in turbulence is reduced from 143.92 to 116.76, because the turbulence will increase the drag of the airfoil. According to the Fig. 4.11, the lift coefficient of the CQU-DTU-A18 airfoil begins to decrease from AOA of 5° with the maximum lift coefficient slightly decreased from 1.87 to 1.82. Meanwhile, the lift-to-drag ratio drops to 12.767 from 150.09. Similarly, as shown in Fig. 4.12, the difference in the lift coefficient of airfoil CQU-DTU-A21 between laminar and turbulent flow conditions is not obvious. However, the lift-to-drag ratio drops to 113.10 from 130.10. In general, the aerodynamic performance of this airfoil series is good in the condition of turbulent flow.

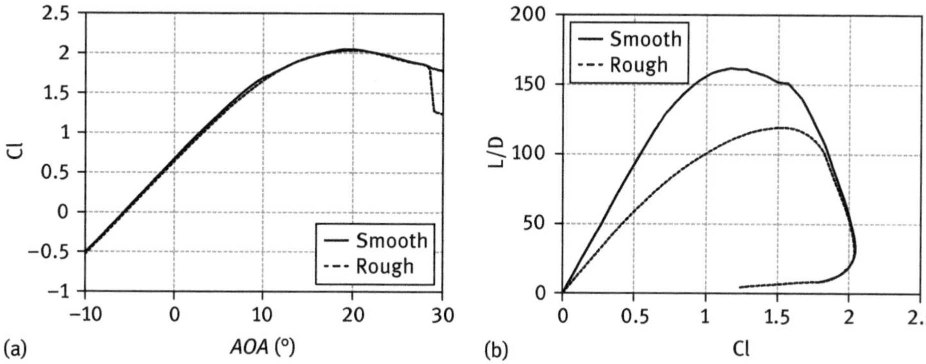

Fig. 4.7: Aerodynamic characteristics of CQU-DTU-A15 airfoil at $Re = 6.0 \times 10^6$.

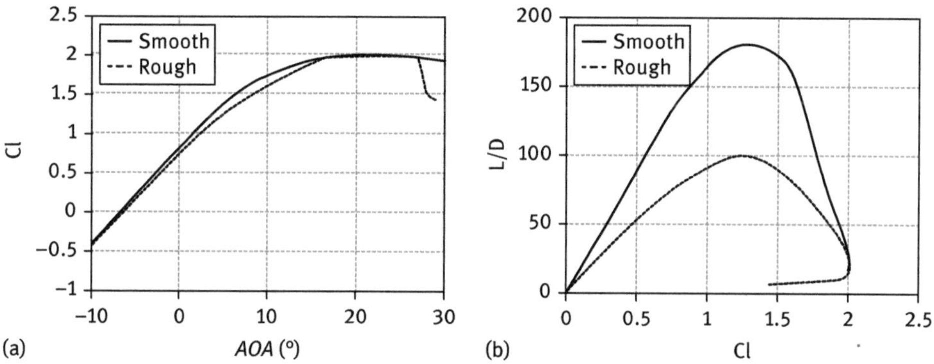

Fig. 4.8: Aerodynamic characteristics of CQU-DTU-A18 airfoil at $Re = 6.0 \times 10^6$.

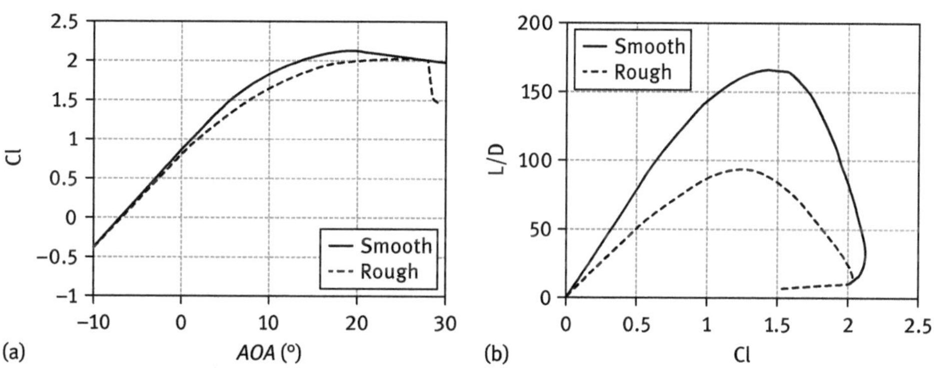

Fig. 4.9: Aerodynamic characteristics of CQU-DTU-A21 airfoil at $Re = 6.0 \times 10^6$.

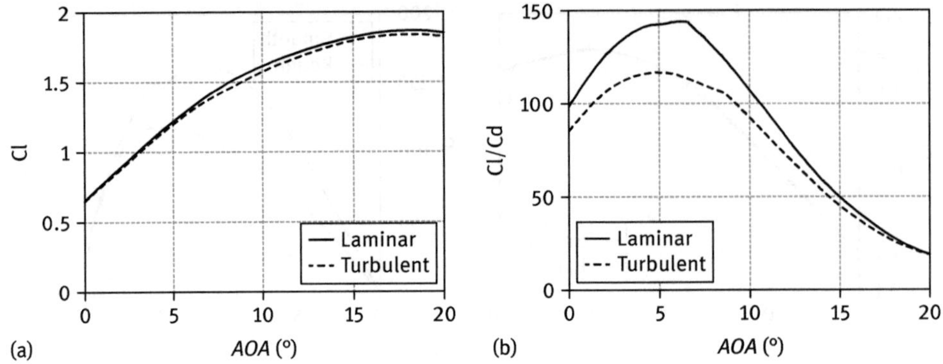

Fig. 4.10: Comparison of performance of CQU-DTU-A15 airfoil under turbulent and laminar flow.

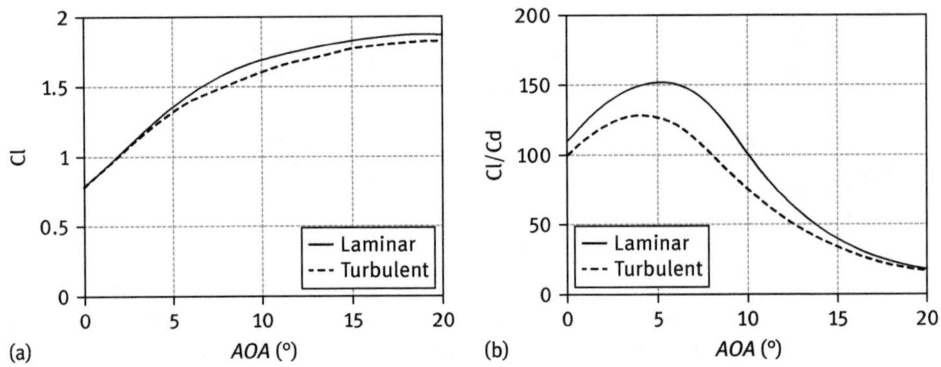

Fig. 4.11: Comparison of performance of CQU-DTU-A18 airfoil under turbulent and laminar flow.

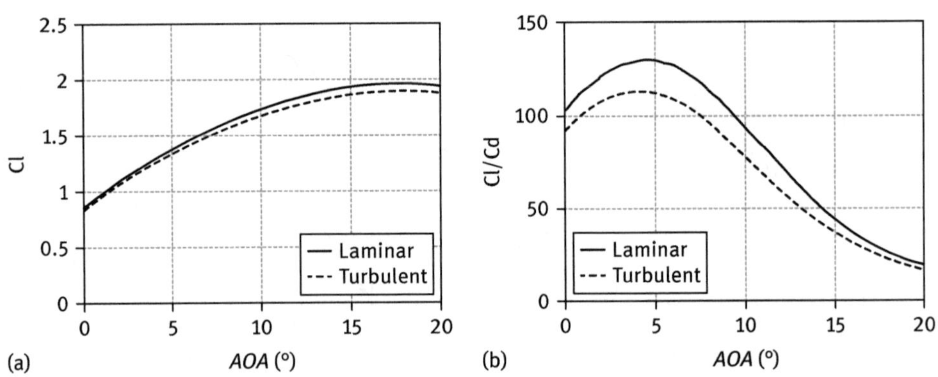

Fig. 4.12: Comparison of performance of CQU-DTU-A21 airfoil under turbulent and laminar flow.

4.3.7 Comparative analysis of the performance of optimized airfoils

In order to illustrate further the aerodynamic performance of the optimized airfoil series, the airfoils were compared with the commonly used Risø, DU, NACA and FFA airfoil series [121, 122]. Under Reynolds number of $Re = 1.6 \times 10^6$, the comparison of lift coefficient and drag coefficient for CQU-DTU-A15 and NACA 63215 airfoils are shown in Fig. 4.13. The maximum relative thickness of the two airfoils are both 15 %. Compared with the NACA 63215 airfoil, the lift coefficient of the CQU-DTU-A15 airfoil is higher and continues to increase until AOA reaches 18°. However, stall will occur at AOA of 16° for the NACA 63215 airfoil. Moreover, the minimum drag coefficient of the two airfoils is similar. However, the drag coefficient of the NACA 63215 airfoil has a sudden jump which will lead to the lift-to-drag ratio decreasing rapidly. It is sure that the CQU-DTU-A15 airfoil must have a higher lift-to-drag ratio.

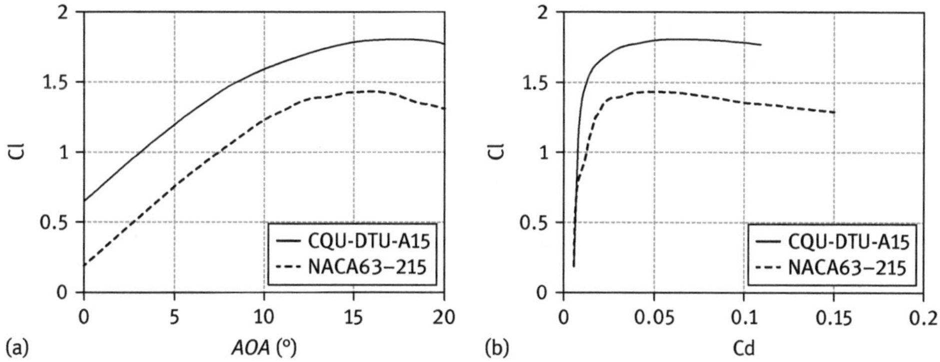

Fig. 4.13: Comparison of performance of CQU-DTU-A15 and NACA 63215 airfoils.

Under Reynolds number of $Re = 1.6 \times 10^6$, the comparison of lift coefficient and drag coefficient for CQU-DTU-A18 and NACA 64418 airfoils is shown in Fig. 4.14. The maximum relative thickness of the two airfoils are both 18 %. The variation trend of lift coefficients of the two airfoils is consistent. However, in the whole range of AOA, the lift coefficient of the CQU-DTU-A18 airfoil is larger than that of the NACA 64418 airfoil by about 0.5. Under the same Reynolds number, the comparison of lift coefficient and drag coefficient for CQU-DTU-A18 and Risø-A1-18 airfoils is shown in Fig. 4.15. Of course, the maximum relative thickness of the two airfoils are both 18 %. It can be seen that the minimum drag coefficient of the two airfoils are similar, but the CQU-DTU-A18 airfoil has a higher lift coefficient.

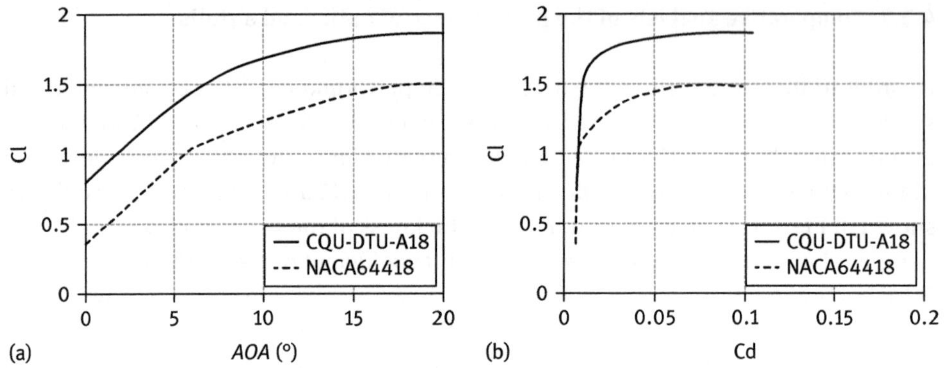

Fig. 4.14: Comparison of performance of CQU-DTU-A18 and NACA 64418 airfoils.

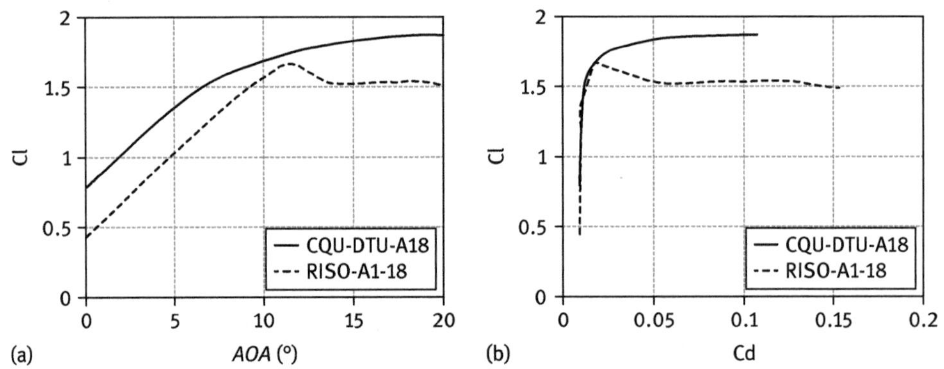

Fig. 4.15: Comparison of performance of CQU-DTU-A18 and Risø-A1-18 airfoils.

With Reynolds number of $Re = 1.6 \times 10^6$, the comparison of lift coefficient and drag coefficient for CQU-DTU-A21 and Risø-A1-21 airfoils are shown in Fig. 4.16. The comparison of lift coefficient and drag coefficient under the same Reynolds number for CQU-DTU-A21 and DU 93-W-210 airfoils are shown in Fig. 4.17. With Reynolds number of $Re = 1.8 \times 10^6$, the comparison for CQU-DTU-A21 and FFA-W3-211 airfoils are shown in Fig. 4.18. As can be seen from these figures, the lift characteristic of the CQU-DTU-A21 airfoil is obviously improved compared with those of the commonly used airfoils. And the lift-to-drag ratio is also increased.

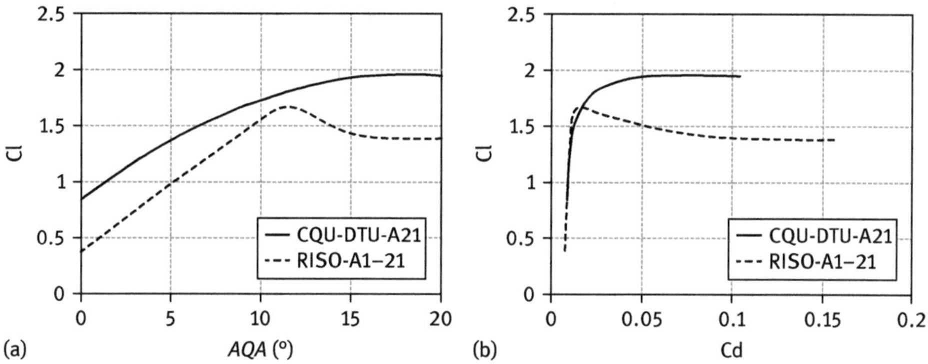

Fig. 4.16: Comparison of performance of CQU-DTU-A21 and Risø-A1-21 airfoils.

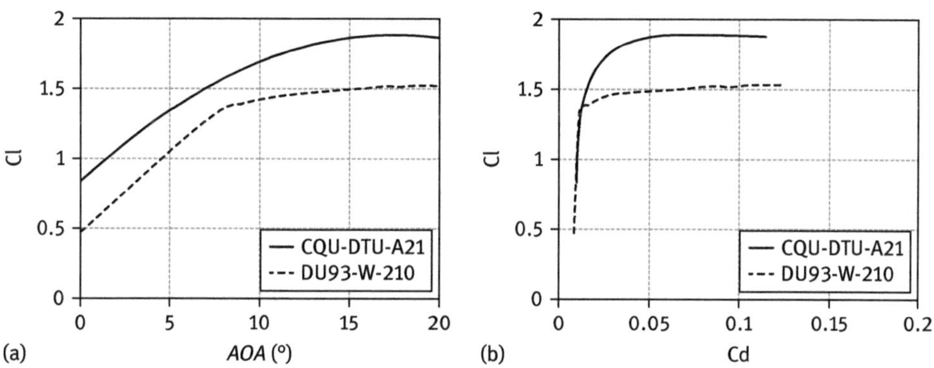

Fig. 4.17: Comparison of performance of CQU-DTU-A21 and DU 93-W-210 airfoils.

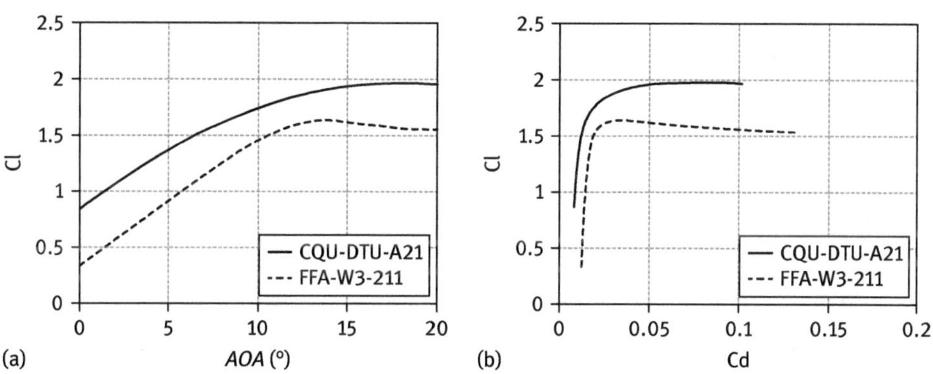

Fig. 4.18: Comparison of performance of CQU-DTU-A21 and FFA-W3-211 airfoils.

4.4 Multiobjective optimization of the wind turbine airfoils

4.4.1 Design variables

Based on the common characteristics of the airfoil profile, Chapter 3 gives the shape function equation $\rho(\theta)$ of the airfoil based on the Taylor series theory

$$\rho(\theta) = C_0 + C_1\theta + C_2\theta^2 + C_3\theta^3 + \cdots + C_k\theta^k + \cdots . \tag{4.9}$$

Meanwhile, the governing equation of shape function is:

$$\begin{cases} C_0 = 1, \\ \sum_{k=1}^{n} 2^k\pi^k C_k = 0, \\ \sum_{k=1}^{n} \pi^k C_k = \varepsilon, \\ \sum_{k=0}^{n} \frac{(3^k - 1)\pi^k}{2^k} C_k = -\Delta. \end{cases} \tag{4.10}$$

When the shape function takes the format of first-order polynomial $\rho(\theta) = C_0 + C_1\theta$, taking $\rho(\theta)$ into the shape function, the following parameters are obtained: $C_0 = 1$, $C_1 = 0$.

The offset of the quasi-circle along the x- and y-directions is 0 in the governing equations of shape function, that is, $\varepsilon = 0s$. Putting $C_0 = 1$, $C_1 = 0$ into the shape function (4.9) and carrying out the conformal transformation, flat shapes graphics are obtained (as shown in Fig. 4.19).

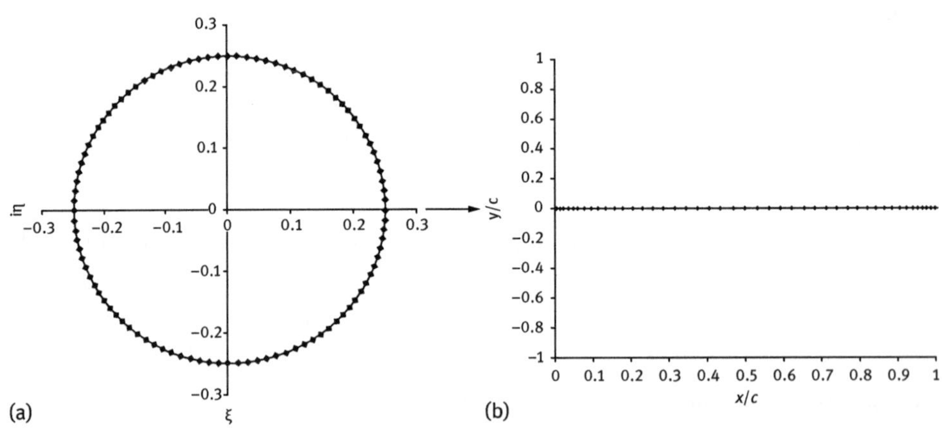

Fig. 4.19: The first-order transformation of an airfoil.

When the shape function takes the format of a second-order polynomial $\rho(\theta) = C_0 + C_1\theta + C_2\theta^2$, taking $\rho(\theta)$ into the shape function, the following parameters are obtained:

$$C_0 = 1, \quad C_1 = \frac{2\varepsilon}{\pi}, \quad C_2 = -\frac{\varepsilon}{\pi^2}.$$

The offset of the quasi-circle along the y-direction is 0 in the governing equations of shape function, that is $\Delta = 0$. Arbitrary values of C_2 and C_1 were selected to meet the above requirements. Putting

$$C_1 = \frac{2}{5\pi}, \quad C_2 = -\frac{1}{5\pi^2},$$

into the shape function (4.9) and carrying out the conformal transformation, the shapes obtained are those of symmetric airfoils (shown in Fig. 4.20).

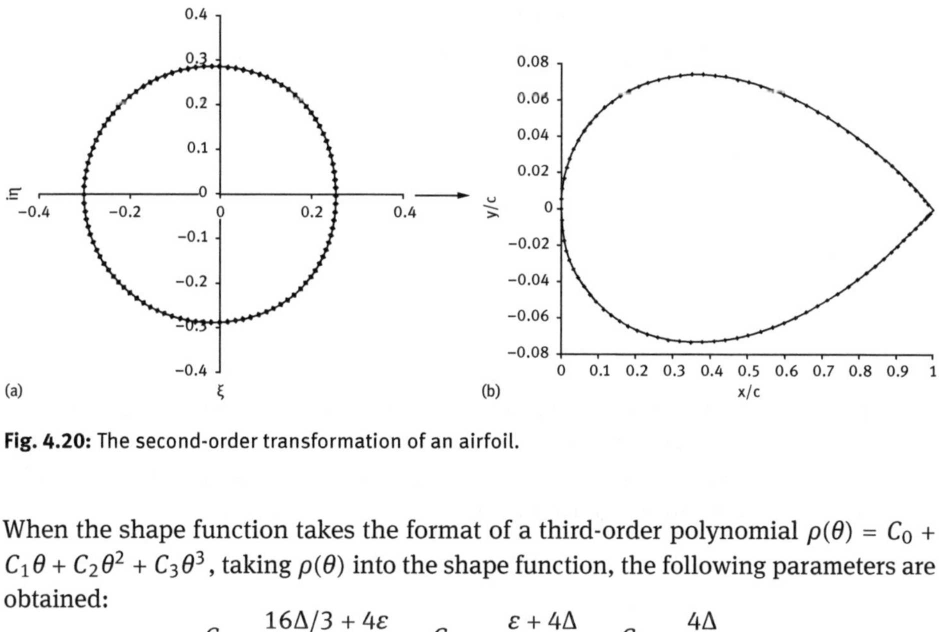

Fig. 4.20: The second-order transformation of an airfoil.

When the shape function takes the format of a third-order polynomial $\rho(\theta) = C_0 + C_1\theta + C_2\theta^2 + C_3\theta^3$, taking $\rho(\theta)$ into the shape function, the following parameters are obtained:

$$C_1 = \frac{16\Delta/3 + 4\varepsilon}{2\pi}, \quad C_2 = -\frac{\varepsilon + 4\Delta}{\pi^2}, \quad C_3 = \frac{4\Delta}{3\pi^3}.$$

Arbitrary values of C_1, C_2 and C_3 were selected to meet the above requirements. Taking

$$C_1 = \frac{5}{6\pi}, \quad C_2 = -\frac{19}{20\pi^2}, \quad C_3 = \frac{4}{15\pi^3},$$

into the shape function (4.9) and carrying out the conformal transformation, the shapes obtained are those of nonsymmetric ordinary airfoils (Fig. 4.21).

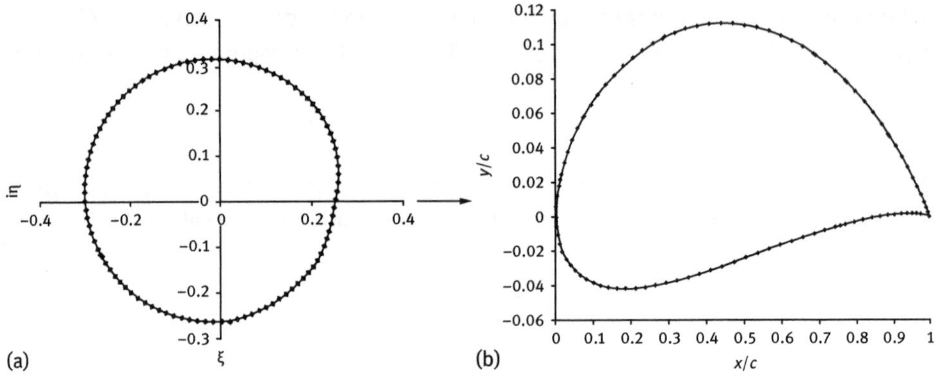

Fig. 4.21: The third-order transformation of an airfoil.

When the shape function is in polynomial of order larger than 3, symmetrical or asymmetrical airfoils can be designed according to the design requirements. Based on the analysis of general shape functions, considering the design space coverage characteristics and the degree of freedom for the profile, the second to ninth coefficients of the shape function are selected as optimization design variables:

$$X = (x_1, x_2, x_3, x_4, x_5, x_6, x_7, x_8) = (C_1, C_2, C_3, C_4, C_5, C_6, C_7, C_8). \tag{4.11}$$

4.4.2 Objective function

The main design goal is to get the maximum power coefficient with other factors such as the lift coefficient and the structural and acoustic requirements taken as design constraints. Under the corresponding Reynolds number and Maher number of normal operating conditions, the maximum lift-to-drag ratio at the design AOA and under smooth airfoil condition is taken as one of the objective functions

$$f_1(X) = \max(c_l/c_d), \tag{4.12}$$

where c_l, c_d are the lift coefficient and drag coefficient of the airfoil in smooth surface condition. The free transition is set to simulate the smooth surface condition.

Meanwhile, the influence of surface roughness on wind turbine performance should be reduced as far as possible. The maximum lift-to-drag ratio of an airfoil at design AOA and under the rough condition is taken as the second objective function

$$f_2(X) = \max(c_l'/c_d'), \tag{4.13}$$

where c_l', c_d' are lift coefficient and drag coefficient under rough conditions. The fixed transition was set to simulate the worst rough conditions, with the suction surface fixed transition at 1 % chord location, the pressure surface fixed transition at 10 % chord location [4, 6, 110].

The system level (total) objective function is set as:

$$f(X) = \lambda_1 \cdot f_1(X) + \lambda_2 \cdot f_2(X), \tag{4.14}$$

where λ_1, λ_2 are the weight factors for lift-to-drag ratio under free transition conditions and fixed transition conditions, $\lambda_1, \lambda_2 \in [0, 1]$ and $\lambda_1 + \lambda_2 = 1$.

In terms of the free transition condition, the Orr–Sommerfeld equation (i.e., e^9 method) is adopted to predict the transition position [96]. The basic idea of this method is as follows. Whether the transition occurs was judged through calculating the disturbance degree of boundary layer flow. The comprehensive disturbance amplitude ratio was calculated though integrating parameters along the surface downstream of a point. When the ratio is above e^9 (approximately 8100), it is decided that transition occurs at this point [46].

The typical amplitude coefficient is calculated as

$$\ln(A/A_0) = \tilde{n} = \frac{\mathrm{d}\tilde{n}}{\mathrm{d}Re_\theta}(H)[Re_\theta - Re_{\mathrm{crit}}(H)], \tag{4.15}$$

where the slope of the amplitude ratio and the critical Reynolds number Re_{crit} are

$$\frac{\mathrm{d}\tilde{n}}{\mathrm{d}Re_\theta} = 0.01\{(2.4H - 3.7 + 2.5\tanh[1.5(H - 3.1)]) + 0.25\}^{1/2}, \tag{4.16}$$

$$\lg Re_{\mathrm{crit}} = \left(\frac{1.45}{H-1} - 0.489\right)\tanh\left(\frac{20}{H-1} - 12.9\right) + \frac{3.295}{H-1} + 0.44, \tag{4.17}$$

where Re_θ is the local Reynolds number.

For similar flow, the boundary layer shape parameter H is constant and local Reynolds number Re_θ is determined directly by airfoil coordinates x. For nonsimilar flow, local Reynolds number Re_θ needs a series of empirical formula in order to realize the transformation between x and Re_θ. So the relationship between amplitude ratio and the airfoil coordinates is obtained:

$$\frac{\mathrm{d}\tilde{n}}{\mathrm{d}x} = \frac{\mathrm{d}\tilde{n}}{\mathrm{d}Re_\theta}\frac{\mathrm{d}Re_\theta}{\mathrm{d}x} = \frac{\mathrm{d}\tilde{n}}{\mathrm{d}Re_\theta}\frac{1}{2}\left(\frac{x}{U_1}\frac{\mathrm{d}U_1}{\mathrm{d}x} + 1\right)\frac{\rho U_1 \theta^2}{U_1 x}\frac{1}{\theta}. \tag{4.18}$$

Empirical correction:

$$\frac{\rho U_1 \theta^2}{U_1 x} = l(H) = \frac{(6.54H - 14.07)}{H^2}, \tag{4.19}$$

$$\frac{x}{U_1}\frac{\mathrm{d}U_1}{\mathrm{d}x} = m(H) = \left(0.058\frac{(H-4)^2}{H-1} - 0.068\right)\frac{1}{l(H)}. \tag{4.20}$$

Taking the correction equations (4.19) and (4.20) into (4.18), the relation between the amplitude ratio, the boundary layer shape parameter H and the momentum thickness θ [96] is obtained:

$$\frac{\mathrm{d}\tilde{n}}{\mathrm{d}x}(H, \theta) = \frac{\mathrm{d}\tilde{n}}{\mathrm{d}Re_\theta}(H)\frac{m(H) + 1}{2} l(H)\frac{1}{\theta}. \tag{4.21}$$

When the amplitude coefficient reaches $\tilde{n} = 9$, transition starts. The laminar boundary layer integral equation will be used before the transition point. The turbulent boundary layer integral equation will be used after the transition point.

At the transition location, the continuity of boundary layer flow should be considered. A weight factor was used to consider the laminar and turbulent regimes characteristic of the dynamic shape parameters which will make the boundary layer equations show continuity along the whole airfoil surface [96].

$$H^* = (1 - \gamma_{\text{tr}})H^*_{\text{laminar}} + \gamma_{\text{tr}} H^*_{\text{turbulent}}, \tag{4.22}$$

where the weight factor is

$$\gamma_{\text{tr}} = \begin{cases} 0 & i < i_{\text{tr}}, \\ \dfrac{\tilde{n}_i - 9}{(\mathrm{d}\tilde{n}/\mathrm{d}x)_i} \dfrac{1}{x_i - x_{i-1}} & i = i_{\text{tr}}, \\ 1 & i > i_{\text{tr}}, \end{cases} \tag{4.23}$$

where i_{tr} is the transition point on the airfoil surface, namely the point of $\tilde{n} = 9$.

4.4.3 Design constraints

Based on the parametric representation research on 30 kinds of commonly used airfoils [106] and ensuring that the design space of the airfoil profile covers an approximate circular space, the boundary constraints of variables were established as

$$X_{\min} \le X \le X_{\max}. \tag{4.24}$$

In order to ensure that the shape of the generated airfoil has the characteristics of rounded leading edge, sharp trailing edge, smooth profile and narrow shape at trailing edge, the design variables must satisfy the constraints of the shape function equation:

$$\begin{cases} C_0 = 1, \\ \sum_{k=1}^{8} 2^k \pi^k C_k = 0, \\ \sum_{k=1}^{8} \pi^k C_k = \varepsilon, \\ \sum_{k=0}^{8} \dfrac{(3^k - 1)\pi^k}{2^k} C_k = -\Delta, \end{cases} \tag{4.25}$$

where ε, Δ are the offsets of the quasi-circle which will be used to generate an asymmetric and cambered airfoil.

In order to ensure the design of an airfoil with a relatively high lift coefficient [5], the lift coefficient under design AOA of the airfoil is restricted

$$c_{\text{l}} \ge c_{\text{l,min}}, \tag{4.26}$$

where $c_{l,min}$ represents the minimum lift coefficient required under design AOA and free transition condition.

The roughness sensitivity constraint of the airfoil is also needed. Even if the leading edge of the airfoil is subjected to external contamination or has certain manufacturing errors, the airfoil would still show good aerodynamic performance. At the design AOA, the difference of lift coefficient under free transition and fixed transition conditions should be in a certain range Δc_l. And the size of Δc_l changes according to the different design requirements:

$$c_l - c_l' \leq \Delta c_l. \tag{4.27}$$

The airfoil thickness is the most important structural requirement. The main power production area of the wind turbine blade is the outer part (generally at 70–90 % span location). For most of the wind turbine blade, the airfoils have the maximum relative thickness of 0.12–0.20 in this area. So the relative thickness constraint of the airfoil is set as

$$\frac{th}{c} = t \in [0.12, 0.20]. \tag{4.28}$$

Another important structural parameter is the chordwise location of the maximum thickness L_{max}. Considering the torque characteristics and airfoil compatibility [11] with other airfoils, the constraints for chordwise location of the maximum thickness is

$$0.24 \leq L_{max} \leq 0.35. \tag{4.29}$$

The rotor blade tip is the main area producing the aerodynamic noise. In order to control the noise, the blade tip airfoil should have sharp trailing edge characteristics [5]:

$$y_{u,0.99} - y_{l,0.99} \leq 0.01, \tag{4.30}$$

where $y_{u,0.99}$, $y_{l,0.99}$ represent the y-coordinate values of the upper and lower airfoil when x-coordinate is 0.99.

4.4.4 Multiobjective genetic algorithm

Genetic algorithm (GA) is a kind of searching method that imitates biological evolution based on stochastic theory. In recent years, the GA method has been widely used in various fields of science. It demonstrates strong ability in solving some complex optimization problems [123–133]. A multiobjective optimization problem can be transformed into a single object problem through adding the objective function $f_i(X)$ with the weight factor [133], which can be expressed as

$$f(X) = \sum_{i=1}^{n} \lambda_i \cdot f_i(X), \tag{4.31}$$

where λ_i is the weight factor of corresponding $f_i(X)$, which represents the importance of each object and should follow the condition of $\sum_{i=1}^{n} \lambda_i = 1$.

Then the genetic algorithm for single-objective optimizations can be used to solve this multiobjective optimization problem. An improved genetic algorithm was adopted to solve the optimization problem. The combination of numerical optimization and flow solver RFOIL software are shown in Fig. 4.22. The GA parameters are as follows: the population size is 30, the maximum generation number is 200, the initial crossover probability is 0.7 and the mutation probability is 0.1.

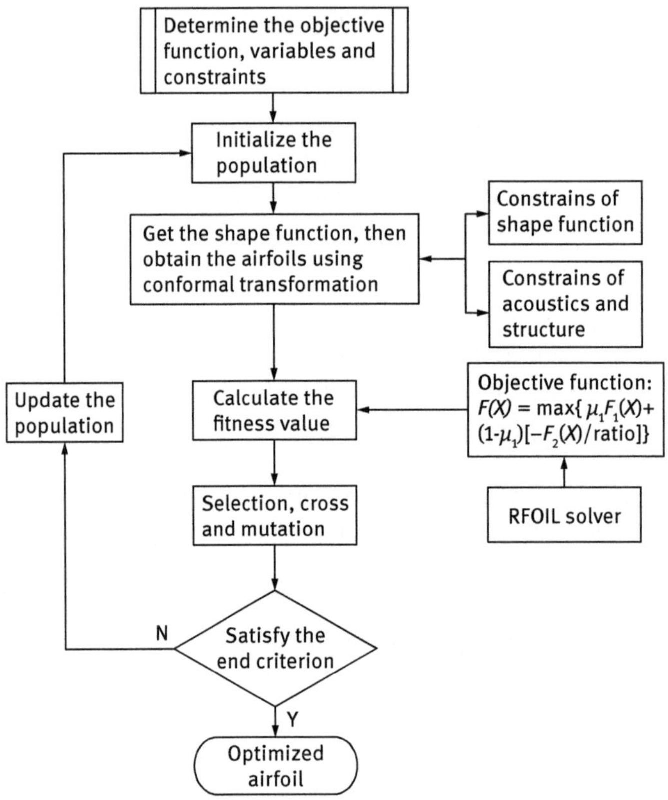

Fig. 4.22: Flow chart for airfoil optimization.

4.4.5 WT series wind turbine airfoils of high performance

According to the actual requirements of most wind farms, the high performance airfoil series (high lift-to-drag ratio, high lift coefficient, low leading edge roughness sensitivity) were designed. The relative thicknesses of 0.15, 0.17, 0.18 and 0.20 were selected as constraints for optimization design. The weight factors of free transition condition and fixed transition condition are set as: $\lambda_1 \in [0.4, 0.6]$, $\lambda_2 = 1 - \lambda_1$. The optimization objective at system level is: $f(X) = \lambda_1 \cdot f_1(X) + \lambda_2 \cdot f_2(X)$. Four kinds of wind turbine airfoils were generated: WT150, WT170, WT180 and WT201, in which WT (Wind Turbine) represents the wind turbine dedicated airfoils and the following number is 1000 times the relative thickness of airfoil. Tab. 4.3 presents the coefficients of the shape function of these airfoils. Results show that this algorithm has fast convergence speed and has good global optimization ability. The application of the improved genetic algorithm can push the whole population to noninferior solutions simultaneously. Thus the corresponding global optimal solution in the entire design space was figured out.

Tab. 4.3: Airfoil shape function coefficients.

Name	Coefficients			
	C_1	C_2	C_3	C_4
WT150	0.131796	−0.060058	−0.150282	0.154312
WT170	0.101061	−0.0626973	−0.159258	0.166905
WT180	0.135179	0.169777	−0.650805	0.526446
WT201	0.136947	0.168366	−0.680133	0.555068
	C_5	C_6	C_7	C_8
WT150	0.0149111	−0.066557	−0.00167887	7.47892e−5
WT170	−0.0697439	0.0148122	−0.00157114	6.59562e−5
WT180	−0.199565	0.0399258	−0.00407183	0.000166823
WT201	−0.210776	0.0421114	−0.00428201	0.000174758

4.4.5.1 WT180 airfoil

Selecting the third line of optimization results in Tab. 4.3 and putting the coefficients into the airfoil shape function, the shape and coordinates of the WT180 airfoil (shown in Fig. 4.23) can be obtained through the conformal transformation.

The WT180 airfoil has good structural characteristics. The maximum thickness is $t = 0.179\,998$ and the chordwise position of the maximum thickness is $x/c = 0.294$, which is compatible with other wind turbine airfoils. The maximum camber is $cam/c = 0.030\,005$ and its chordwise position is $x/c = 0.4688$. The airfoil has a closed trailing edge which is helpful to reduce the aerodynamic noise of the wind turbine.

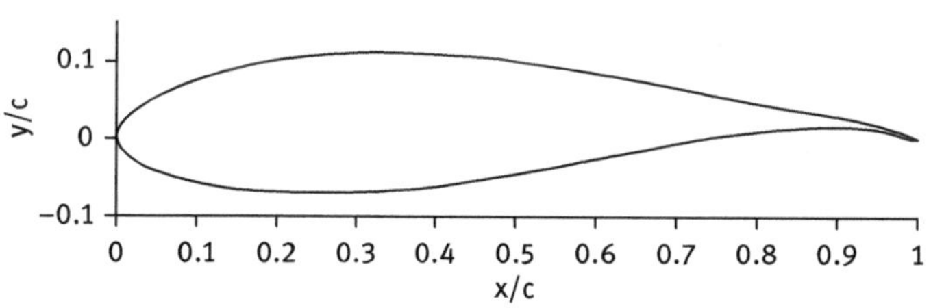

Fig. 4.23: WT180 airfoil profile.

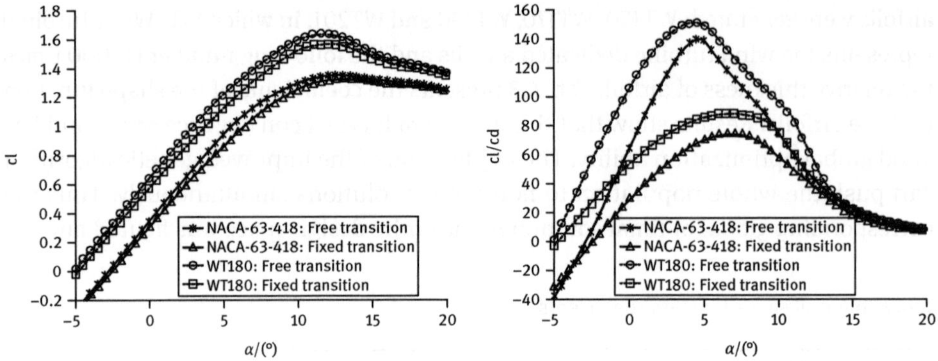

(a) Lift coefficient (b) Lift to drag ratio

Fig. 4.24: Comparison of aerodynamic characteristics between WT180 and NACA 63418.

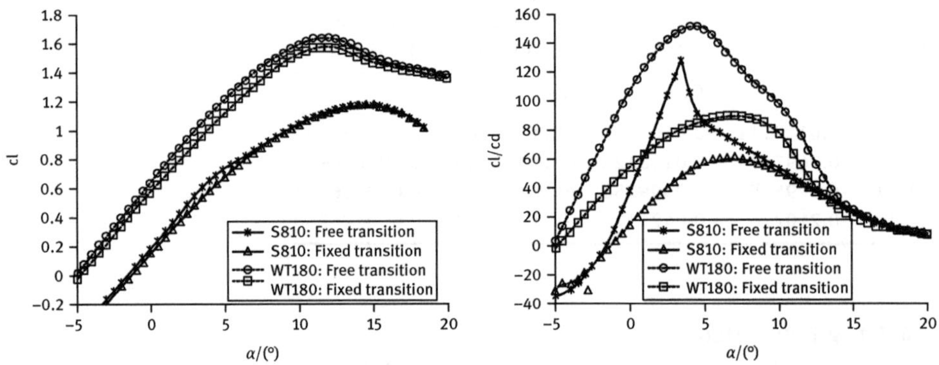

(a) Lift coefficient (b) Lift to drag ratio

Fig. 4.25: Comparison of aerodynamic characteristics between WT180 and S810.

Tab. 4.4: Comparison of aerodynamic parameters of airfoils with relative thickness 18 %.

Airfoil name	**Free transition**		**Fixed transition** $X_{tr,s} = 0.01; X_{tr,p} = 0.1$	
	$c_{l,max}$	L/D_{max}	$c_{l,max}$	L/D_{max}
NACA 63418	1.3521	140.302	1.3211	77.3897
S810	1.1871	126.674	1.1736	60.9084
WT180	1.6341	151.556	1.5647	91.4793

The comparison of aerodynamic performance under the same conditions ($Re = 3.0 \times 10^6$, $Ma = 0.2$) for WT180, NACA 63418 and S810 [7] airfoils is shown in Fig. 4.24 and 4.25 respectively. The maximum lift-to-drag ratio of 151.556 is reached under free transition conditions at the AOA of $\alpha = 4.5°$. The maximum lift-to-drag ratio of 91.4793 is reached under fixed transition conditions at the AOA of $\alpha = 6.5°$. The new airfoil has higher lift coefficient and lift-to-drag ratio (free transition and fixed transition) in the main AOA range and has better characteristics during design and off-design operating conditions.

The comparison of lift coefficient and lift-to-drag ratio under the same conditions for these three airfoils is shown in Tab. 4.4. Compared with NACA 632-418 airfoil, the maximum lift-to-drag ratio of WT180 under free transition and fixed transition increased by 8 % and 18.2 % respectively. Compared with S810 airfoil, the maximum lift-to-drag ratio of WT180 under free transition and fixed transition increased by 19.6 % and 50.2 % respectively. The new airfoil shows better performance compared with the other two airfoils, which validates this optimization model.

4.4.5.2 WT150 airfoil

Selecting the first line of optimization results in Tab. 4.3 and putting the coefficients into the airfoil shape function, the shape and coordinates of WT150 airfoil (shown in Fig. 4.26) can be obtained through the conformal transformation.

The maximum thickness of the WT150 airfoil is t = 0.15 and the chordwise position of the maximum thickness is $x/c = 0.254$, which is compatible with other wind turbine airfoils. The maximum camber is $cam/c = 0.037385$ and its chordwise position is $x/c = 0.443$. The airfoil has a closed trailing edge which is helpful to reduce the aerodynamic noise of the wind turbine.

The RFOIL software was used to analyze the aerodynamic performance of the WT150 airfoil and the results are shown in Fig. 4.27. It can be seen that the sensitivity to roughness of the WT150 airfoil is low. It has higher lift coefficient and lift-to-drag ratio and has good characteristics during off-design operating conditions. The maximum lift coefficients of WT150 under free transition and fixed transition are 1.7748 and 1.6504 respectively. The maximum lift-to-drag ratios of WT150 under free transition and fixed transition are 158.079 and 97.6994 respectively.

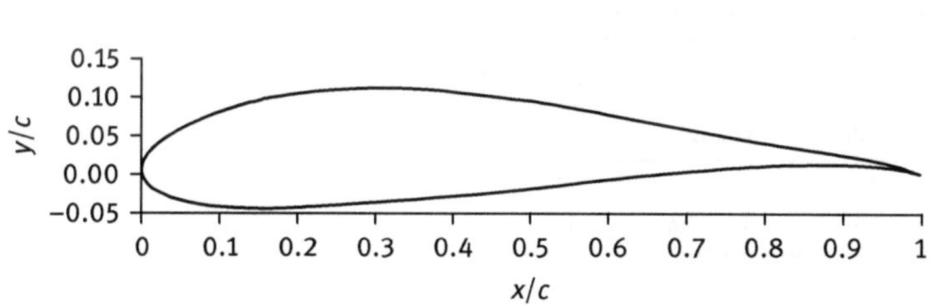

Fig. 4.26: WT150 airfoil profile.

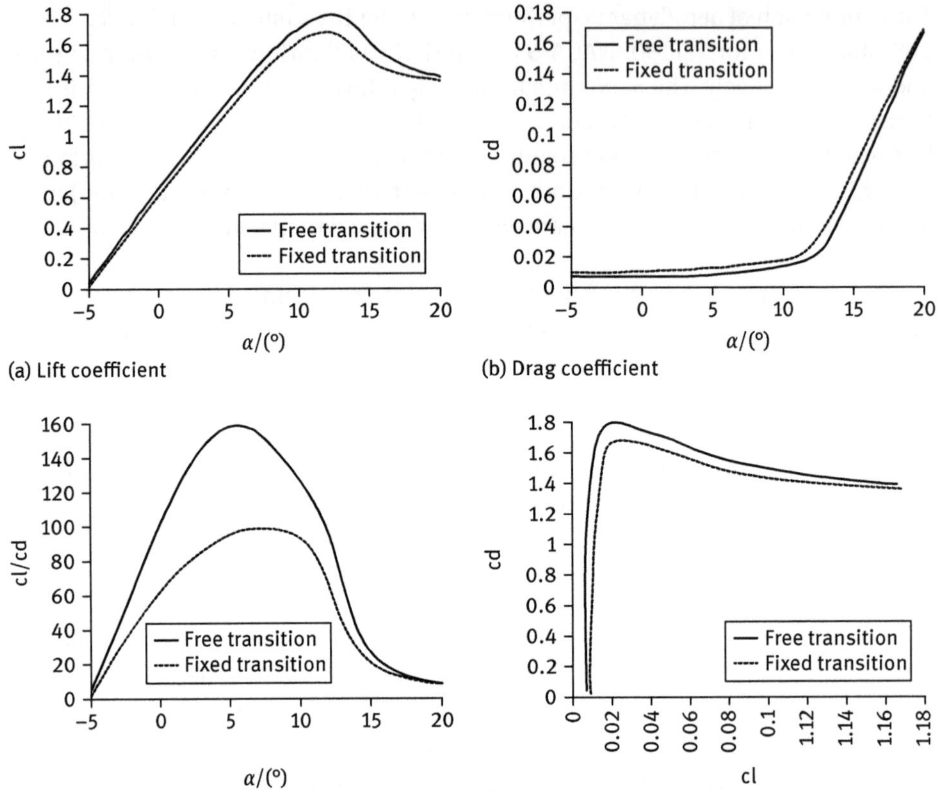

(a) Lift coefficient

(b) Drag coefficient

(c) Lift to drag ratio

(d) Lift to drag curve

Fig. 4.27: Comparison of aerodynamic characteristics of WT150.

The comparison of aerodynamic performance under the same conditions for WT150 and NACA 64415 airfoils is shown in Fig. 4.28. The new airfoil has higher lift coefficient and lift-to-drag ratio (free transition and fixed transition) in the main AOA range and has better characteristics during design and off-design operating conditions.

Tab. 4.5: Comparison of aerodynamic parameters of airfoils with relative thickness 15 %.

Airfoil name	Free transition		Fixed transition $X_{tr,s} = 0.01; X_{tr,p} = 0.1$	
	$c_{l,max}$	L/D_{max}	$c_{l,max}$	L/D_{max}
NACA 64415	1.4054	133.934	1.3883	86.476
WT150	1.7748	158.079	1.6504	97.6994

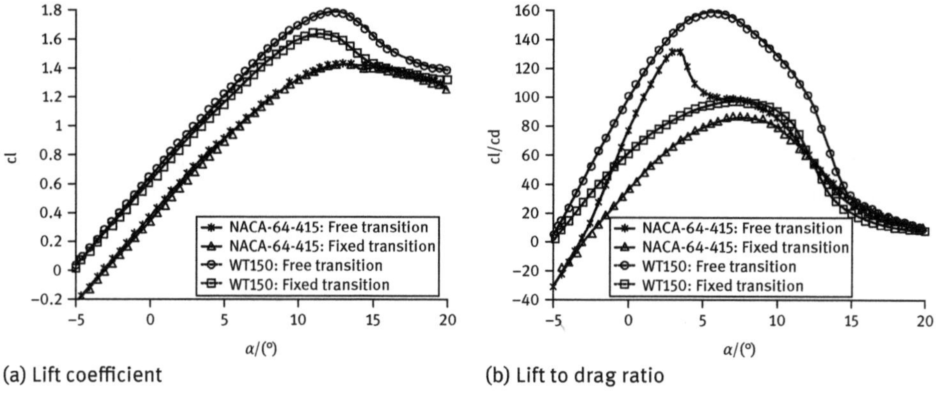

Fig. 4.28: Comparison of aerodynamic characteristics between WT150 and NACA 64415.

The comparison of lift coefficient and lift-to-drag ratio under the same conditions for these two airfoils is shown in Tab. 4.5. Compared with the NACA 64415 airfoil, the maximum lift-to-drag ratio of WT150 under free transition and fixed transition increased by 18 % and 12.9 % respectively. The new airfoil shows better performance compared with the other two airfoils, which validates this optimization model.

4.4.5.3 WT170 airfoil

Selecting the second line of optimization results in Tab. 4.3 and putting the coefficients into the airfoil shape function, the shape and coordinates of the WT170 airfoil (shown in Fig. 4.29) can be obtained through the conformal transformation.

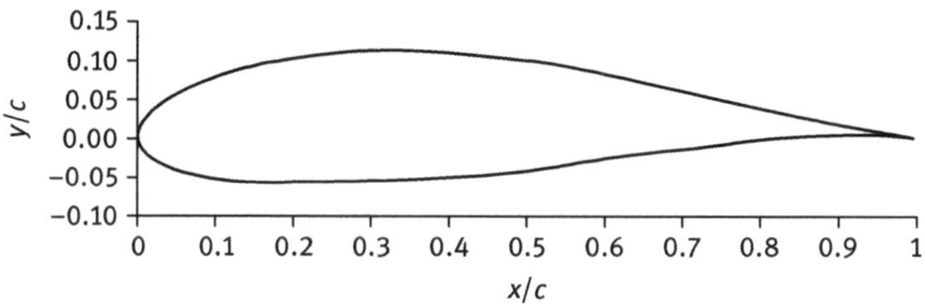

Fig. 4.29: WT170 airfoil profile.

The maximum thickness of the WT170 airfoil is $t = 0.17$ and the chordwise position of the maximum thickness is $x/c = 0.289$, which is compatible with other wind turbine airfoils. The maximum camber is $cam/c = 0.029445$ and its chordwise position is $x/c = 0.460$. The airfoil has a closed trailing edge which is helpful to reduce the aerodynamic noise of the wind turbine.

The RFOIL software was used to analyze the aerodynamic performance of the WT170 airfoil and the results are shown in Fig. 4.30. It can be seen that the sensitivity to roughness of the WT170 airfoil is low. It has higher lift coefficient and lift-to-drag ratio and has good characteristics during off-design operating conditions. The maximum lift coefficients of WT170 under free transition and fixed transition are 1.5817 and 1.4887 respectively. The maximum lift-to-drag ratios of WT170 under free transition and fixed transition are 147.829 and 88.4434 respectively.

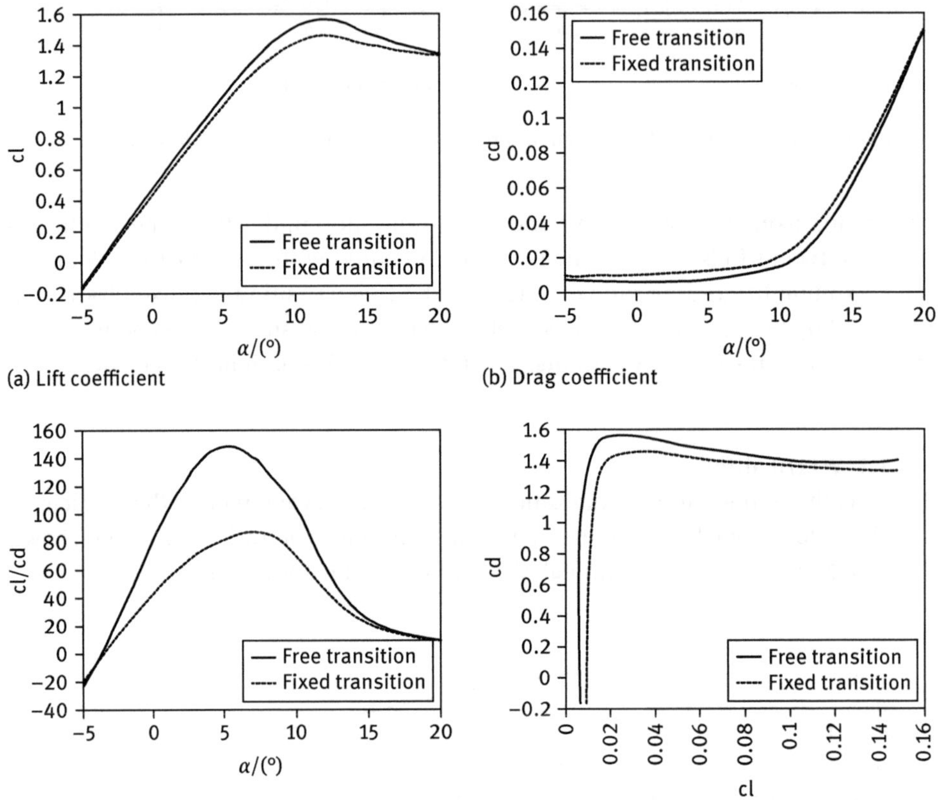

(a) Lift coefficient

(b) Drag coefficient

(c) Lift to drag ratio

(d) Lift to drag curve

Fig. 4.30: Comparison of aerodynamic characteristics of WT170.

The comparison of lift coefficient and lift-to-drag ratio under the same conditions for these two airfoils is shown in Tab. 4.6. Compared with the NACA 63417 airfoil, the maximum lift-to-drag ratio of WT170 under free transition and fixed transition increased by 8.5 % and 7.0 % respectively as shown in Fig. 4.31. The new airfoil shows better performance compared with the other two airfoils, which validates this optimization model.

Tab. 4.6: Comparison of aerodynamic parameters of airfoils with relative thickness 17 %.

Airfoil name	Free transition		Fixed transition $X_{tr,s} = 0.01$; $X_{tr,p} = 0.1$	
	$c_{l,max}$	L/D_{max}	$c_{l,max}$	L/D_{max}
NACA 63417	1.3763	136.704	1.3501	79.8299
WT170	1.5438	148.287	1.4292	85.3645

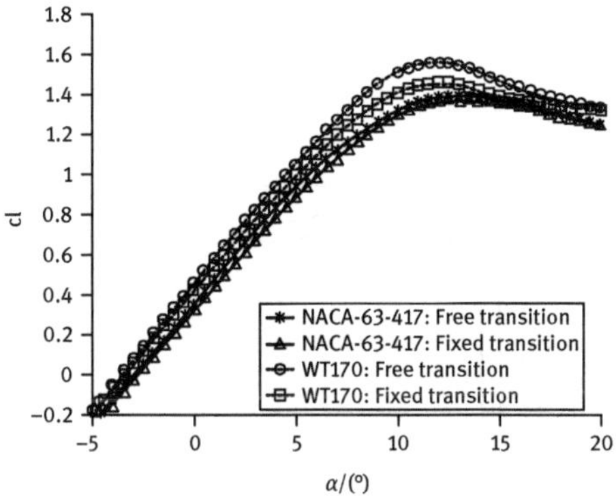

(a) Lift coefficient

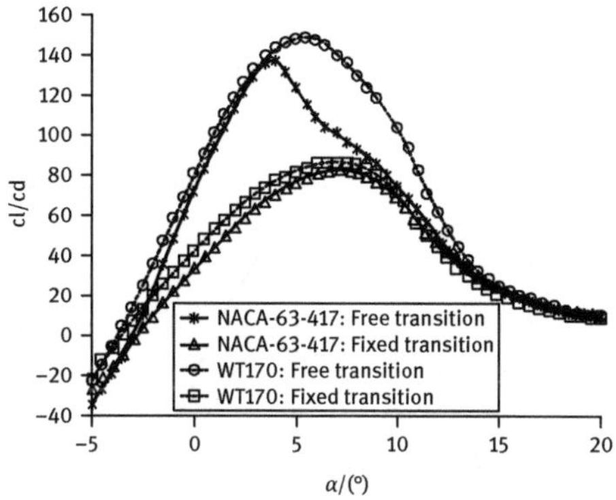

(b) Lift to drag ratio

Fig. 4.31: Comparison of aerodynamic characteristics between WT170 and NACA 63417.

4.4.5.4 WT201 airfoil

Selecting the fourth line of optimization results in Tab. 4.3 and putting the coefficients into the airfoil shape function, the shape and coordinates of WT201 airfoil (shown in Fig. 4.32 can be obtained through the conformal transformation.

The maximum thickness of the WT210 airfoil is $t = 0.21$ and the chordwise position of the maximum thickness is $x/c = 0.294$. The maximum camber is $cam/c = 0.030005$ and its chordwise position is $x/c = 0.688$.

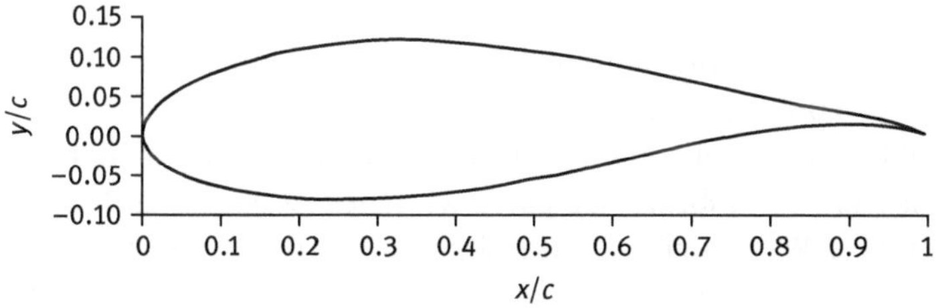

Fig. 4.32: WT201 airfoil profile.

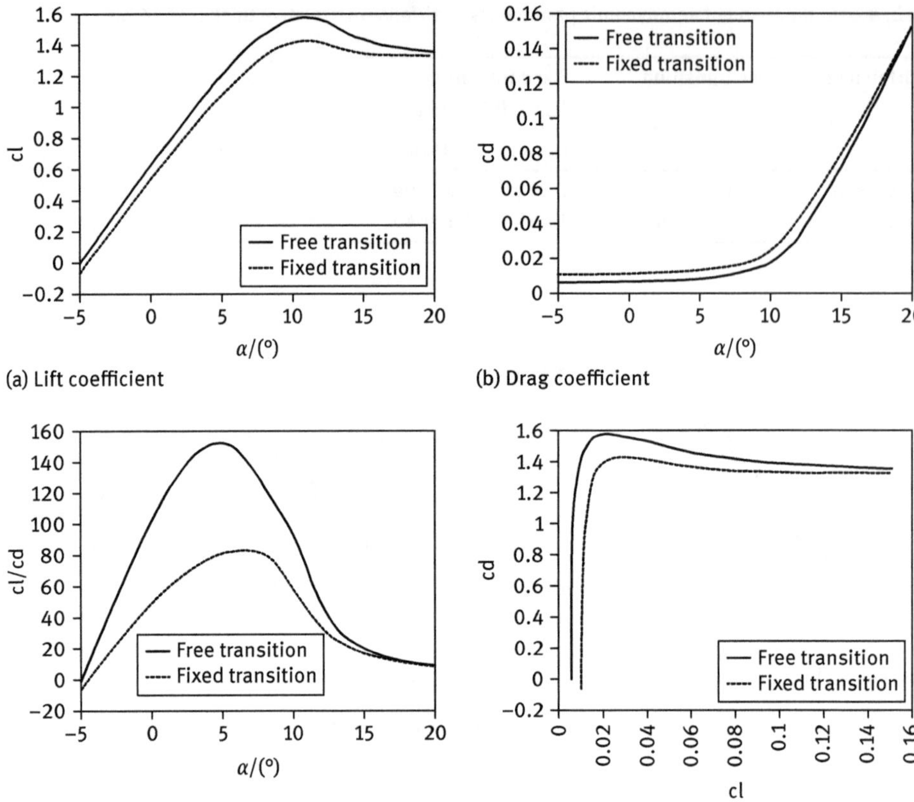

(a) Lift coefficient

(b) Drag coefficient

(c) Lift to drag ratio

(d) Lift to drag curve

Fig. 4.33: Comparison of aerodynamic characteristics of WT201.

The RFOIL software was used to analyze the aerodynamic performance of the WT201 airfoil and the results are shown in Fig. 4.33. It can be seen that the sensitivity to roughness of the WT201 airfoil is low. It has higher lift coefficient and lift-to-drag ratio and has good characteristics during off-design operating conditions. The maximum lift coefficients of WT201 under free transition and fixed transition are 1.5689 and 1.4211 respectively. The maximum lift-to-drag ratios of WT170 under free transition and fixed transition are 152.65 and 83.0673 respectively.

The comparison of lift coefficient and lift-to-drag ratio under the same conditions for these two airfoils is shown in Tab. 4.7. Compared with the NACA 63421 airfoil, the maximum lift-to-drag ratio of WT201 under free transition and fixed transition increased by 5.2 % and 32.7 % respectively as shown in Fig. 4.34. The new airfoil shows better performance compared with the other two airfoils, which validates this optimization model.

Tab. 4.7: Comparison of aerodynamic parameters of airfoils with relative thickness 20 %.

Airfoil name	Free transition		Fixed transition $X_{tr,s} = 0.01; X_{tr,p} = 0.1$	
	$c_{l,max}$	L/D_{max}	$c_{l,max}$	L/D_{max}
NACA 63421	1.2668	144.971	1.2088	62.5668
WT201	1.5689	152.65	1.4211	83.0673

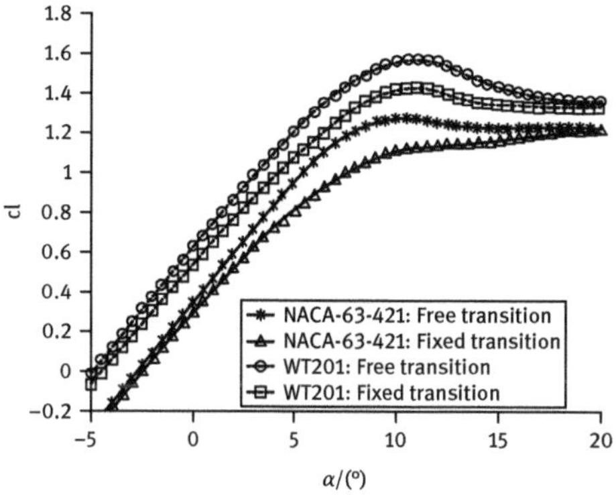

(a) Lift coefficient

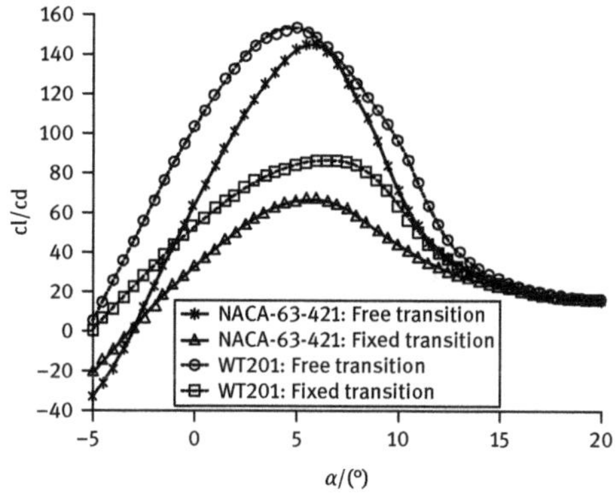

(b) Lift to drag ratio

Fig. 4.34: Comparison of aerodynamic characteristics between WT201 and NACA 63421.

4.4.6 WTH series wind turbine airfoils with high lift-to-drag ratio

For the wind farm environment where the wind resources are stable and blades are not exposed to pollution (such as specially surface-treated blades on offshore wind turbines), airfoils with high lift-to-drag ratio under free transition conditions are required. Then the expectation of high lift-to-drag under the fixed transition conditions is lower. Thus the design of an airfoil series with high lift-to-drag ratio and high lift coefficient under free transition condition was completed. Relative thickness of 0.12–0.19 was selected as constraint for optimization design. The weight factors of free transition condition and fixed transition condition are set as: $\lambda_1 \in [0.8, 1.0]$, $\lambda_2 = 1 - \lambda_1$. Three kinds of wind turbine airfoils were generated: WTH122, WTH150 and WTH188, in which WTH (Wind Turbine High L/D) represents the wind turbine dedicated airfoils of high lift-to-drag ratio and the following number is 1000 times the relative thickness of the airfoil. Randomly taking the WTH122 airfoil for analysis, the airfoil profile is shown in Fig. 4.35.

The maximum thickness of the WTH122 airfoil is $t = 0.122$ and the chordwise position of the maximum thickness is $x/c = 0.343$, which is compatible with other wind turbine airfoils. The maximum camber is $cam/c = 0.0776$ and its chordwise position is $x/c = 0.487$. The airfoil has a closed trailing edge which is helpful to reduce the aerodynamic noise of the wind turbine.

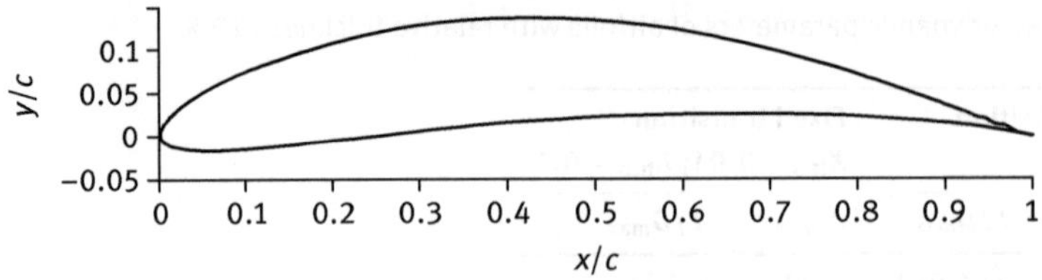

Fig. 4.35: WTH122 airfoil profile.

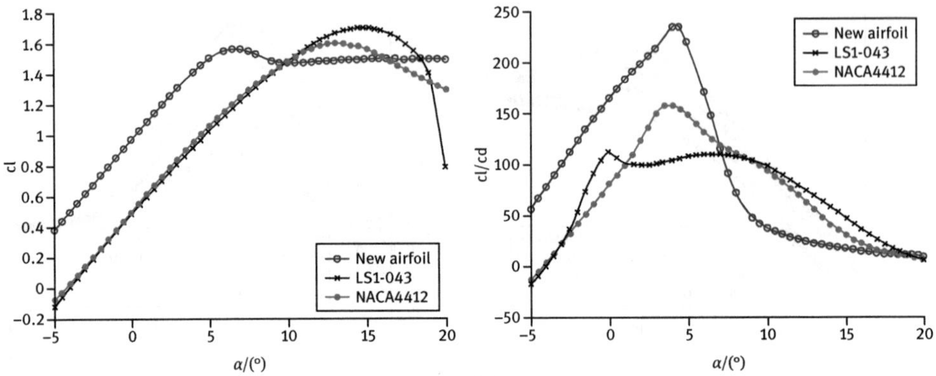

(a) Lift coefficient

(b) Lift to drag ratio

Fig. 4.36: Comparison of aerodynamic characteristics between WTH122 and traditional airfoils.

The comparison of aerodynamic performance under the same conditions ($Re = 3.0 \times 10^6$, $Ma = 0.25$) for WTH122, LS1-0413 and NACA 4412 [7] airfoils is shown in Fig. 4.36. The maximum lift coefficient is 1.5582. The maximum lift-to-drag ratio of 239.01 is reached under the fixed transition condition at AOA of $\alpha = 4°$.

Tab. 4.8: Comparison of aerodynamic parameters of airfoils with relative thickness 12 %.

Airfoil name	$c_{l,max}$	L/D_{max}
LS1-0413	1.6917	111.726
NACA 4412	1.5977	147.054
WTH122	1.5582	239.01

The comparison of lift coefficient and lift-to-drag ratio under the same conditions for these three airfoils is shown in Tab. 4.8. The maximum lift-to-drag ratio of WTH122 increased respectively by 114 % (compared with LS1-0413) and 62.6 % (compared with NACA 4412). The new airfoil shows better performance compared with the other two airfoils, which validates this optimization model.

Tab. 4.9: Aerodynamic parameters of WTH airfoil series.

Airfoil name	$C_{l,max}$	L/D_{max}	at $\alpha =$	Max. thickness	at $x =$	Max. camber	at $x =$
WTH122	1.5582	239.01	4°	0.1223189	0.343	0.077643	0.487
WTH150	1.6709	190.883	4.5°	0.149264	0.313	0.078728	0.488
WTH188	1.8004	159.645	6.5°	0.187542	0.288	0.075425	0.427

The main aerodynamic and structural parameters of this airfoil series, with relative thickness ranging from 0.12 to 0.19, are shown in Tab. 4.9. Compared with the traditional airfoils, the maximum lift-to-drag ratio of the WTH airfoil series has been greatly improved.

4.4.7 WTI series wind turbine airfoils with low roughness sensitivities

For the wind farm environment where the air contains significant amounts of dust, dirt and staining elements, the wind turbine blade can be easily polluted and becomes rough. So airfoils with high lift-to-drag ratio under fixed transition conditions are required. Then the expectation of high lift-to-drag under free transition conditions is lower. Thus the design of an airfoil series with low roughness sensitivities was completed. Relative thickness of 0.15–0.20 was selected as constraint for optimization design. The weight factors of free transition condition and fixed transition condition are set as: $\lambda_1 \in [0, 0.2]$, $\lambda_2 = 1 - \lambda_1$. Seven kinds of wind turbine airfoils were generated: WTI150, WTI151, WTI154, WTI180, WTI181, WTI181-2 and WTI197, in which WHI (Wind Turbine Insensitive) represents the wind turbine dedicated airfoils of low roughness sensitivities and the following number is 1000 times the relative thickness of airfoil. WTI181-2 means that there are many airfoils with the same relative thickness, and "-2" means the second airfoil among them. Randomly taking the WTI180 airfoil for analysis, the airfoil profile is shown in Fig. 4.37.

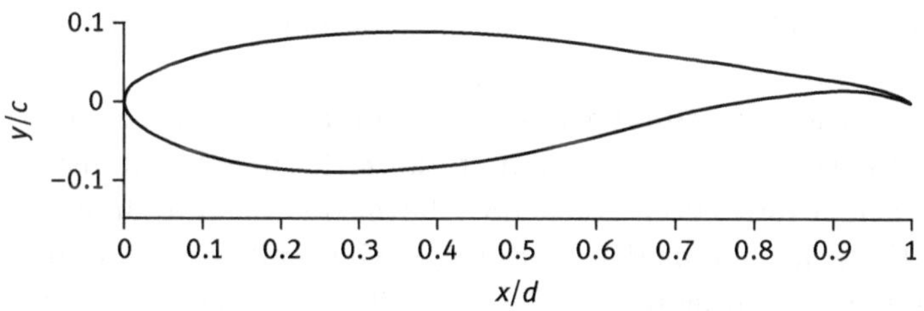

Fig. 4.37: WTI180 airfoil profile.

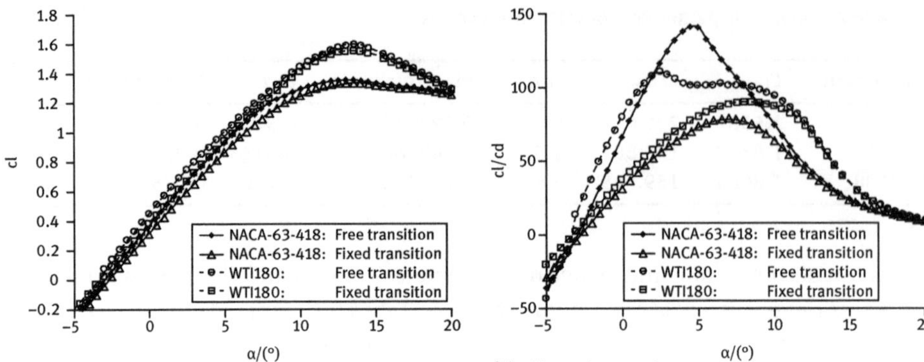

(a) Lift coefficient (b) Lift to drag ratio

Fig. 4.38: Comparison of aerodynamic characteristics between WTI180 and traditional airfoils.

Tab. 4.10: Comparison of aerodynamic parameters of airfoils with relative thickness 18 %.

Airfoil name	Free transition		Fixed transition $X_{tr,s} = 0.01; X_{tr,p} = 0.1$	
	$c_{l,max}$	L/D_{max}	$c_{l,max}$	L/D_{max}
NACA 63418	1.3521	140.302	1.3211	77.3897
WTI180	1.6339	110.223	1.6076	91.8991

The maximum thickness of the WTI180 airfoil is $t = 0.18$ and the chordwise position of the maximum thickness is $x/c = 0.320$. The maximum camber is $cam/c = 0.023\,274$ and its chordwise position is $x/c = 0.821$.

The comparison of lift coefficient and lift-to-drag ratio under the same conditions for two airfoils is shown in Tab. 4.10. Compared with the NACA 63418 airfoil, the maximum lift-to-drag ratio of WTI180 increased by 18.7 %. The new airfoil shows better performance compared with the other two airfoils, which validates this optimization model. The WTI180 airfoil is insensitive to roughness conditions, with the maximum lift coefficient decreasing from 1.6339 to 1.6075 (drop of 1.6 %) and with the maximum lift-to-drag ratio decreasing from 110.223 to 91.8991 (drop of 16 %).

The main aerodynamic and structural parameters of this airfoil series, with relative thickness ranging from 0.15 to 0.20, are shown in Tab. 4.11. Under free transition condition, the lift coefficient varies from 1.63 to 1.83, the maximum lift-to-drag ratio varies from 102 to 137. Under fixed transition condition, the lift coefficient varies from 1.61 to 1.79, the maximum lift-to-drag ratio varies from 89 to 101. This airfoil series is insensitive to roughness conditions, with the maximum decrease of lift coefficient smaller than 4.7 %.

Tab. 4.11: Aerodynamic parameters of WHI airfoil series.

Airfoil name	Free transition		Fixed transition $X_{tr,s} = 0.01; X_{tr,p} = 0.1$		Max. thickness	at x =	Max. camber	at x =
	$c_{l,max}$	L/D_{max}	$c_{l,max}$	L/D_{max}				
WHI150	1.8235	136.868	1.7873	100.839	0.150320	0.262	0.036038	0.644
WHI151	1.7791	136.148	1.7438	100.35	0.150700	0.262	0.025100	0.583
WHI154	1.8455	129.851	1.7913	100.551	0.154419	0.250	0.025067	0.675
WHI180	1.6339	110.223	1.6076	91.8991	0.180240	0.320	0.023569	0.821
WHI181	1.8268	107.558	1.7787	90.853	0.181089	0.266	0.039666	0.800
WHI181-2	1.715	102.421	1.6913	92.6011	0.181197	0.294	0.020421	0.866
WHI197	1.6914	115.083	1.611	88.9914	0.197497	0.280	0.027016	0.812
NACA 64415	1.4589	131.52	1.4439	89.5692	0.149934	0.344	0.020522	0.531
NACA 63418	1.3893	139.676	1.3617	80.0824	0.180002	0.340	0.020254	0.550
NACA 63421	1.2935	144.466	1.2298	66.3805	0.210212	0.340	0.019905	0.550

4.5 Design of airfoils with medium relative thickness

With increasing rotor blade radius, the design of thicker airfoils is necessary. For an MW-level wind turbine, the maximum Reynolds number on the rotor can reach up to $8.0–10.0 \times 10^6$ and it generally appears in the middle span of the blade [110, 134]. When the rotor is rotating, the pressure gradient, centrifugal force and Coriolis force on the blade surface will impact airfoil aerodynamic performance [109]. Especially for airfoils of medium thickness, the difference between two-dimensional (2D) aerodynamic characteristics and actual performance is large. So, three-dimensional (3D) rotational effects must be considered when designing airfoils with medium thickness.

4.5.1 Geometric characteristics analysis of medium thickness airfoils

Several representative wind turbine airfoils [106] with relative thickness of about 24 %, which are widely used in the design of large wind turbine blades, were selected for geometric characteristics analysis. The shapes of these airfoils are shown in Fig. 4.39.

(1) The FFA-W3-W241 airfoil is a wind turbine airfoil developed by the Swedish Institute of Aeronautics.
(2) DU 91-W2-250 is a wind turbine airfoil developed by Delft University of the Netherlands.
(3) The NACA 63425 airfoil is an aircraft airfoil developed by the National Advisory Committee for Aeronautics (NACA).
(4) The S814 airfoil is a wind turbine airfoil developed by D. Somers from the National Renewable Energy Laboratory (NREL).

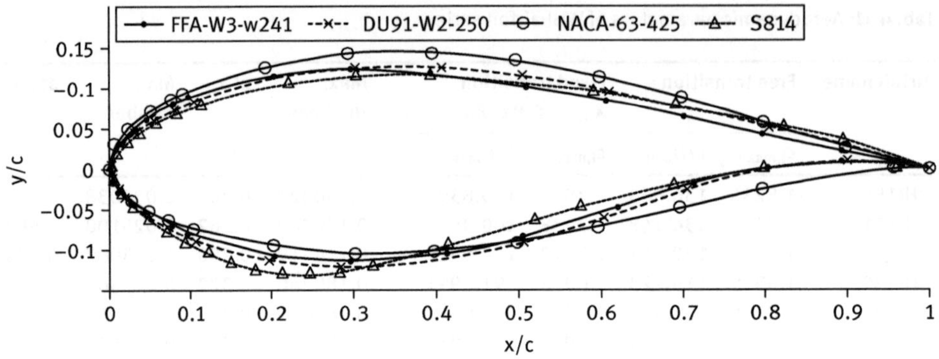

Fig. 4.39: Four airfoils with relative thickness from 23.8 % to 25 %.

The airfoil with the lowest thickness is FFA-W3-W241 ($t/c = 0.238$), the airfoil with the highest thickness is NACA 63425 ($t/c = 0.250$). The differences between NACA 63425 (a traditional aviation airfoil) and the other three dedicated wind turbine airfoils are large. The NACA 63425 airfoil has the maximum upper surface thickness and the minimum lower surface thickness among the four airfoils. The thickness of the trailing edge of NACA 63425 is greater than those of the other airfoils. The three dedicated wind turbine airfoils are similar in shape. They all have smaller upper surface thickness and large lower surface thickness, with the S814 airfoil having the minimum upper surface thickness and the maximum lower surface thickness. The chordwise position of maximum thickness of the three airfoils is closer to the leading edge compared to the NACA airfoil, and that ($x/c = 0.262$) of the S814 airfoil is the closest to the aerodynamic center. The pressure surfaces of the three airfoils are all S-shaped, which can not only increase the lift but also prevent flow separation under small inflow angle. The detailed parameters are shown in Tab. 4.12.

Tab. 4.12: Geometric characteristics of the four airfoils.

Airfoil name	Maximum thickness	Chordwise position	Maximum camber	Chordwise position
FFA-W3-W241	0.237775	0.297	0.020294	0.749
DU 91-W2-250	0.248434	0.326	0.026918	0.780
NACA 63425	0.250038	0.353	0.018236	0.577
S814	0.241664	0.262	0.030868	0.737

4.5.2 Aerodynamic characteristics of airfoils with medium thickness

In order to calculate the aerodynamic characteristics under smooth and rough conditions, free transition was adopted to simulate smooth airfoil condition under viscous condition. The fixed transition was set to simulate the worst rough conditions, with the suction surface fixed transition at 1 % chord location ($X_{tr,s} = 0.01$), the pressure surface fixed transition at 10 % chord ($X_{tr,p} = 0.1$) [4, 6, 110]. Considering the actual working conditions of large wind turbines, the Reynolds number was set as $Re = 3 \times 10^6$, Mach number as $Ma = 0.16$.

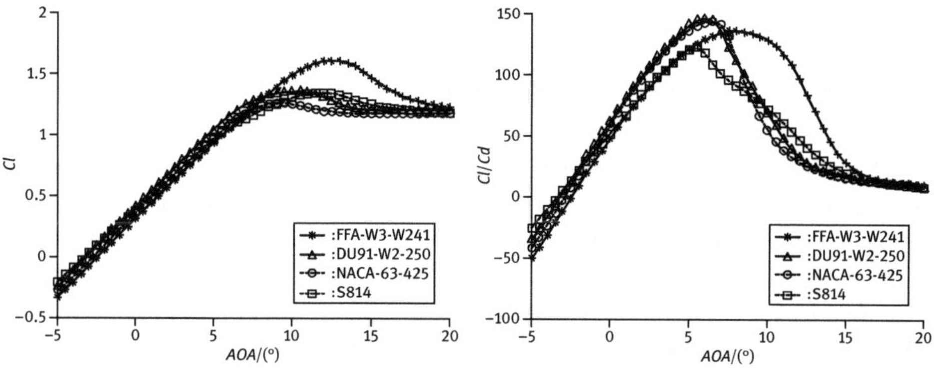

(a) Lift characteristics (b) Lift to drag ratio characteristics

Fig. 4.40: Aerodynamic performance of airfoils under free transition condition.

As shown in Fig. 4.40, the aerodynamic performance of the four airfoils varies little under free transition conditions. The maximum lift coefficient of the FFA-W3-W241 airfoil is 1.6, which is larger than those of the other three airfoils. The slope of the lift coefficient curve of the four airfoils is close before stall. The variation in lift-to-drag ratio for NACA 63425 and DU 91-W2-250 is very close, with the maximum lift-to-drag ratio of 146 and 143 respectively. The maximum lift-to-drag ratio of FFA-W3-W241 is lower by 11 than that of DU 91-W2-250. The maximum lift-to-drag ratio of S814 is the lowest, reaching 121. Among the four airfoils, the performance of FFA under the airfoil design condition is the best. According to Fig. 4.41 (b), the airfoil has high lift-to-drag ratio in a large AOA range, which ensures that the wind turbine rotor can maintain a high power coefficient over a wide range of wind speeds.

As shown in Fig. 4.41, the roughness sensitivity of the NACA 63425 airfoil is worse than that of the other three airfoils, because the initial design objective of NACA airfoils is not for wind turbines. The separation emerges on the airfoil trailing edge at an AOA of 5° and gradually moves towards the leading edge, resulting in increased lift and sharply declining drag. In the whole AOA range, the lift-to-drag ratio is very low with the maximum value of 39. The other three wind turbine airfoils have low leading edge

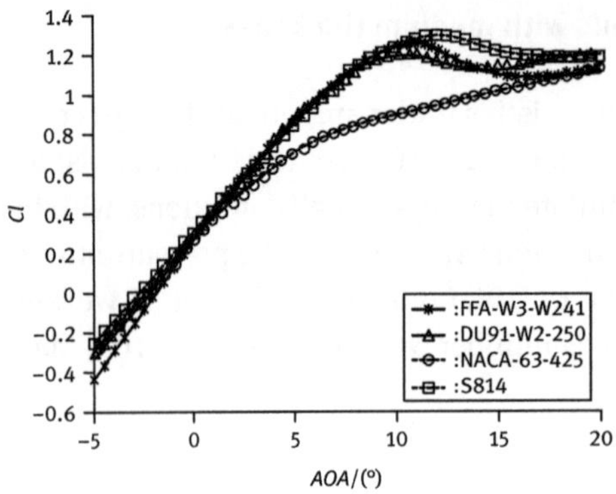

(a) Lift characteristic

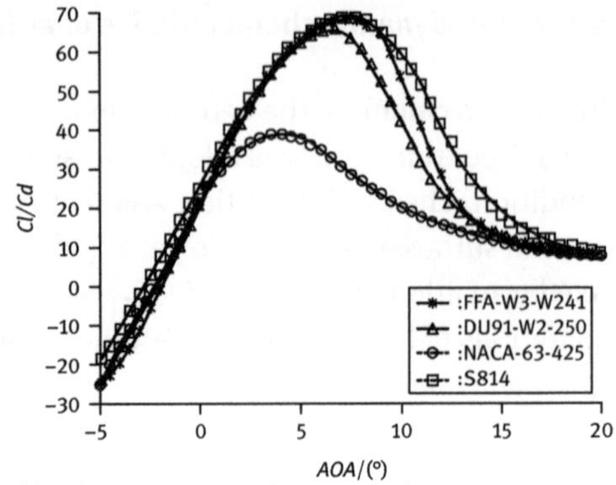

(b) Lift to drag ratio characteristic

Fig. 4.41: Aerodynamic performance of airfoils under fixed transition condition.

roughness sensitivity. Under fixed transition condition, the maximum lift coefficient is in the range of 1.21–1.30 and the maximum lift-to-drag ratio is in the range of 65.1–69.8. The change in aerodynamic performance between smooth and rough conditions of S814 is the smallest among the three wind turbine airfoils. And the maximum lift coefficient decreases by 0.04 ($\Delta c_{l,max} = 0.04$) and the maximum lift-to-drag ratio decreases by 53 ($\Delta L/D_{max} = 53$). Detailed aerodynamic parameters are shown in Tab. 4.13.

Tab. 4.13: Aerodynamic characteristics of the four airfoils.

Airfoil name	Smooth condition		Rough condition	
	$c_{l,max}$	L/D_{max}	$c_{l,max}$	L/D_{max}
FFA-W3-241	1.6019	134.747	1.2512	69.7508
DU 91-W2-250	1.3509	146.112	1.2096	65.0962
NACA 63425	1.2605	143.162	1.1286	38.8079
S814	1.3409	121.304	1.2966	67.8684

4.5.3 The design of a new airfoil with medium thickness

4.5.3.1 The design requirements

For the airfoils used at the middle span of the blade, the aerodynamic and structural properties must be comprehensively considered. An airfoil with relative thickness of about 24 % is generally arranged at spanwise range of 40–60 % ($r/R = 0.4$–0.6). The design guidelines should synthesize the requirements from the inner part and outer part of blade [5, 6], which should meet the following requirements:

(1) Relatively higher (considering the 3D rotational effects) maximum lift coefficient is needed so as to reduce chord and, thereby, the blade load.

(2) In order to ensure maximum energy capture, a good performance under nondesign conditions is needed, which means higher lift-to-drag ratio in a large AOA (considering 3D rotational effects).
(3) The chordwise position of maximum thickness should be controlled between 24 and 35 % to ensure compatibility with other airfoils.
(4) A lower roughness sensitivity (considering 3D rotational effects) is needed. In other words, the maximum lift-to-drag ratio and the maximum lift coefficient are slightly affected by the rough condition. Even if the leading edge of the airfoil is polluted or under certain conditions of manufacturing error, good aerodynamic performance can still be guaranteed.

4.5.3.2 The optimum model

Relative thickness of 0.24–0.25 was selected as constraint for optimization design. Removing the constraint for noise, the rest of the constraints are the same as above. In addition, the 3D rotational effects are considered ($c/r = 0.2$). The weight factors of free transition condition and fixed transition condition are set as: $\lambda_1 \in [0.4, 0.6]$, $\lambda_2 = 1 - \lambda_1$. The optimization objective at system level is: $f(X) = \lambda_1 \cdot f_1(X) + \lambda_2 \cdot f_2(X)$. The wind turbine airfoil generated was the WT247.

4.5.3.3 Performance of the WT247 airfoil

The coefficients of the shape function of the WT247 airfoil are listed in Tab. 4.14.

Taking the coefficients in Tab. 4.12 into the airfoil shape function and after conformal transformation, the airfoil shape and coordinates can be obtained as shown in Fig. 4.42.

As shown in Fig. 4.42, the relative thickness of the WT247 airfoil is 24.7 %. The chordwise position of maximum thickness is 27.0 % ($x/c = 0.270$), which is closer to the aerodynamic center compared with FFA-W3-241. The new airfoil has larger upper surface thickness, larger camber and larger leading edge radius. Meanwhile, the lower surface at the trailing edge of the new airfoil is S-shaped, which is beneficial for a larger lift coefficient.

Tab. 4.14: Coefficients of shape function of airfoil WT247.

	Coefficient			
	C_1	C_2	C_3	C_4
WT247	−0.024044	0.685269	−1.20551	0.782323
	C_5	C_6	C_7	C_8
WT247	−0.257323	0.0461754	−0.00431411	0.00016443

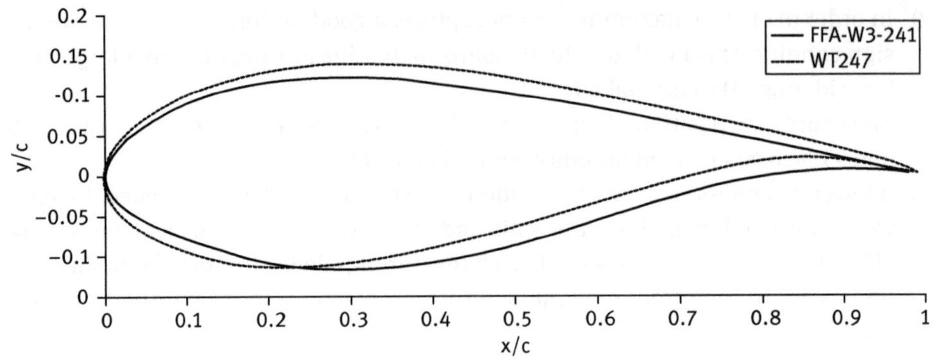

Fig. 4.42: The shape of WT247 and FFA-W3-241.

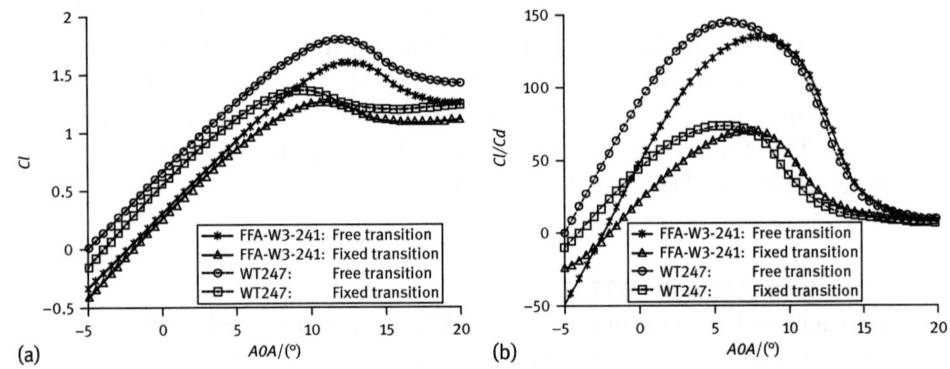

Fig. 4.43: Comparison of aerodynamic characteristics.

The airfoil aerodynamic characteristics under conditions of $Re = 3.0 \times 10^6$, $Ma = 0.16$ and with AOA range from −5° to 20° are shown in Fig. 4.43. In the whole AOA range, the lift coefficients of new airfoil under free and fixed transition conditions are higher than those of FFA-W3-241. Under free transition conditions, the maximum lift coefficient and the maximum lift-to-drag ratio reach 1.80 and 145 respectively; under the fixed transition conditions, the maximum lift coefficient and the maximum lift-to-drag ratio are 1.36 and 74 respectively. The new airfoil has high lift-to-drag ratio over a wide range of AOA so that the rotor can maintain a high power coefficient over a wide range of wind speeds. Tab. 4.15 presents the main geometric and aerodynamic parameters of the new airfoil.

The pressure distribution of the airfoil under three AOAs (with the maximum lift coefficient, maximum drag coefficient and maximum lift-to-drag ratio) and under conditions of $Re = 3.0 \times 10^6$, $Ma = 0.16$ are shown in Fig. 4.44. The maximum lift-to-drag ratio reaches 145.037 at AOA of 6.1°, which is very close to the design AOA. The maximum lift coefficient reaches 1.7977 at AOA of 12.1°. At this time, the transition point

Tab. 4.15: The geometric and aerodynamic parameters of the new airfoil.

Maximum thickness	Chordwise location of maximum thickness	Maximum camber	Chordwise location of maximum camber
0.247415	0.270	0.038052	0.725
Smooth condition		**Rough condition**	
$c_{l,max}$	L/D_{max}	$c_{l,max}$	L/D_{max}
1.7977	145.037	1.365	73.9405

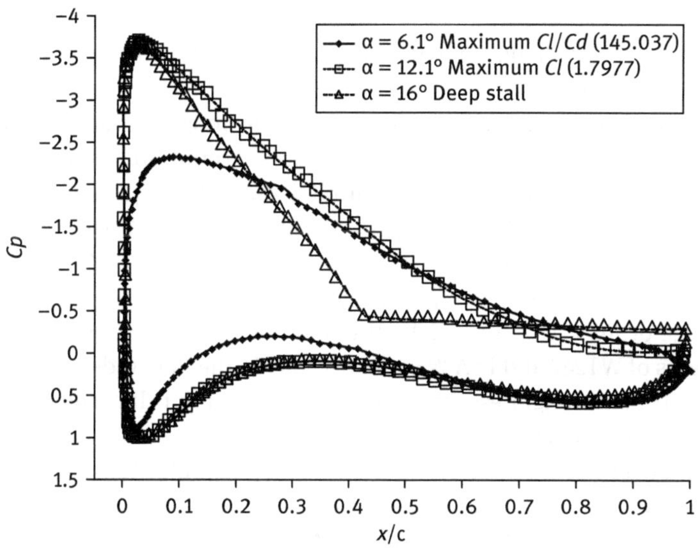

Fig. 4.44: Pressure distribution at three specific AOA.

is located at the leading edge which results in a low roughness sensitivity. In the deep stall region, the pressure on the pressure side of the airfoil was almost unchanged and the pressure on the suction side is obviously different. It is apparent that separation occurs on the suction side, which leads to a decrease in the lift coefficient.

4.5.4 The effects of turbulence, Reynolds number and blade rotation

The Reynolds number, turbulence intensity and rotational effects will affect the airfoil boundary layer condition and flow separation, which will then change the blade aerodynamic characteristics. Therefore, it is necessary to study the performance of the airfoil in these conditions.

4.5.4.1 The influence of Reynolds number

For MW-level wind turbines, the maximum Reynolds number on the rotor can reach up to $8.0–10.0 \times 10^6$ and it generally appears in the middle span of the blade [110, 134]. The variation in the maximum lift coefficient and the maximum lift-to-drag ratio of WT247 and FFA-W3-241, with the Reynolds number in the range of $1.0–8.0 \times 10^6$, is shown in Fig. 4.45. The maximum lift coefficients and the maximum lift-to-drag ratios of the two airfoils increase with increasing Reynolds number. Under free transition conditions, the maximum lift coefficient of the WT247 airfoil increases from 1.6434 ($Re = 1.0 \times 10^6$) to 1.8982 ($Re = 8.0 \times 10^6$). The change in the maximum lift coefficient is larger ($\Delta c_{l,max} = 0.1541$) when the Reynolds number is in the range of $1.0–3.0 \times 10^6$ compared to when the Reynolds number is in the range 3.0×10^6 to 8.0×10^6 ($\Delta c_{l,max} = 0.1007$). Similarly, the change of maximum lift-to-drag ratio is larger in the Reynolds range of $1.0–3.0 \times 10^6$.

As shown in Fig. 4.45, with increasing Reynolds number, the roughness sensitivity of the airfoil gradually decreases. Under the fixed transition condition, the rate of decrease of the maximum lift coefficient of the FFA airfoil is 28.5 % ($\Delta c_{l,max} = 0.4266$) at $Re = 1.0 \times 10^6$, and it decreases to 14.4 % ($\Delta c_{l,max} = 0.2408$) at $Re = 8.0 \times 10^6$. The rate of decrease of the maximum lift coefficient of WT247 drops from 28.5 % ($\Delta c_{l,max} = 0.4777$) at $Re = 1.0 \times 10^6$ to 18.9 % ($\Delta c_{l,max} = 0.3581$) at $Re = 8.0 \times 10^6$. In the Reynolds number range of $Re = 1.0–3.0 \times 10^6$, the roughness sensitivities for the maximum lift coefficients of WT247 and FFA-W3-241 are close. In the Reynolds number range of $Re = 3.0–8.0 \times 10^6$, the roughness sensitivities for the maximum lift coefficient of FFA-W3-241 is slightly better than that of WT247. Within the whole Reynolds number range, the maximum lift coefficient and the maximum lift-to-drag ratio of WT247 are larger than those of FFA-W3-241, both under free transition and the fixed transition condition, which shows the superior characteristics of WT247.

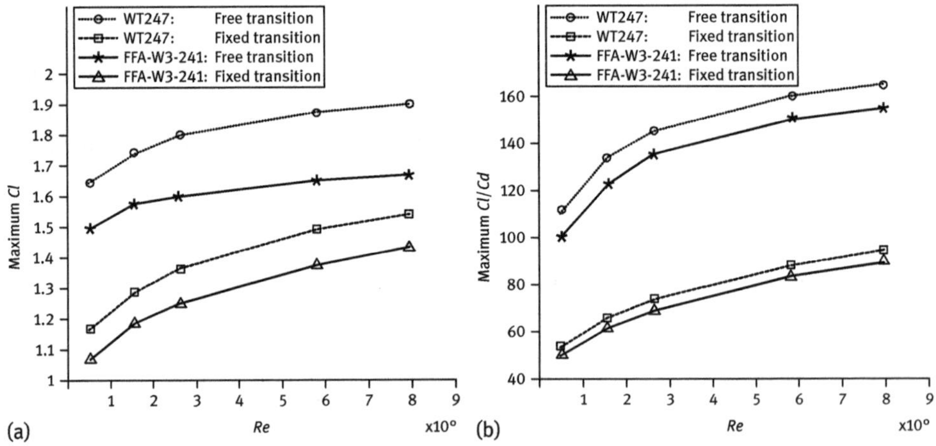

Fig. 4.45: Influence of Reynolds numbers on airfoil performance.

4.5.4.2 Turbulent flow conditions

Under the operating conditions for wind turbine blades, the inflow turbulence level will influence the aerodynamic performance of the airfoil. In RFOIL software, the degree of inflow turbulence is specified through the parameter N_{crit}. The range of N_{crit} parameters [99] in typical cases are shown below.

- Glider: $N_{crit} = 11–14$;
- Wind tunnel with lower turbulence level: $N_{crit} = 10–12$;
- Ordinary wind tunnel: $N_{crit} = 9$;
- Wind tunnel with higher turbulence level: $N_{crit} = 4–8$.

$N_{crit} = 4$ was selected here to calculate the aerodynamic performance with higher turbulence conditions, as shown in Fig. 4.46. The maximum lift coefficient declines from 1.7977 to 1.6978 ($\Delta c_{l,max} = 0.0999$), the maximum lift-to-drag force reduces from 14.5 to 124.5 ($\Delta L/D_{max} = 20.5$). Thus, the new airfoil is not sensitive to turbulent conditions which satisfies the working requirements of wind turbine airfoils.

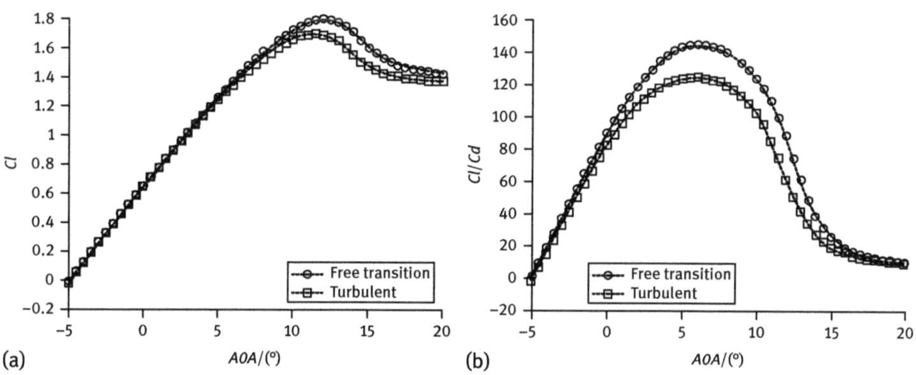

Fig. 4.46: Aerodynamic performance under free transition and turbulent flow.

4.5.4.3 Three-dimensional rotational effects

The influences of 3D rotational effects on airfoil aerodynamic performance (free transition and the transition condition $c/r = 0.2$) are shown in Fig. 4.47. The boundary layer thickness becomes thinner with the rotation effects, so that boundary layer stability is enhanced. The maximum lift coefficient of the airfoil under the free transition condition increases by 0.18. Under the fixed transition condition, the maximum lift coefficient at AOA of 20° increases by a large margin of 0.58.

With 3D rotational effects and the fixed transition condition, the lift coefficient maintained a continuous increase in an AOA range from –5° to 20°, which reduces the gap to that of the free transition condition. The roughness sensitivity of the airfoil is reduced by rotational effects. According to [4, 5], rotational effects can basically elimi-

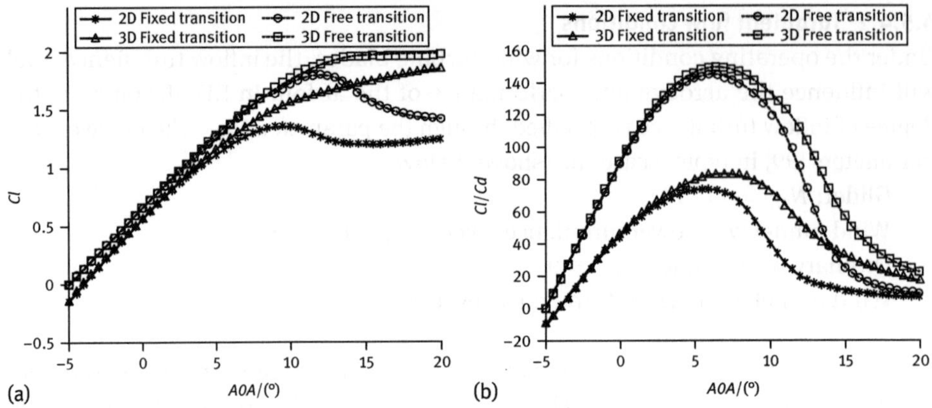

Fig. 4.47: Aerodynamic performance with and without 3D rotational effects.

nate the roughness sensitivity of the airfoils located near the blade root. Meanwhile, compared with two-dimensional airfoil aerodynamic results, the lift-to-drag ratio is greatly improved in the AOA range of 5° to 20°.

4.6 Design of airfoils based on noise

4.6.1 Acoustic theory for wind turbines

The noise generated from wind turbines can be divided into three groups according to the mechanism of aerodynamic noise: low frequency noise, inflow turbulence noise and noise generated by airfoils.

Low frequency noise originates from the pressure pulses caused by interaction between inflow variations (due to tower shadow, wind shear, wake effects, etc.) and rotating blades. It is periodic and discrete noise with frequency of an integral multiple of blade passing frequency [92].

Inflow turbulence noise is broadband noise caused by the vortices originating from the interaction between blade and inflow turbulence. It is related to rotor speed, the profile of the airfoil section and the turbulence intensity of inflow.

The noise generated by the airfoil is mainly broadband noise, and it also includes tonal noise. Airfoil-generated noise is divided into the following groups [135]:

(1) Trailing edge noise. This is formed by the interaction between the turbulent boundary layer and the blade trailing edge, has the frequency range of 750–2000 Hz and is the main high frequency noise produced by the wind turbine.
(2) Blade tip noise. This is produced by 3D blade tip effects and is one of the main noise sources of wind turbines. In addition, the vortex generator at the blade tip and the sudden movement of the blade surface will also cause the impulse noise.

(3) Noise caused by separation. The unsteady flow on blade surface when the blade is at stall, makes the radiation of broadband noise increase.
(4) Blunt trailing edge noise. This is produced by the vortex caused by blade trailing edge thickness, and is a kind of tonal noise.
(5) Laminar vortex noise. This is also a kind of tonal noise, which is caused by the laminar vortex formed by the holes and gaps on the blade surface.

4.6.2 The measurement of noise

Noise can be objectively measured by SPL (sound pressure level), sound intensity (sound intensity level), acoustic power (sound power level) and frequency (spectrum) etc. It can be also subjectively felt by sound loudness, tone and timbre.

4.6.2.1 Sound pressure

The sound pressure p indicates the sound intensity, which is the difference between the local air pressure caused by sound waves and the average atmospheric pressure (the unit is Pa). The sound pressure level (SPL) L_p can be expressed as

$$L_p = 10\lg\left(\frac{p}{p_0}\right)^2, \tag{4.32}$$

where p is sound pressure, p_0 is reference sound pressure ($p_0 = 2 \times 10^{-5}$ Pa). The reference pressure is the sound pressure at noise frequency of 1000 Hz (the limit value that can be heard by normal human ears). Therefore, it is also known as the threshold pressure. The unit of sound pressure level L_p is dB.

When considering the change of sound pressure with time at a certain point in the sound field, the root mean square value is adopted, which is also called the equivalent sound pressure p_{eq} expressed as

$$p_{eq} = \sqrt{\frac{1}{T}\int_0^T p^2\,dt}, \tag{4.33}$$

where p is the instantaneous sound pressure, T is the measuring time (generally as long as 30–180 s).

Then, the equivalent sound pressure level can be expressed as

$$L_{eq} = 10\lg\left(\frac{1}{T}\int_0^T \frac{p^2}{p_0^2}\,dt\right). \tag{4.34}$$

4.6.2.2 Sound intensity

Sound intensity I is defined as the sound energy per unit time and per unit area (perpendicular to the wave propagation direction), with the unit of W/m^2. Sound intensity level L_I can be expressed as

$$L_I = 10 \lg \frac{I}{I_0}, \tag{4.35}$$

where I is the sound intensity, I_0 is the reference sound intensity, $I_0 = 10^{-12}\ W/m^2$. The reference sound intensity corresponds to the reference pressure and, therefore, is also known as threshold sound intensity. The unit of sound intensity level L_I is dB.

When the sound is transmitted through a spherical sound wave, the sound intensity on the spherical surface (the distance to the sound source is r) is

$$I = \frac{W}{4\pi r^2}. \tag{4.36}$$

When the sound wave propagates only in a half space, the sound intensity on the spherical surface (the distance to the sound source is r) is

$$I = \frac{W}{2\pi r^2}. \tag{4.37}$$

4.6.2.3 Sound power

Sound power W is the sound energy (emitted, reflected, transmitted or received) per unit time, and the unit is W. The sound power level L_W can be expressed as

$$L_W = 10 \lg \frac{W}{W_0}, \tag{4.38}$$

where W is the sound power, W_0 is reference sound power, $W_0 = 10^{-12}$ W. The reference sound power corresponds to the sound pressure reference, which is also called the threshold power. The unit of sound power level L_W is dB.

4.6.2.4 Weighted sound pressure level

The sensitivity of the human ear to sound is not only related to sound pressure, but also to frequency. Sounds with the same sound pressure level and different frequencies may sound different. Therefore, the weighted pressure level is adopted to carry out the subjective assessment of noise. The equivalent A-weighted sound pressure level $L_{A,eq}$ can be expressed as

$$L_{A,eq} = 10 \lg \left(\frac{1}{T} \int_0^T \frac{p_A^2}{p_0^2} \, dt \right), \tag{4.39}$$

where p_A is A-weighted instantaneous sound pressure, p_0 is the reference sound pressure, $p_0 = 2 \times 10^{-5}$ Pa. The unit of equivalent A-weighted sound pressure level is

Tab. 4.16: Weighted relative response values of different frequencies.

1/3 octave center frequency (Hz)	A-weighted response value (dB)
200	−10.9
250	−8.6
315	−6.6
400	−4.8
500	−3.2
630	−1.9
800	−0.8
1000	0
1250	0.6
1600	1.0
2000	1.2
2500	1.3
3150	1.2
4000	1.0
5000	0.5

dB (A). For different frequencies, the A-weighted relative response value (partial) is shown in Tab. 4.16.

In addition, the weighted sound power level is adopted to carry out subjective noise assessments. The calculation method is similar to the A-weighted sound pressure level.

4.6.2.5 Frequency spectrum

The noise spectrum is adopted to describe the characteristics of noise, that is, the noise signal is divided into frequency bands and then accumulated in the frequency band. The analysis of the noise spectrum can provide a basis for noise control.

The sound frequency range that the human ear can hear is roughly from 20 Hz to 20 kHz. Usually the 1 Hz bandwidth is not adopted, instead, the octave bandwidth or the 1/3 octave bandwidth are utilized. Octave refers to bandwidth in which the upper frequency limit is twice that of the lower frequency limit. So the ratio of the two frequencies is 2 : 1. One-third octave bandwidth is obtained through dividing the octave bandwidth into three sections. The upper frequency limit of each section is $\sqrt[3]{2}$ times that of the lower frequency limit.

4.6.3 The acoustics model of the airfoil

Brooks, Pope and Marcolini from NREL proposed a semi-empirical formula for noise calculation for wind turbine airfoils, which is obtained by summarizing the experimental data of the NACA 0012 airfoil. However, airfoil shapes and the inflow conditions on the wind turbine blade vary with blade spanwise locations which lead to certain errors when predicting the boundary layer parameters according to the empirical formula [136–138]. The formula has many intrinsic limitations for the calculation of airfoil noise. Therefore, the formula is to be improved so as to be suitable for different airfoils under various conditions. In the following, the BPM (Brooks, Pope and Marcolini) semi-empirical formula for noise calculation of the NACA 0012 airfoil will be discussed.

4.6.3.1 Trailing edge noise of the turbulent boundary layer (TBL-TE)

When the turbulent boundary layer, attached to the blade surface, flows through the trailing edge, TBL-TE can be generated. This is one of the main noise sources of large wind turbines. Under certain AOA and Reynolds number, transition will happen at some point, turning the laminar flow into turbulent flow. Turbulent flow on the pressure side and the suction surfside of the trailing edge will cause fluctuating pressure which will then produce noise. Under a small AOA, the pressure and suction side are the main noise sources [139–141]. As shown in Fig. 4.48, the influence of trailing edge shape on noise is relatively large.

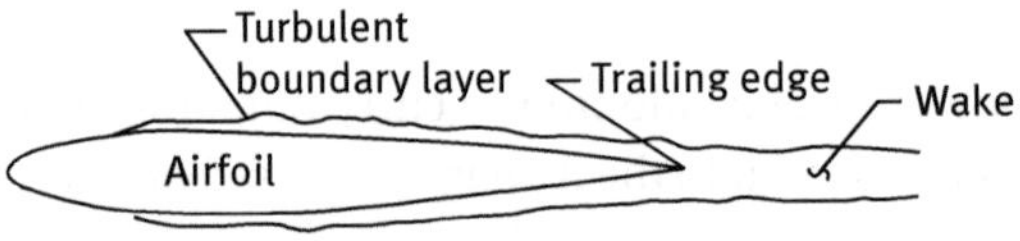

Fig. 4.48: Trailing edge noise of the turbulent boundary layer.

The TBL-TE is composed of the noise SPL_p generated by the turbulent boundary layer on the pressure side, and the noise SPL_s generated on the suction side [135]:

$$SPL_{TBLTE} = 10\lg\{10^{SPL_s/10} + 10^{SPL_p/10}\}, \tag{4.40}$$

$$SPL_s = 10\lg\left(\frac{\delta_s^* M^5 \Delta l \bar{D}_h}{r^2}\right) + A\left(\frac{St_s}{St_1}\right) + (W_1 - 3), \tag{4.41}$$

$$SPL_p = 10\lg\left(\frac{\delta_p^* M^5 \Delta l \bar{D}_h}{r^2}\right) + A\left(\frac{St_p}{St_1}\right) + (W_1 - 3) + \Delta W_1, \tag{4.42}$$

where δ_p^* is the relative thickness of the boundary layer on the pressure side of the airfoil and δ_s^* is the relative thickness of the boundary layer on the suction side of the airfoil, both are associated with AOA and Reynolds number Re of the airfoil; M is the inflow Mach number, which is the ratio of relative inflow speed U and sound speed c_0; St is the Strouhal number; $\bar{D}_h$ is the direction function of high frequency sound; r is the distance from the sound source to observers; A is the frequency spectrum shape function; W_1 is amplitude function; ΔW_1 is sound pressure level correction function; Δl is the spanwise length of airfoil section.

4.6.3.2 Noise caused by separation (SEP)

With increasing AOA, the boundary layer will separate and the turbulent vortex on the suction surface of the airfoil is larger than that under low AOA. When the vortex moves into the wake, separation noise is produced. When the AOA increases to a certain value, large-scale separation of boundary layer (deep stall) occurs and the whole suction surface is under unsteady flow [142, 143], which is shown in Fig. 4.49. Under this condition, SEP is the main noise.

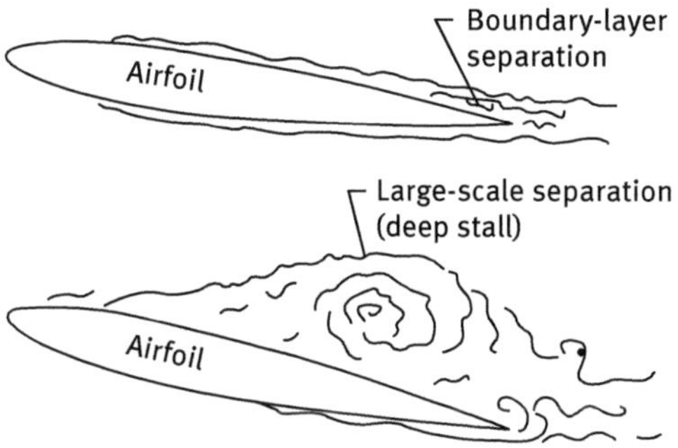

Fig. 4.49: Separation noise.

The calculation equation of separation noise is as follows [135]:

$$\mathrm{SPL}_{\mathrm{SEP}} = 10\lg\left(\frac{\delta_{\mathrm{s}}^{*} M^{5} \Delta l \bar{D}_{\mathrm{h}}}{r^{2}}\right) + B\left(\frac{St_{\mathrm{s}}}{St_{2}}\right) + W_{2}, \tag{4.43}$$

where B is the frequency spectrum shape function; W_2 is the amplitude function.

In order to facilitate the analysis, the TBL-TE and SEP are generally considered together. When the AOA satisfies $\alpha \leq 12.5°$, the equations (4.40)–(4.43) are adopted to calculate the noise. When the AOA satisfies $\alpha > 12.5°$, the following equations are used:

$$\mathrm{SPL}_{\mathrm{p}} = -\infty, \tag{4.44}$$

$$\mathrm{SPL}_{\mathrm{s}} = -\infty, \tag{4.45}$$

$$\mathrm{SPL}_{\mathrm{SEP}} = 10\lg\left(\frac{\delta_{\mathrm{s}}^{*} M^{5} \Delta l \bar{D}_{\mathrm{h}}}{r^{2}}\right) + A'\left(\frac{St_{\mathrm{s}}}{St_{2}}\right) + W_{2}. \tag{4.46}$$

In equation (4.46), $\bar{D}_{\mathrm{h}}$ is the direction function of high frequency; A and A' are the frequency spectrum shape function, which is related to the Reynolds number R_{c} (based on chord) and is three times its actual value.

The equations for calculating the Strouhal number on the pressure and the suction side surfaces are respectively:

$$St_{\mathrm{p}} = \frac{f\delta_{\mathrm{p}}^{*}}{U}, \tag{4.47}$$

$$St_{\mathrm{s}} = \frac{f\delta_{\mathrm{s}}^{*}}{U}, \tag{4.48}$$

where $\delta_p^* = \delta_p^*(\alpha, R_c)$ and $\delta_s^* = \delta_s^*(\alpha, R_c)$ represent the boundary layer displacement thickness on the pressure side and the suction side of the airfoil, which are related to the AOA and Reynolds number R_c; U is the inflow velocity; f is 1/3 octave band.

The equations of St_1 and St_2 are:

$$St_1 = 0.02M^{-0.6}; \tag{4.49}$$

$$St_2 = St_1 \times \begin{cases} 1 & (\alpha < 1.33°), \\ 10^{0.0054(\alpha-1.33)^2} & (1.33° \le \alpha \le 12.5°), \\ 4.72 & (12.5° < \alpha). \end{cases} \tag{4.50}$$

The frequency spectrum shape functions A and B are as follows:

$$A(a) = A_{\min}(a) + A_R(a_0)[A_{\max}(a) - A_{\min}(a)], \tag{4.51}$$

$$B(b) = B_{\min}(b) + B_R(b_0)[B_{\max}(b) - B_{\min}(b)], \tag{4.52}$$

where the curves $A_{\max}(a)$ and $A_{\min}(a)$ are defined as

$$A_{\min}(a) = \begin{cases} \sqrt{67.552 - 886.788a^2} - 8.219 & (a < 0.204), \\ -32.665a + 3.981 & (0.204 \le a \le 0.244), \\ -142.795a^3 + 103.656a^2 & (0.244 < a); \\ \quad - 57.757a + 6.006 & \end{cases} \tag{4.53}$$

$$A_{\max}(a) = \begin{cases} \sqrt{67.552 - 886.788a^2} - 8.219 & (a < 0.13), \\ -15.901a + 1.098 & (0.13 \le a \le 0.244), \\ -4.669a^3 + 3.491a^2 & (0.244 < a); \\ \quad - 16.699a + 1.149 & \end{cases} \tag{4.54}$$

where a represents the ratio of the Strouhal number

$$a = |\log(St/St_{\text{peak}})|, \tag{4.55}$$

where $St = St_p$ or St_s, $St_{\text{peak}} = St_1$, St_2 or $(St_1 + St_2)/2$.

$A_R(a_0)$ is an interpolation coefficient, which is defined as

$$A_R(a_0) = \frac{-20 - A_{\min}(a_0)}{A_{\max}(a_0) - A_{\min}(a_0)}; \tag{4.56}$$

$a_0(R_c)$ is a function related to Reynolds number

$$a_0(R_c) = \begin{cases} 0.57 & (R_c < 9.52 \times 10^4), \\ (-9.57 \times 10^{-13}) & (9.52 \times 10^4 \le R_c \le 8.57 \times 10^5), \\ \quad \times (R_c - 8.57 \times 10^5)^2 + 1.13 & \\ 1.13 & (8.57 \times 10^5 < R_c). \end{cases} \tag{4.57}$$

Similarly, the equation for frequency spectrum shape function B is

$$B_{\min}(b) = \begin{cases} \sqrt{16.888 - 886.788b^2} - 4.109 & (b < 0.13), \\ -83.607b + 8.138 & (0.13 \le b \le 0.145), \\ -817.81b^3 + 355.21b^2 & (0.145 < b); \\ \quad - 135.024b + 10.619 & \end{cases} \tag{4.58}$$

$$B_{\max}(b) = \begin{cases} \sqrt{16.888 - 886.788b^2} - 4.109 & (b < 0.1), \\ -31.33b + 1.854 & (0.1 \le b \le 0.145), \\ -80.541b^3 + 44.174b^2 & (0.145 < b); \\ \quad - 39.381b + 2.344 & \end{cases} \tag{4.59}$$

where b represents the ratio of Strouhal numbers:

$$b = \left|\log\left(\frac{St_s}{St_2}\right)\right|; \tag{4.60}$$

$B_R(b_O)$ is the interpolation coefficient, which is defined as

$$B_R(b_O) = \frac{-20 - B_{\min}(b_O)}{B_{\max}(b_O) - B_{\min}(b_O)}; \tag{4.61}$$

$b_O(R_c)$ is a function related to the Reynolds number

$$b_O(R_c) = \begin{cases} 0.30 & (R_c < 9.52 \times 10^4), \\ (-4.48 \times 10^{-13}) & (9.52 \times 10^4 \le R_c \le 8.57 \times 10^5), \\ \quad \times (R_c - 8.57 \times 10^5) + 0.56 & \\ 0.56 & (8.57 \times 10^5 < R_c). \end{cases} \tag{4.62}$$

The amplitude functions W_1 and W_2 are defined as

$$W_1 = \begin{cases} -4.31\log(R_c) + 156.3 & (R_c < 2.47 \times 10^5), \\ -9.01\log(R_c) + 181.6 & (2.47 \times 10^5 \le R_c \le 8.0 \times 10^5), \\ 128.5 & (8.0 \times 10^5 < R_c); \end{cases} \tag{4.63}$$

$$W_2 = W_1 + \begin{cases} -1000 & (\alpha < \gamma_O - \gamma), \\ \sqrt{\beta^2 - (\beta/\gamma)^2(\alpha - \gamma_O)^2} + \beta_O & (\gamma_O - \gamma \le \alpha \le \gamma_O + \gamma), \\ -12 & (\gamma_O + \gamma < \alpha). \end{cases} \tag{4.64}$$

The coefficients in equation (4.64) are defined as

$$\begin{cases} \gamma = 27.094M + 3.31, \\ \gamma_O = 23.43M + 4.651, \\ \beta = 72.65M + 10.74, \\ \beta_O = -34.19M - 13.82, \end{cases} \tag{4.65}$$

where M is Mach number, $M = U/c_0$; ΔW_1 is the sound pressure level correction function for the pressure side of the airfoil which is shown below

$$\Delta W_1 = \begin{cases} \alpha[1.43\log(R_{\delta_p^*}) - 5.29] & (R_{\delta_p^*} \le 5000), \\ 0 & (5000 < R_{\delta_p^*}), \end{cases} \tag{4.66}$$

where $R_{\delta_p^*}$ is the Reynolds number based on the displacement thickness of the pressure surface.

4.6.3.3 Laminar vortex noise (LBL-VS)

As shown in Fig. 4.50, vortex noise is mainly caused by the vortex shedding from the trailing edge and the unsteady flow originating from the upstream laminar boundary layer. When the vortex is shedding from the trailing edge, the inducted pressure will spread upstream which leads to fluctuation in the boundary layer; when the unstable boundary layer reaches the trailing edge, the vortex will shed. Therefore, laminar boundary layer vortex shedding noise is generated through this periodic process [144–146]. Laminar vortex noise is a kind of discrete noise, which can be ignored for large wind turbines.

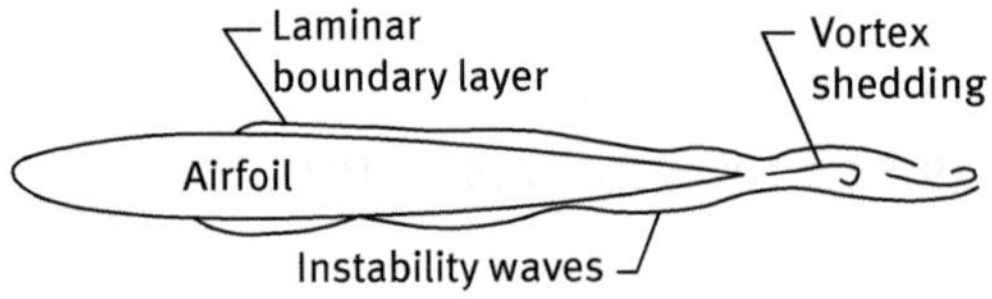

Fig. 4.50: Laminar vortex noise.

The noise prediction equation is [135]:

$$\mathrm{SPL_{LBLVS}} = 10\lg\left(\frac{\delta_p M^5 \Delta l \bar{D}_h}{r^2}\right) + G_1\left(\frac{St'}{St'_{peak}}\right) + G_2\left(\frac{Re}{Re_0}\right) + G_3(\alpha), \tag{4.67}$$

where δ_p represents the boundary layer thickness on the pressure side of the airfoil; G_1, G_2 and G_3 are the empirical functions based on Strouhal number, Reynolds number and angle of attack, respectively. In fact, they are generally neglected according to the design condition of the airfoil.

The Strouhal number St' can be expressed as

$$St' = \frac{f\delta_p}{U}; \tag{4.68}$$

St'_{peak} is the maximum Strouhal number

$$St'_{peak} = St' \times 10^{-0.04\alpha}, \tag{4.69}$$

where St'_{peak} is defined as

$$St' = \begin{cases} 0.18 & (R_c \le 1.3 \times 10^5), \\ 0.001756 R_c^{0.3931} & (1.3 \times 10^5 \le R_c \le 4.0 \times 10^5), \\ 0.28 & (4.0 \times 10^5 < R_c). \end{cases} \quad (4.70)$$

Function G_1 defines the shape of the spectrum:

$$G_1(e) = \begin{cases} 39.8 \log(e) - 11.12 & (e \le 0.5974), \\ 98.409 \log(e) + 2.0 & (0.5974 < e \le 0.8545), \\ -5.076 + \sqrt{2.484 - 506.25[\log(e)]^2} & (0.8545 < e \le 1.17), \\ -98.409 \log(e) + 2.0 & (1.17 < e \le 1.674), \\ -39.8 \log(e) - 11.12 & (1.674 < e), \end{cases} \quad (4.71)$$

where $e = St'/St'_{\text{peak}}$.

The peak of the spectrum is determined by the function G_2

$$G_2(d) = \begin{cases} 77.852 \log(d) + 15.328 & (d \le 0.3237), \\ 65.188 \log(d) + 9.125 & (0.3237 < d \le 0.5689), \\ -114.052[\log(d)]^2 & (0.5689 < d \le 1.7579), \\ -65.188 \log(d) + 9.125 & (1.7579 < d \le 3.0889), \\ -77.852 \log(d) + 15.328 & (3.0889 < d), \end{cases} \quad (4.72)$$

where $d = R_c/(R_c)_0$

$$(R_c)_0 = \begin{cases} 10^{0.215\alpha + 4.978} & (\alpha \le 3.0), \\ 10^{0.120\alpha + 5.263} & (3.0 < \alpha); \end{cases} \quad (4.73)$$

G_3 is the function of angle of attack:

$$G_3 = 171.04 - 3.03\alpha. \quad (4.74)$$

4.6.3.4 Blade tip noise (TIP)

Because of 3D flow effects, a tip vortex will be formed at the blade tip. Interaction between the tip vortex and the trailing edge at the tip can cause noise. The generation mechanism for this tip noise is similar to that for the previous described trailing edge noise. What is more, if tip flow is separated, additional noise will occur [147, 148]. As shown in Fig. 4.51, tip vortex noise is closely related to the geometry of the blade tip. The tip shape of an old wind turbine blade is square and thick, so the noise is loud. Currently, this noise can be reduced through good design.

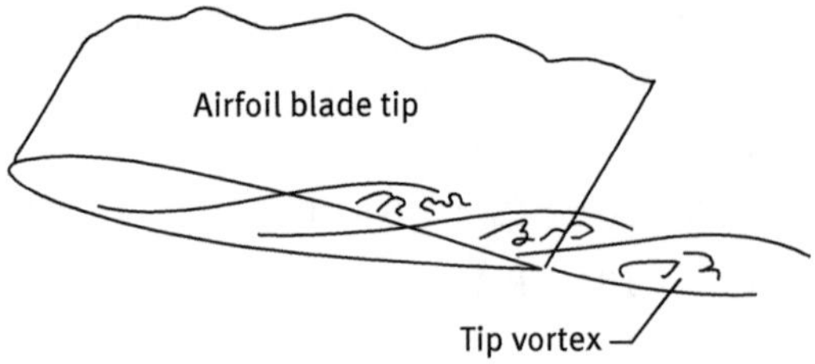

Fig. 4.51: Blade tip noise.

The calculation equations for the vortex noise [135] are as follows:

$$\mathrm{SPL_{TIP}} = 10\lg\left(\frac{M_{\max}^3 M^2 l_{\mathrm{tip}}^2 \bar{D}_{\mathrm{h}}}{r^2}\right) - 30.5(\lg St'' + 0.3)^2 + 126; \tag{4.75}$$

l_{tip} is the spanwise length of the blade tip. For a round blade tip, it is

$$l_{\mathrm{tip}} = 0.008\alpha_{\mathrm{tip}}c. \tag{4.76}$$

For a flat tip, it is

$$l_{\mathrm{tip}}/c = \begin{cases} 0.0230 + 0.0169\alpha'_{\mathrm{tip}} & (0^\circ \le \alpha'_{\mathrm{tip}} \le 2^\circ), \\ 0.0378 + 0.0095\alpha'_{\mathrm{tip}} & (2^\circ < \alpha'_{\mathrm{tip}}), \end{cases} \tag{4.77}$$

where α_{tip} is the geometry AOA at tip; α'_{tip} is the modified AOA at tip, which is based on tip load characteristics (mainly according to the ideal conditions with large spanwise length, free bending and torsion, constant flow [135])

$$\alpha'_{\mathrm{tip}} = \left[\left(\frac{\partial L'/\partial y}{(\partial L'/\partial y)_{\mathrm{ref}}}\right)_{y\to\mathrm{tip}}\right]\alpha_{\mathrm{tip}}, \tag{4.78}$$

where L' is the lift force per unit span at spanwise positon of y.

The Strouhal number St'' is

$$St'' = \frac{fl}{U_{\max}}, \tag{4.79}$$

where $U_{\max} = c_0 M_{\max}$, $M_{\max}$ is the maximum velocity of fluid in the trailing edge stall region.

$$M_{\max}/M \approx (1 + 0.036\alpha_{\mathrm{tip}}), \tag{4.80}$$

where M is the inflow speed in the tip region.

4.6.3.5 Blunt trailing edge noise (TEB-VS)

As shown in Fig. 4.52, when the trailing edge thickness increases to a certain value, vortices will shed from the blade trailing edge which leads to the Karman vortex street. The periodic loads will appear on both the top and bottom of the blade trailing edge.

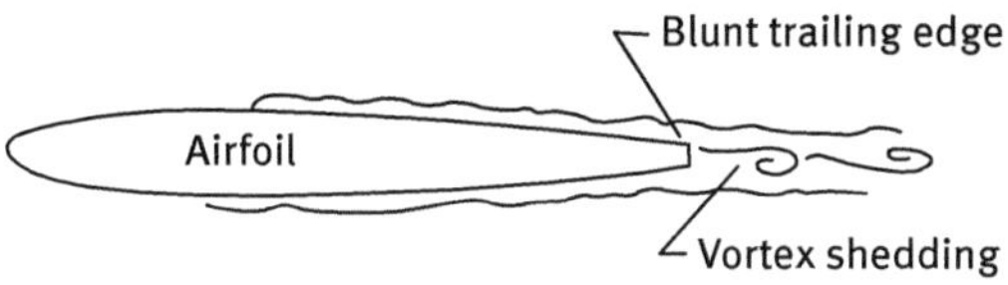

Fig. 4.52: Blunt trailing edge noise.

Then, discrete blunt trailing edge noise is generated which depends on the shape, thickness and local Reynolds number on the blade's trailing edge [140]. This kind of noise can be reduced by decreasing the thickness of the trailing edge.

The equation of blunt trailing edge noise [135] is as follows:

$$\mathrm{SPL}_{\mathrm{TEBVS}} = 10\lg\left(\frac{hM^{5.5}\Delta l\bar{D}_{\mathrm{h}}}{r^2}\right) + G_4\left(\frac{h}{\delta^*_{\mathrm{avg}}}, \Psi\right) + G_5\left(\frac{h}{\delta^*_{\mathrm{avg}}}, \Psi, \frac{St'''}{St'''_{\mathrm{peak}}}\right), \tag{4.81}$$

where h is the thickness of the blunt trailing edge; Ψ is the dullness angle of the trailing edge; δ^*_{avg} represents the average relative thickness of the boundary layer; G_4 and G_5 are the frequency spectrum shape function.

The Strouhal number is

$$St'' = \frac{fh}{U}; \tag{4.82}$$

St''_{peak} is the maximum Strouhal number

$$St''_{\mathrm{peak}} = \begin{cases} \dfrac{0.212 - 0.0045\Psi}{1 + 0.235(h/\delta^*_{\mathrm{avg}})^{-1} - 0.0132(h/\delta^*_{\mathrm{avg}})^{-2}} & (0.2 \le h/\delta^*_{\mathrm{avg}}), \\ 0.1(h/\delta^*_{\mathrm{avg}}) + 0.095 - 0.00243\Psi & (h/\delta^*_{\mathrm{avg}} < 0.2), \end{cases} \tag{4.83}$$

where δ^*_{avg} is the average displacement thickness of the boundary layer

$$\delta^*_{\mathrm{avg}} = \frac{\delta^*_{\mathrm{p}} + \delta^*_{\mathrm{s}}}{2}. \tag{4.84}$$

The peak value of sound pressure spectrum shape for blunt trailing edge noise is defined by the function G_4

$$G_4(h/\delta^*_{\mathrm{avg}}, \Psi) = \begin{cases} 17.5\log(h/\delta^*_{\mathrm{avg}}) + 157.5 - 1.114\Psi & (h/\delta^*_{\mathrm{avg}} \le 5), \\ 169.7 - 1.114\Psi & (5 < h/\delta^*_{\mathrm{avg}}). \end{cases} \tag{4.85}$$

The shape of the frequency spectrum shape function is defined by G_5, which includes the interpolation calculation from $\Psi = 0°$ to $\Psi = 14°$

$$G_5\left(\frac{h}{\delta^*_{\mathrm{avg}}}, \Psi, \frac{St'''}{St'''_{\mathrm{peak}}}\right) = (G_5)_{\Psi=0°} + 0.0714\Psi[(G_5)_{\Psi=14°} - (G_5)_{\Psi=0°}]. \tag{4.86}$$

Here $(G_5)_{\Psi=14°}$ is defined as

$$(G_5)_{\Psi=14°} = \begin{cases} m\eta + k & (\eta < \eta_0), \\ 2.5\sqrt{1-(\eta/\mu)^2} - 2.5 & (\eta_0 \le \eta < 0), \\ \sqrt{1.5625 - 1194.99\eta^2} - 1.25 & (0 \le \eta < 0.03616), \\ -155.543\eta + 4.375 & (0.03616 \le \eta), \end{cases} \tag{4.87}$$

where

$$\eta = \log(St'''/St'''_{peak}); \tag{4.88}$$

$$\mu = \begin{cases} 0.1221 & (h/\delta^*_{avg} < 0.25), \\ -0.2175(h/\delta^*_{avg}) + 0.1755 & (0.25 \le h/\delta^*_{avg} < 0.62), \\ -0.0308(h/\delta^*_{avg}) + 0.0596 & (0.62 \le h/\delta^*_{avg} < 1.15), \\ 0.0242 & (1.15 \le h/\delta^*_{avg}); \end{cases} \tag{4.89}$$

$$m = \begin{cases} 0 & (h/\delta^*_{avg} \le 0.02), \\ 68.724(h/\delta^*_{avg}) - 1.35 & (0.02 < h/\delta^*_{avg} \le 0.5), \\ 308.475(h/\delta^*_{avg}) - 121.23 & (0.5 < h/\delta^*_{avg} \le 0.62), \\ 224.811(h/\delta^*_{avg}) - 69.35 & (0.62 < h/\delta^*_{avg} \le 1.15), \\ 1583.28(h/\delta^*_{avg}) - 1631.59 & (1.15 < h/\delta^*_{avg} \le 1.2); \\ 268.344 & (1.2 < h/\delta^*_{avg}) \end{cases} \tag{4.90}$$

$$\eta_0 = -\sqrt{\frac{m^2\mu^4}{6.25 + m^2\mu^2}}; \tag{4.91}$$

$$k = 2.5\sqrt{1 - \left(\frac{\eta_0}{\mu}\right)^2} - 2.5 - m\eta_0. \tag{4.92}$$

The term $(G_5)_{\Psi=0^\circ}$ can be obtained through substituting h/δ^*_{avg} in equation (4.90) as $(h/\delta^*_{avg})'$, which is shown as

$$(h/\delta^*_{avg})' = 6.724(h/\delta^*_{avg})^2 - 4.019(h/\delta^*_{avg}) + 1.107. \tag{4.93}$$

From the equations for the five kinds of noise, it is found that the trailing edge noise of the turbulent boundary layer, noise of separation, and blunt trailing edge noise are associated with the airfoil boundary layer parameters. What is more, they are the main types of airfoil noise. In the BPM model, the airfoil boundary layer parameters are based on experimental data obtained from the NACA 0012 airfoil [135]. In order to obtain noise characteristics under different conditions, the calculation method for the airfoil boundary layer parameters had been improved.

The previously mentioned XFOIL software can solve the coupled flow and boundary layer equations. With the transition prediction method based on the laminar flow stability e^N theory, XFOIL can deal with free transition and fixed transition flow and can solve aerodynamic parameters (including the parameters of airfoil boundary layer) at various flow conditions. The calculation results are accurate and reliable. XFOIL was used to solve the boundary layer parameters instead of using the empirical equations in the BPM model, thus broadening the scope of the original noise model and improving the accuracy of the calculation. The airfoil boundary layer parameters can be expressed as a function of Reynolds number, AOA, and turbulence intensity:

$$\delta = \delta(Re, \alpha, tur). \tag{4.94}$$

Entering the corresponding parameters of Reynolds number *Re*, angle of attack α and turbulence intensity *tur* into XFOIL, the airfoil boundary parameter files can be obtained. Then through reading the boundary layer parameters, the noise value can be calculated with the noise model above.

4.6.3.6 The direction function of noise

In the 3D coordinate system (shown in Fig. 4.53), the thin flat plate represents the airfoil plane which is moving with speed U along the x-direction and the observer is stationary.

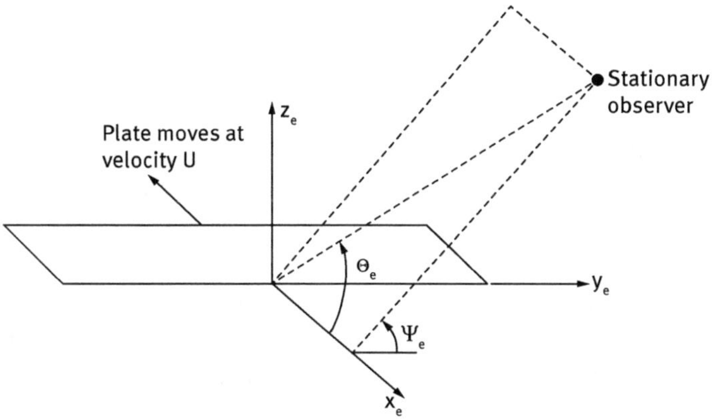

Fig. 4.53: The direction of sound.

For high frequency noise, the direction function is

$$\bar{D}_h(\Theta_e, \Phi_e) \approx \frac{2\sin^2(0.5\Theta_e)\sin^2(\Phi_e)}{(1 + M\cos\Theta_e)[1 + (M - M_c)\cos\Theta_e]^2}, \tag{4.95}$$

where M represents the velocity of an airfoil; M_c represents the fluid velocity at the trailing edge; Θ_e and Φ_e represent the direction angle as shown in Fig. 4.53.

For low frequency noise, the direction function is

$$\bar{D}_l(\Theta_e, \Phi_e) \approx \frac{\sin^2(\Theta_e)\sin^2(\Phi_e)}{(1 + M\cos\Theta_e)^2}. \tag{4.96}$$

4.6.4 Comparison of noise calculations

In order to validate the established calculation model, computations for the S809 and FFA-W3-241 airfoils were carried out and compared with NAFNoise software from NREL [149]. As the self-generated noise of the airfoil mainly comes from trailing edge

noise and separation noise [136–138], only these two kinds of noise were considered. The initial calculation conditions of the two models are the same and the detailed parameters are as below.
Sound speed: $c_0 = 340$ m/s;
Spanwise length: $d = 1$ m;
Viscous coefficient: $\mu = 1.5 \times 10^{-5}$;
Chord: $C = 0.4$ m;
Air density: $\rho = 1.225$ kg/m^3;
AOA: $\alpha = 6°$;
Inflow velocity: $V_0 = 70$ m/s;
Bluntness of trailing edge: 0.0002 m;
Observation distance: $r = 1.2$ m;
Angle of bluntness: $\Psi = 12.5°$;
Viewing angle: $\theta = 90°$.

The calculation results of the airfoil noise in 1/3 octave frequency spectrum are shown in Fig. 4.54 and 4.55. As space in the two figures is limited, trailing edge noise is abbreviated to TEN.

There is a certain gap in the noise prediction on the pressure side at the trailing edge between the new model and NAFNoise. Especially for the FFA-W3-241 airfoil, the results from the new model are mainly in the middle and high frequency region (200–800 Hz) and the results from NAFNoise are mainly in the high frequency region

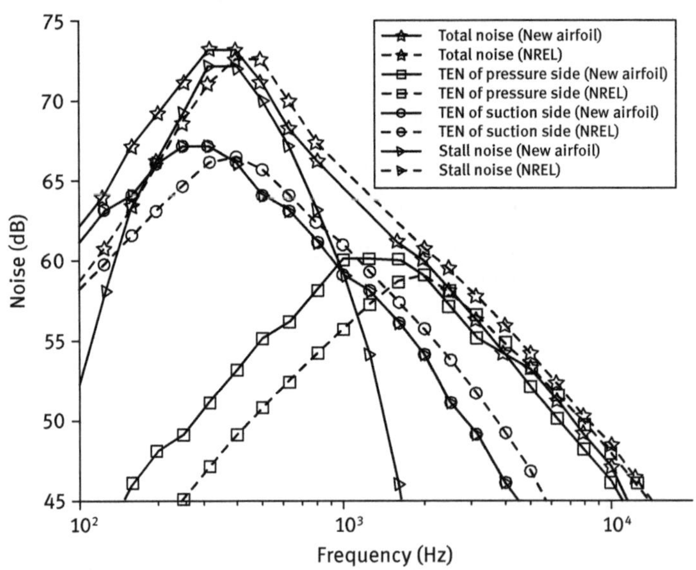

Fig. 4.54: Noise curve of S809.

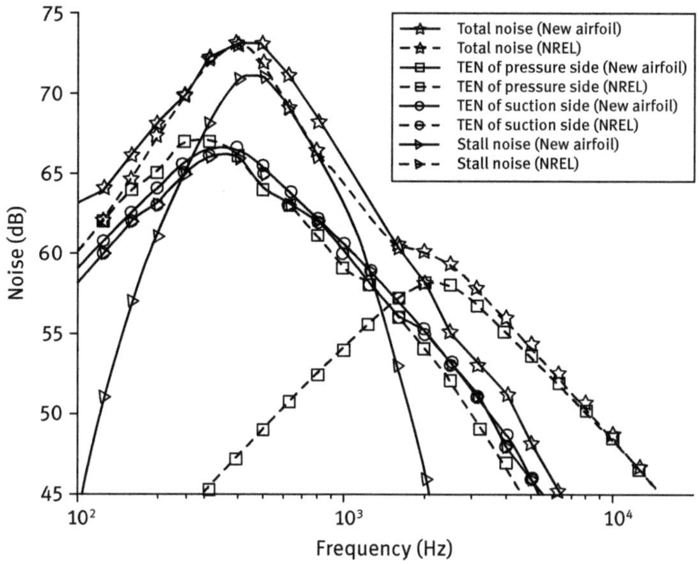

Fig. 4.55: Noise curve of FFA-W3-241.

(1200–4000 Hz). The prediction of suction surface noise with the two methods is close. For separation noise, the main band area of the predicted noise is close (concentrated in the high frequency region of 200–800 Hz), but there is a deviation between the noise values. The differences may be caused by the different calculation methods for the trailing edge boundary layer thickness. Generally, the two calculation methods are basically consistent which indicates validation of this new model.

4.6.5 Influence of geometric parameters of airfoils on noise

In Section 2.6, the influences of airfoil geometric parameters on aerodynamic performance were discussed. Referring to design methods for other airfoils, the corresponding constraints can be given according to the design objects. However, the influence of these geometric parameters on noise is not very clear. In order to design a low noise airfoil, the influence of these parameters should be studied. Based on the constraints normally used in airfoil design, the chordwise location of the maximum relative thickness, the maximum camber and the chordwise location of the maximum camber location were selected to study their relationships.

All the conditions for dynamic and acoustic calculations are consistent with the AOA at 7°, the Reynolds number $Re = 2.0 \times 10^6$ and wind speed of 70 m/s. The calculation cases are divided into free transition and fixed transition conditions. The noise is shown using the A-weighted sound power level.

4.6.5.1 The influence of the chordwise position of maximum thickness

Based on the previously established noise prediction model, four airfoils (NACA 63415, NACA 64415, NACA 65415 and NACA 66415) with relative thickness of 15 % were selected to study the influences of chordwise location of maximum thickness on noise.

The four airfoils have the same maximum camber and maximum camber location, with the maximum thickness position gradually shifting towards the trailing edge direction. As shown in Tab. 4.17, with the chordwise location of the maximum thickness moving towards the trailing edge, the maximum lift coefficient gradually decreases; the differences between the maximum lift coefficients under free transition and fixed transition is decreasing, implying the roughness sensitivity is getting lower; the maximum lift-to-drag ratio under the free transition condition is increasing, the trend is opposite to that under the fixed transition condition; the noise value increases under both free and fixed transition conditions and the increment increases gradually.

Tab. 4.17: The impact of chordwise position of maximum thickness on noise.

Airfoil name	Maximum thickness location	Maximum camber	Maximum camber location	Maximum lift coefficient	Maximum lift-to-drag ratio	Noise (dB)
NACA 63415	0.344	0.02	0.53	1.62(1.61)	131.3(81.8)	90.9(95.9)
NACA 64415	0.374	0.02	0.53	1.59(1.59)	133.7(80.3)	91.5(96.2)
NACA 65415	0.405	0.02	0.53	1.53(1.53)	134.1(75)	92.5(97.1)
NACA 66415	0.457	0.02	0.53	1.51(1.51)	143(67.2)	96.0(99.3)

Note: the value under fixed transition conditions is in parentheses.

4.6.5.2 The influence of maximum camber

In order to study the influence of chordwise location of maximum thickness on airfoil noise, several airfoils (NACA 63415, NACA 64415, NACA 65415 and NACA 66415, divided into two groups) with relative thickness of 15 % were selected for analysis.

The airfoils in each group have the same maximum relative thickness location and maximum camber location, with the maximum camber increasing gradually. As shown in Tab. 4.18, as the maximum camber increases, the maximum lift coefficient gradually increases both under free and fixed transition conditions; the difference between the maximum lift coefficients under free transition and fixed transition increases, which implies that roughness sensitivity is increasing; the maximum lift-to-drag ratio under both free and fixed transition conditions is increasing; the noise value increases under both free and fixed transition conditions and the increment is almost the same and can be ignored

Tab. 4.18: The impact of maximum camber on noise.

Airfoil name	Maximum thickness location	Maximum camber	Maximum camber location	Maximum lift coefficient	Maximum lift-to-drag ratio	Noise (dB)
NACA 63215	0.36	0.01	0.52	1.50(1.50)	104.8(77.0)	93.1(96.6)
NACA 63415	0.36	0.02	0.52	1.57(1.56)	130.8(80.7)	93.5(97.1)
NACA 63615	0.36	0.03	0.52	1.68(1.67)	153.9(83.3)	93.8(97.8)
NACA 64215	0.36	0.01	0.52	1.44(1.43)	106.5(76.5)	93.3(96.5)
NACA 64415	0.36	0.02	0.52	1.54(1.53)	132.3(80.7)	93.7(97.1)
NACA 64615	0.36	0.03	0.52	1.64(1.63)	153.6(83.6)	94.1(97.8)

Note: the value under fixed transition conditions is in parentheses.

4.6.5.3 The influence of the location of maximum curvature

In order to study the influence of chordwise location of maximum camber on airfoil noise, several airfoils (NACA 63415, NACA 64415, NACA 65415 and NACA 66415) with relative thickness of 15 % were selected for analysis.

The airfoils have the same maximum relative thickness location and maximum camber, with the location of maximum camber gradually moving towards the trailing edge. As shown in Tab. 4.19, with increasing maximum camber, the maximum lift coefficient firstly decreases to the minimum value (when the chordwise location of maximum camber is 0.4) and then increases; the differences between the maximum lift coefficients under free transition and fixed transition (representing roughness sensitivity) follow the same trend as that of the lift coefficient; the varying trend of the maximum lift-to-drag ratio is opposite to that of the lift coefficient; the varying trend of noise under the free transition condition is the same as that of the lift coefficient, however, the trend under the fixed transition condition is always increasing with increasing position of the maximum chord.

Tab. 4.19: The impact of maximum camber location on airfoil noise.

Airfoil name	Maximum thickness location	Maximum camber	Maximum camber location	Maximum lift coefficient	Maximum lift-to-drag ratio	Noise (dB)
NACA 4115	0.3	0.04	0.1	1.97(1.67)	92.8(76.5)	92.4(97.0)
NACA 4215	0.3	0.04	0.2	1.88(1.67)	130.5(79.3)	91.3(97.0)
NACA 4315	0.3	0.04	0.3	1.75(1.66)	147.9(79.8)	90.6(97.1)
NACA 4415	0.3	0.04	0.4	1.74(1.69)	151.0(78.6)	90.8(97.5)
NACA 4515	0.3	0.04	0.5	1.79(1.75)	149.3(77.3)	91.6(98.0)
NACA 4615	0.3	0.04	0.6	1.84(1.81)	129.4(76.3)	92.6(98.4)

Note: the value under fixed transition conditions is in parentheses.

4.6.6 Design of wind turbine airfoils with high efficiency and low noise

The design of wind turbine airfoils is a basic but important task for designing optimal wind turbine rotors. Employing efficient airfoils with high lift coefficient and high lift-to-drag ratio can reduce the cost of wind turbine blades and therefore reduce the cost of energy. On the other hand, noise from wind turbines can give annoyance to people living nearby, which becomes a barrier for the further development of (at least onshore) wind energy. Therefore, designing highly efficient wind turbines and at the same time reducing their noise emission are the design goals of future wind turbine rotors. Through the analysis in the above section, it is known that airfoil noise is closely related to the thickness of the trailing edge boundary layer on the upper and lower sides of the airfoil. Airfoil noise can be controlled through controlling the boundary layer thickness.

4.6.6.1 Objective function

However, the design characteristics of wind turbine airfoils are a trade-off between several possible conflicting requirements such as high lift, high lift-to-drag ratio and low noise emission; therefore, the choice of an objective function which can weight desirable features of an airfoil is very important in the design procedure. As discussed above, designing wind turbines of high efficiency and low noise is a design goal of future wind turbine rotors. So, the lift-to-drag ratio and the noise value are taken into the optimization objective function using the following equations:

$$f(x) = \max(\mathrm{RLD}/\mathrm{SPL}), \tag{4.97}$$

$$\mathrm{RLD} = \frac{1}{m-n+1}\sum_{i=n}^{m} C_\mathrm{l}(i)/C_\mathrm{d}(i), \tag{4.98}$$

$$\mathrm{SPL} = \frac{1}{m-n+1}\sum_{i=n}^{m} (\mathrm{SPL}_\mathrm{TBLTE}(i) + \mathrm{SPL}_\mathrm{SEP}(i)), \tag{4.99}$$

where C_l represents the lift coefficient of the airfoil, C_d represents the drag coefficient, RLD represents the average lift-to-drag ratio in the design AOA range, SPL represents the average noise value of the airfoil in the design range of AOA range, (n, m) represents the design AOA range. Here the airfoil design AOA range of 5–10° is selected.

4.6.6.2 Design variables and constraints

For the integrated expression of an airfoil, the first six coefficients of equation (3.26) are selected as design variables:

$$X = (x_1, x_2, x_3, x_4, x_5, x_6). \tag{4.100}$$

The range of the design variables is determined according to the shape of the airfoil. If the value of a variable exceeds a certain range, the integrated representation will no longer have the shape characteristics of the airfoil. So the scope of the variable is constrained:

$$X_{\min} \le X \le X_{\max}. \tag{4.101}$$

In addition to requirements for the shape, the other constraints are mainly aimed at the structure of the airfoil. Normally, the airfoil thickness is the most important requirement on structural characteristics. The main power production zone of a wind turbine blade is the outer part (generally within 70–90 % span). The maximum relative thickness of airfoil in the region is generally between 0.15 and 0.21. Relative thickness of 0.21 is selected as the design constraint:

$$t/c = 0.21, \tag{4.102}$$

where t is the maximum thickness of the airfoil, c is the chord of airfoil.

Another important structural parameter of the airfoil is the chordwise position of the maximum relative thickness which is constrained as

$$0.25 \le x/c \le 0.35. \tag{4.103}$$

The constraint for the maximum camber of the airfoil is

$$0.3 \le cam/c \le 0.4. \tag{4.104}$$

The constraint for the chordwise position of maximum camber is

$$0.4 \le cam/c \le 0.6. \tag{4.105}$$

In addition, in the practical engineering application, considering the manufacturing process and the structural strength of the blade, the thickness of the trailing edge of the airfoil is usually thickened. In view of the fact that the trailing edge processing of an airfoil after completion of the design will have a certain influence on the aerodynamic performance, the trailing edge thickness is constrained during the optimization process. Considering the airfoil thickness and leaf expansion to the location, the constraint conditions for the airfoil trailing edge thickness is

$$te/c = 0.005. \tag{4.106}$$

4.6.6.3 Optimization results

Based on the integrated expression and noise prediction method, for the establishment of the above optimization model, the following parameters were selected as the initial conditions: Reynolds number of $Re = 2.0 \times 10^6$, relative inflow speed of $V_0 = 70$ m/s, distance between noise source and observer of $r = 1$ m, observation angle of 90°, airfoil chord of $c = 1$ m, span length of 1 m, maximum relative thickness of 0.21. After optimization, the new wind turbine airfoil CQU-DTU-T21 is obtained, which is shown in Fig. 4.56.

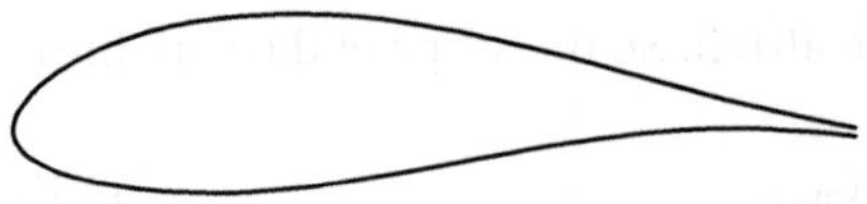

Fig. 4.56: Profile of CQU-DTU-T21 airfoil.

In order to verify the performance of the airfoil, the noise and aerodynamic characteristics of the new airfoil were calculated and compared with the commonly used 21 % wind turbine airfoils (DU 93-W-210 and FFA-W3-211). The geometric properties of the three airfoils are shown in Tab. 4.20.

Tab. 4.20: The geometric characteristics of three airfoils.

Airfoil name	Maximum thickness	Chordwise location of maximum relative thickness	Maximum camber	Chordwise location of maximum camber
FFA-W3-W211	0.21	0.324	0.0217	0.714
DU 93-W-210	0.207	0.341	0.0274	0.711
CQU-DTU-T21	0.21	0.288	0.0391	0.517

The variation of noise and the ratio of RLD (average lift-to-drag ratio) to SPL (average noise value) with Reynolds number of $Re = 2.0 \times 10^6$ and the AOA range of –10° to 20° are shown in Fig. 4.57 and 4.58 respectively. As shown in Fig. 4.57, the noise of the new airfoil is quieter than that of FFA-W3-211 with AOA in the range of 0–10° (the maximum reduction is 1 dB at AOA of 0°) and is slightly larger than that of FFA-W3-211 with AOA in the range of 11–20° (the maximum increment is 0.7 dB at AOA of 12°). Over the whole AOA range, the performance of FFA-W3-211 is close to the new airfoil. Compared with the DU 93-W-210, the noise of the new airfoil is quieter with AOA in the range of 0–14° (the maximum reduction is 4.5 dB at AOA of 10°). From the comparison with the two dedicated wind turbine airfoils, the new airfoil has good noise performance. Fig. 4.58 shows the RLD/SPL of the three airfoils. The new airfoil has higher RLD/SPL over almost the whole AOA range. This demonstrates that the new type not only has low noise characteristics, but also has higher lift-to-drag ratio.

The lift coefficient and lift-to-drag ratio (RLD) of three airfoils with Reynolds number of $Re = 2.0 \times 10^6$ and AOA in the range of –10° to 20° are shown in Fig. 4.59 and 4.60 respectively. Over the whole range of AOA, the new airfoil has a higher lift coefficient and lift-to-drag ratio. For the new airfoil, the maximum lift coefficient is 1.89 and the maximum lift-to-drag ratio is 145.8, which is higher than those of DU 93-W-210 (the maximum lift coefficient is 1.38 and the maximum lift-to-drag ratio is 140.2) and FFA-W3-211 (the maximum lift coefficient is 1.53 and the maximum lift-to-drag ratio is 140.5).

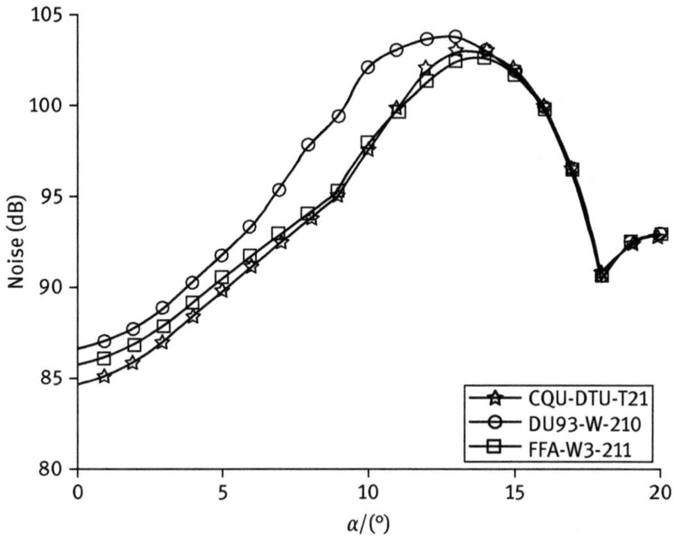

Fig. 4.57: Noise of three airfoils.

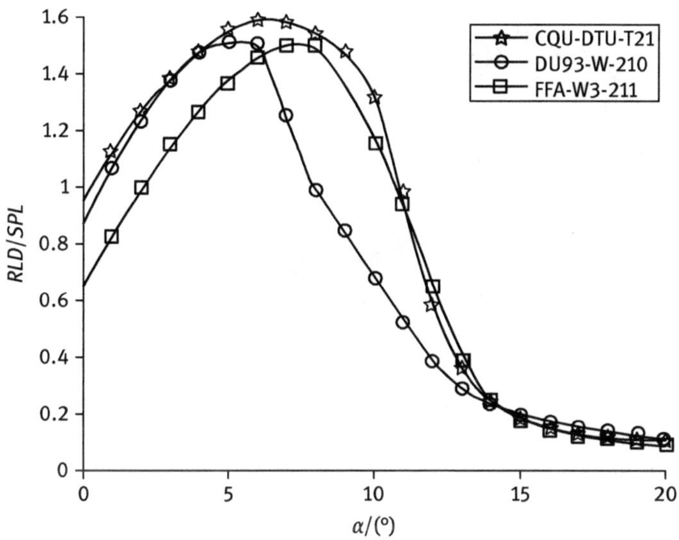

Fig. 4.58: RLD/SPL of three airfoils.

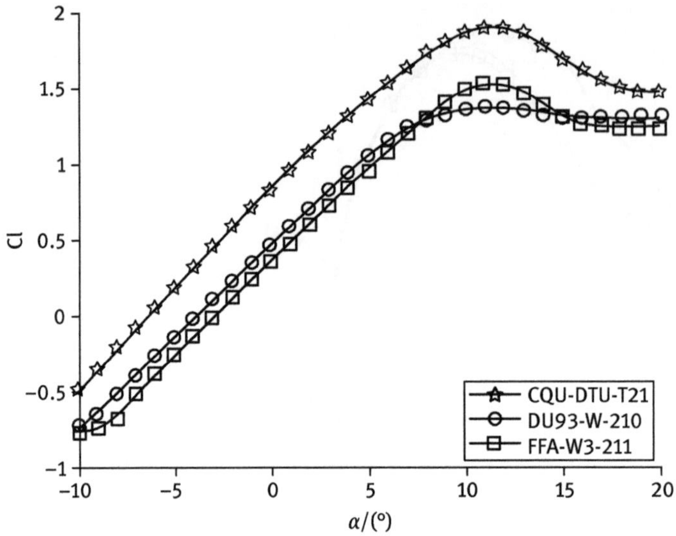

Fig. 4.59: Lift coefficient of three airfoils.

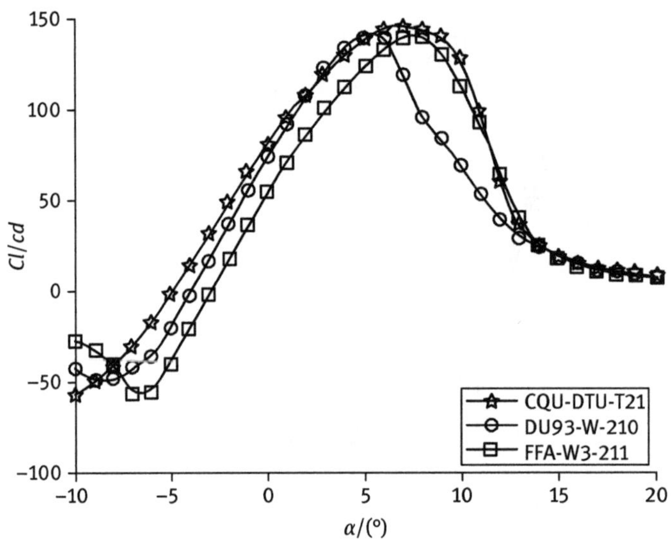

Fig. 4.60: Lift-to-drag ratio (RLD) of three airfoils.

4.7 Airfoil design based on a 2D power coefficient

Higher power efficiency, C_p, is the one of the main goals of wind turbine blade design. In blade element theory, the blade is divided into a number of segments along the span, which are called blade elements. According to this theory, the power coefficient of each blade element can be expressed as

$$C_p = \frac{dF_{driving} r\omega}{\frac{1}{2}\rho V_0^3 dA}, \tag{4.107}$$

where $dF_{driving}$ represents the local tangential force, r is the radius position of the blade element, ω is the rotor speed, ρ is the air density, and V_0 is the wind speed far upstream, $dA = 2\pi r\, dr$ is the rotor areas corresponding to each element.

The power coefficient of the blade can be obtained by integrating the power coefficient of each blade element along the blade span. If the power coefficient of each blade element reaches the largest value, then the power coefficient of the blade is the largest.

Usually, equation (4.107) is expressed as

$$C_p = [(1-a)^2 + x^2(1+a')^2] x c_x \sigma, \tag{4.108}$$

where a is the axial induction factor, a' is the tangential inducing factor, σ is blade solidity, c_x is the tangential force coefficient and c_y is the normal force coefficient. They are expressed by equations (4.109)–(4.113) respectively. x is the local tip speed ratio at blade element $x = X \times r/R$, in which X is the tip speed ratio, and R is the rotor radius:

$$a = \frac{1}{\frac{4\sin^2\varphi F}{\sigma c_y} + 1}, \tag{4.109}$$

$$a' = \frac{1}{\frac{4\sin\varphi\cos\varphi F}{\sigma c_y} - 1}, \tag{4.110}$$

$$\sigma = \frac{2\sin^2\varphi F}{c_y}, \tag{4.111}$$

$$c_x = c_l\left(\sin\varphi - \frac{1}{c_l/c_d}\cos\varphi\right), \tag{4.112}$$

$$c_y = c_l\left(\cos\varphi + \frac{1}{c_l/c_d \sin\varphi}\right), \tag{4.113}$$

where F is the tip-loss correction factor, φ is the local inflow angle at the blade element which can be expressed as

$$\varphi = \arctan\left(\frac{1-a}{x(1+a')}\right). \tag{4.114}$$

Combining the equations (4.108)–(4.114), the axial induction factor a and the tangential inducing factor a' can be obtained through iterations so as to calculate the maximum power coefficient under different lift-to-drag ratios. Results show that the axial induction factor with the maximum power coefficient is $a = \frac{1}{3}$, which is consistent with the results from actuator disc theory. The variation of power coefficient with respect to lift-to-drag ratio is shown in Fig. 4.61.

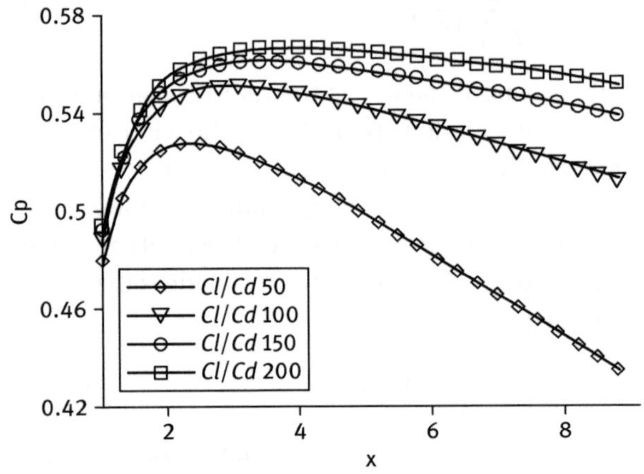

Fig. 4.61: Variations of power coefficient C_p with respect to RLD and x.

As shown in Fig. 4.61, the power efficiency increases with increasing lift-to-drag ratio. When the airfoil RLD increases from 50 to 100, 100 to 150 and 150 to 200, the increment of airfoil power coefficient decreases gradually. Selecting the local speed ratio as the power coefficient increases from 0.5127 to 0.5486 (by 7 %) when the RLD increases from 50 to 100; the power coefficient increases from 0.5486 to 0.5605 (by 2.1 %) when the RLD increases from 100 to 150; the power coefficient increases from 0.5605 to 0.5665 (by 1.07 %) when the RLD increases from 150 to 200.

In addition, the local inflow angle of the blade element φ and the solidity σ can be obtained. According to the design AOA of the airfoil, the local torsion angle can be obtained. It has also been found that when the local ratio is fixed, the term σcl (product of the solidity and the lift coefficient) is a fixed value. All the above parameters determine the blade's aerodynamic shape. If the airfoil is regarded as a 2D airfoil section and taking the Reynolds number, AOA and the spanwise location of the airfoil as design variables, the airfoil design is transformed into the blade design through obtaining the 2D power coefficient.

4.7.1 The optimization model

Airfoils with relative thickness between 15 % and 24 % located on the main power production spanwise zone of the blade (70–90 %), were optimized and obtained. The previously mentioned XFOIL software was adopted in optimization since it can deal with free transition and fixed transition based on the laminar flow stability e^N theory [99]. It was also utilized in the design of the DU airfoil series and the Risø airfoil series. Through coupling the airfoil integrated equation and XFOIL solver, the aerodynamic characteristics (such as lift, drag and boundary layer thickness on pressure side and the suction side of the airfoil) can be obtained directly, which are regarded as the input variables for optimization. Given the local tip speed ratio x at each blade element, the airfoil power coefficient, noise value and other related parameters can be calculated.

4.7.1.1 Objective function

As discussed in the Section 4.6, the design of wind turbine airfoils is a trade-off between several possible conflicting requirements such as aerodynamic performance, structural requirements and low noise emission, etc. With the long-term accumulation of dust, dirt and staining elements, the wind turbine blade can become polluted and rough. And increasing leading edge roughness can directly lead to a substantial decline in the performance of the wind turbine, with a maximum reduction of 30 % shown by some research [150]. The leading edge roughness sensitivity was taken as the main design goal in the design of the Risø airfoil series and the DU airfoil series [5, 6].

Assume that the design life of a wind turbine is N years and in the first M_{th} years it operates under smooth conditions. The power coefficient of the wind turbine during its whole life span can be established:

$$f(x) = \max(\mu_1 C_{\mathrm{p}1} + \mu_2 C_{\mathrm{p}2}), \tag{4.115}$$

where μ_1, μ_2 are weight coefficients for smooth conditions and rough conditions respectively and with the relationship of $\mu_1 + \mu_2 = 1$; $C_{\mathrm{p}1}, C_{\mathrm{p}2}$ are the weighted power coefficients (shown in equations (4.117) and (4.118)) for smooth and rough conditions respectively.

The occurrence of strong winds or turbulence conditions will change the boundary layer flow which will cause the airfoil to operate in off-design conditions. What is more, the impacts of aerodynamic force, inertial force and elastic force on the blades may also lead to off-design conditions. It is required that the wind turbine can achieve a good wind power coefficient under various working conditions within a wide AOA range. Meanwhile, the design lift coefficient should be as close as possible to the maximum lift to ensure smaller solidity which will reduce blade weight and cost. Based on the above considerations, an AOA range of 5–14° is selected in optimization. In addition, the power coefficient has different weight coefficients under different AOAs, with a larger weight coefficient under larger AOA so as to ensure a good stall performance.

The weighted power coefficients C_{p1} and C_{p2} are expressed as:

$$C_{p1} = \sum_{i=5}^{14} \lambda_i C_{p1}^i, \tag{4.116}$$

$$C_{p2} = \sum_{i=5}^{14} \lambda_i C_{p2}^i. \tag{4.117}$$

In equations (4.116) and (4.117), the relationship $\sum_{i=5}^{14} \lambda_i = 1$ should be satisfied. The weight factor λ_i is related to the respective AOA. C_{p1}^i, C_{p2}^i are the power coefficients under each AOA in smooth and rough conditions respectively, which can be obtained from the equations (4.108) and (4.115)–(4.117).

4.7.1.2 Design variables

According to the airfoil integrated expression equation mentioned in Chapter 3, the coefficients of equation (3.26) are selected as design variables. In order to ensure that the expression can describe profiles with the airfoil characteristics, the scope of the variables are constrained:

$$X_{\min} \leq X \leq X_{\max}. \tag{4.118}$$

Through analysis of the influences of integrated expression coefficients on the broadness of design space, degree of freedom of profile and convergence time of optimization, the first six coefficients of equation (3.26) were selected as design variables, namely $X = [a_1, b_1, a_2, b_2, a_3, b_3]$.

4.7.1.3 Design constraints

As mentioned above, the geometric compatibility of airfoils used on the same blade is of great importance. A smooth transition in airfoil thickness can ensure the smoothness of the blade profile so as to minimize the influence of the 3D effects. For the CQU-DTU-B airfoil series, the geometric compatibility is ensured through constraining the chordwise position of maximum thickness in the range of 20–40 %, the maximum camber around $0.04c$, the chordwise position of maximum camber within the range of 50–60 %.

In addition, with increasing rotor size, the noise of the wind turbine attracts more and more attention. Britain, Germany, Denmark and other countries have proposed relevant requirements for the noise of wind turbines [92, 136]. Considering the noise of an airfoil mainly comes from the trailing edge noise and separation noise [151], based on the previously proposed noise calculation method, these two noises were constrained:

$$\text{SPL} < N \quad (\text{dB}). \tag{4.119}$$

The weighted sound power level SPL was adopted in the measurement of noise. The following parameters were selected as design conditions: relative inflow speed of $V_0 = 70$ m/s, distance between noise source and observer of $r = 1$ m, observation angle of 90°, airfoil chord of $c = 1$ m, span length of 1.3 m, and the trailing edge noise and separation noise were calculated at an AOA of 6°. The constraint value N is related to the thickness of the airfoil and is determined through comparing it with that of other wind turbine airfoils.

Finally, considering the actual manufacturing of the blade, a blunt trailing edge should be adopted. Normally, a blunt trailing edge is obtained by cutting the trailing edge directly, which will change the airfoil chord, the corresponding parameters (such as maximum relative thickness and its chordwise location, the maximum camber and its chordwise location) and thus the airfoil's aerodynamic performance. However, the method here for achieving a blunt trailing edge can almost keep the same chord as shown in Fig. 4.62. The pressure surface of the airfoil rotates in a certain clockwise angle without changing the airfoil suction surface and the flow on the suction surface. The rotation of the pressure side of the airfoil has a similar influence as the installation of flap on the original airfoil, which will increase the actual lift and drag of the airfoil. Although the noise also increases, the magnitude is small. Generally speaking, the influence of this method on airfoil performance is small. It also provides references for the design of other airfoils. The trailing edge thicknesses of different airfoils with different relative thickness are different. For the CQU-DTU-B airfoil series, the trailing edge thickness is about 0.5 %c. The thickness of the blunt trailing edge is relative small compared to the chord, thus it will not produce a nonvertical trailing edge line of airfoil, which will not affect the manufacturing process.

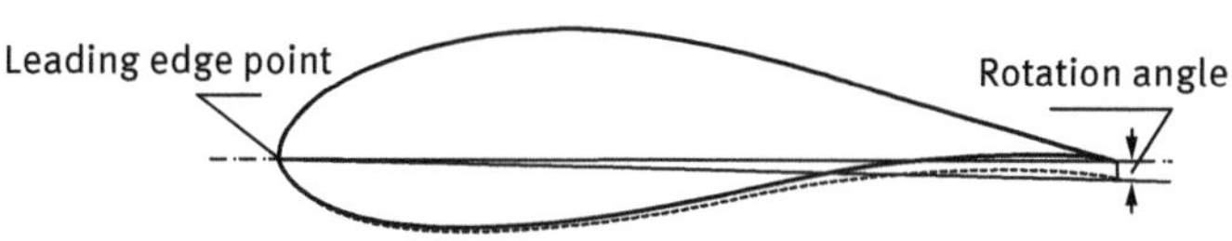

Fig. 4.62: The trailing edge design.

4.7.2 The optimization flow chart

In order to calculate the aerodynamic characteristics under smooth and rough conditions, free transition was adopted to simulate the smooth airfoil condition under the viscous condition. The fixed transition was set to simulate the worst rough conditions, with the fixed transition at 1 % chordwise location on the suction surface and at 10 % chordwise location on the pressure surface ($X_{tr,p} = 0.1$) [6, 110]. The optimization flow chart is shown in Fig. 4.63.

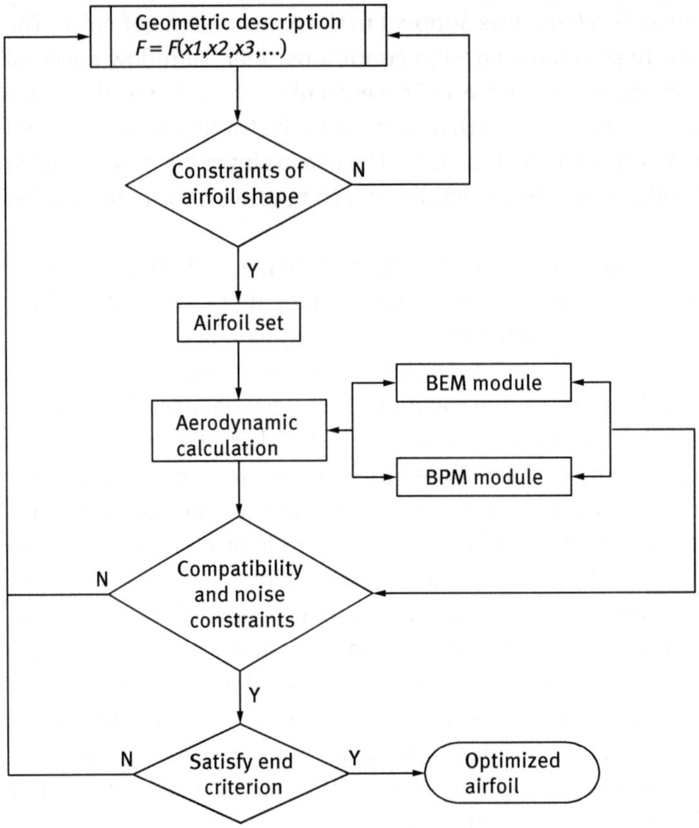

Fig. 4.63: The optimization flow chart

4.7.3 CQU-DTU-B airfoil series

Based on above optimization model, four wind turbine airfoils (CQU-DTU-B15, CQU-DTU-B18, CQU-DTU-B21 and CQU-DTU-B24) with relative thickness of 15 %, 18 % and 21 %, 24 % were obtained, as shown in Fig. 4.64.

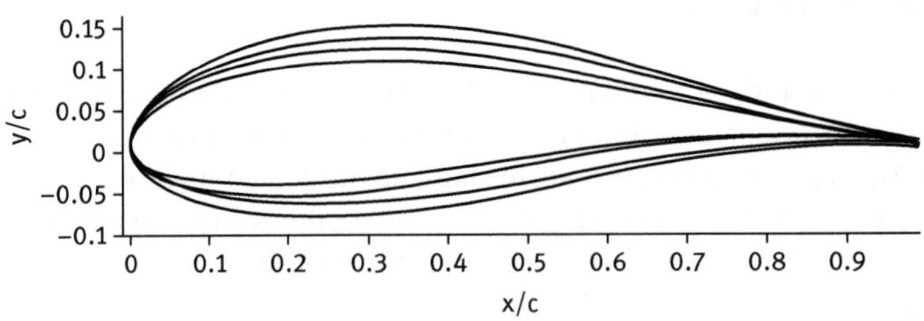

Fig. 4.64: CQU-DTU-B airfoil series.

In the names of airfoil series, CQU-DTU represents Chongqing University and the Technical University of Denmark, B represents the low noise airfoils, which differ from A (the high lift-to-drag ratio airfoils), the following number represents 100 times the relative thickness of the airfoil. It can be seen from the Fig. 4.64 that the profiles of the airfoils change gradually, which can ensure a smooth and continuous blade surface. The main geometric and aerodynamic parameters of the airfoils are listed in Tab. 4.21.

Tab. 4.21: Geometric and aerodynamic parameters of the CQU-DTU-B airfoil series.

	Chordwise location of maximum relative thickness	Thickness of trailing edge	Design AOA	Maximum lift coefficient	Maximum lift-to-drag ratio
CQU-DTU-B15	0.252	0.003	6.0°°°	2.12	164
CQU-DTU-B18	0.254	0.003	7.0°	2.15	171
CQU-DTU-B21	0.288	0.005	6.0°	1.95	174
CQU-DTU-B24	0.290	0.007	6.0°	1.99	178

With increasing maximum thickness, the chordwise position of maximum thickness increases gradually. It could help to distribute the gravity center linearly, thus ensuring the mass is distributed linearly. In addition, the maximum lift coefficient and maximum lift-to-drag ratio are large. And the maximum lift coefficients of the four airfoils are also relatively close, which is beneficial to the uniform distribution of aerodynamic forces on the blade surface.

The lift coefficient and lift-to-drag ratio of the CQU-DTU-B15 airfoil under free transition and the fixed transition conditions are shown in Fig. 4.65. The lift coefficient in the rough condition is almost the same as that in the smooth condition. However, the drag coefficient is increased to some extent with the maximum lift-to-drag ratio reduced from 164 to 113. Generally, this airfoil has low rough sensitivity.

The lift coefficient and lift-to-drag ratio of the CQU-DTU-B18 airfoil under free transition and fixed transition conditions are shown in Fig. 4.66. The lift coefficient of the rough condition is smaller than that of the smooth condition with AOA in the range of 15.0–20.0°, but the lift coefficient is still large. Meanwhile, the drag coefficient increases while the maximum lift-to-drag ratio declined from 171 to 107. The airfoil's roughness sensitivity is higher than that of CQU-DTU-A15.

The lift coefficient and lift-to-drag ratio of the CQU-DTU-B21 airfoil under free transition and fixed transition conditions are shown in Fig. 4.67. The lift coefficient under the rough condition is smaller than that under the smooth condition with AOA in the range of 10.0–20.0°. Meanwhile, the drag coefficient increases strongly and the maximum lift-to-drag ratio declined from 174 to 96. The airfoil's roughness sensitivity is higher than that of CQU-DTU-A15 and CQU-DTU-A18.

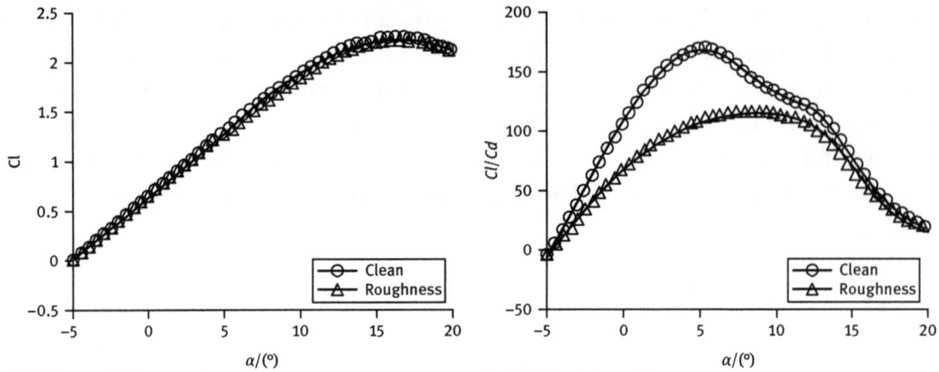

(a) Lift coefficient (b) Lift to drag ratio

Fig. 4.65: Lift and lift-to-drag ratio of CQU-DTU-B15 at Reynolds number of 6.0×10^6.

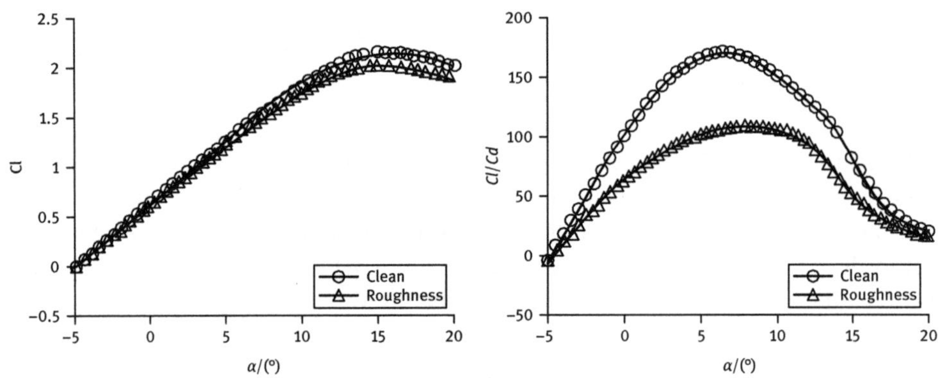

(a) Lift coefficient (b) Lift to drag ratio

Fig. 4.66: Lift and lift-to-drag ratio of CQU-DTU-B18 at Reynolds number of 6.0×10^6.

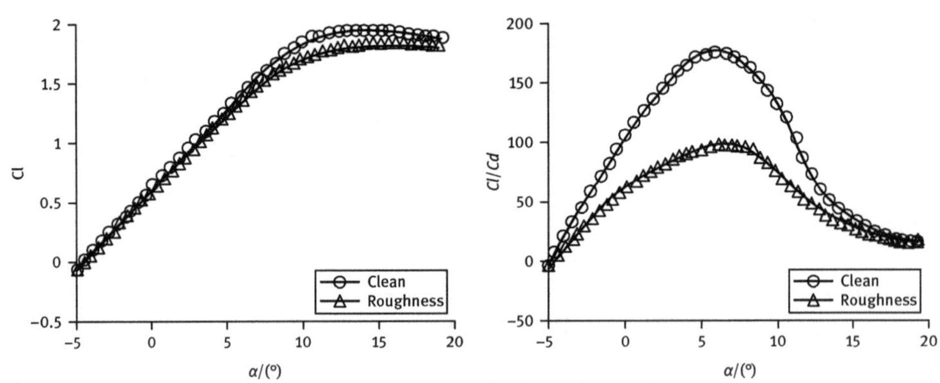

(a) Lift coefficient (b) Lift to drag ratio

Fig. 4.67: Lift and lift-to-drag ratio of CQU-DTU-B21 at Reynolds number of 6.0×10^6.

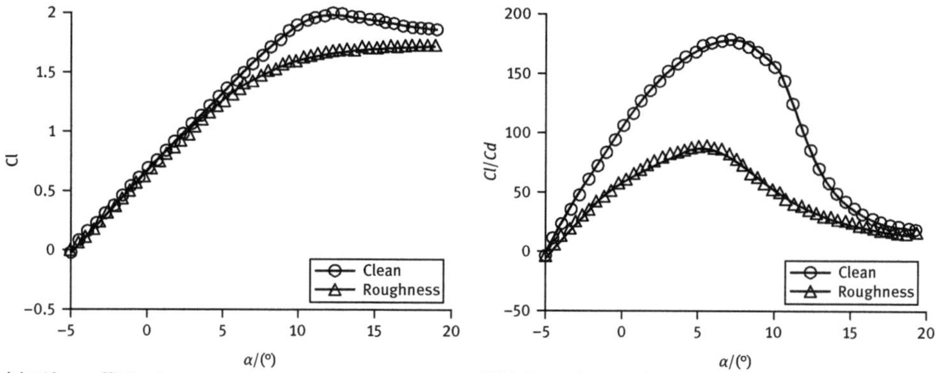

(a) Lift coefficient (b) Lift to drag ratio

Fig. 4.68: Lift and lift-to-drag ratio of CQU-DTU-B24 at Reynolds number of 6.0×10^6.

The lift coefficient and lift-to-drag ratio of the CQU-DTU-B24 airfoil under free transition and fixed transition conditions are shown in Fig. 4.68. The lift coefficient under the rough condition decreases largely while the maximum lift-to-drag ratio declined from 178 to 86. The airfoil's roughness sensitivity is the worst compared with CQU-DTU-A15, CQU-DTU-A18, CQU-DTU-A21 and CQU-DTU-A24. The predicted noise of CQU-DTU-B is listed in Tab. 4.22. With increasing relative thickness, the noise of the airfoil also increases. The airfoil noise in fixed transition conditions increases more strongly than in free transition conditions.

Tab. 4.22: Noise of CQU-DTU-B airfoil series at Reynolds number of 6.0×10^6.

	Wind speed (m/s)	Spanwise length (m)	Chord (m)	AOA (°)	SPL (dB)
CQU-DTU-B15	70	1	1.3	6	92.6(94.7)
CQU-DTU-B18	70	1	1.3	6	92.7(95.4)
CQU-DTU-B21	70	1	1.3	6	93.0(96.7)
CQU-DTU-B24	70	1	1.3	6	94.1(98.5)

Note: the value under fixed transition conditions is in parentheses.

The CQU-DTU-B airfoil series has a higher lift coefficient, which will reduce blade solidity and blade cost. Meanwhile, it has larger maximum lift-to-drag ratio and wider operating AOA range, which can help to capture more wind energy and reduce the cost of energy.

In order to further demonstrate the aerodynamic performance of the optimized airfoil series, CQU-DTU-B18 and CQU-DTU-B21 airfoils were selected to make comparisons with NACA 63418, FFA-W3-211 and DU 93-W-210 airfoils.

The lift coefficient and lift-to-drag ratio of the CQU-DTU-B18 airfoil under free transition and fixed transition conditions are compared with those of the NACA 63418 airfoil, as shown in Fig. 4.69 with Reynolds number of $Re = 2.0 \times 10^6$. The lift coefficient and lift-to-drag ratio are higher than those of NACA 63418, both under smooth and rough conditions in an AOA range of 15.0–20.0°, with their respective maximum values of 1.92 and 149.9 under the free transition condition and the respective maximum values of 1.81 and 86.9 under the free transition condition. The lift-to-drag ratio of CQU-DTU-B18 in fixed transition conditions at AOA of 9° is even greater than that of NACA 63418 under free transition conditions as shown in Fig. 4.69 (b). The aerodynamic characteristics, noise and weighted power coefficient (within the AOA range of 5–14°) are listed in Tab. 4.23. The weighted power coefficient of CQU-DTU-B18 is 0.541, which increases by 2.1 % compared with 0.53 of NACA 63418. CQU-DTU-B18 has lower noise no matter whether under free transition condition or fixed transition condition, as shown in Fig. 4.70 and 4.71. As space in the two figures is limited, trailing edge noise is abbreviated to TEN. The maximum noise of CQU-DTU-B18 is smaller than that of NACA 63418, under the same conditions, with a reduction of 3.1 dB.

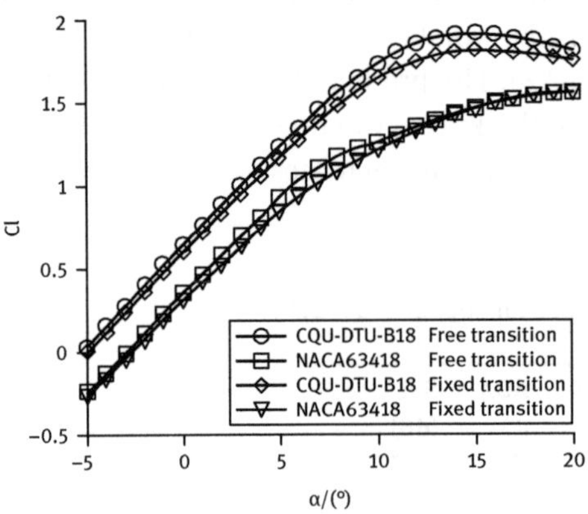

(a) Lift coefficient

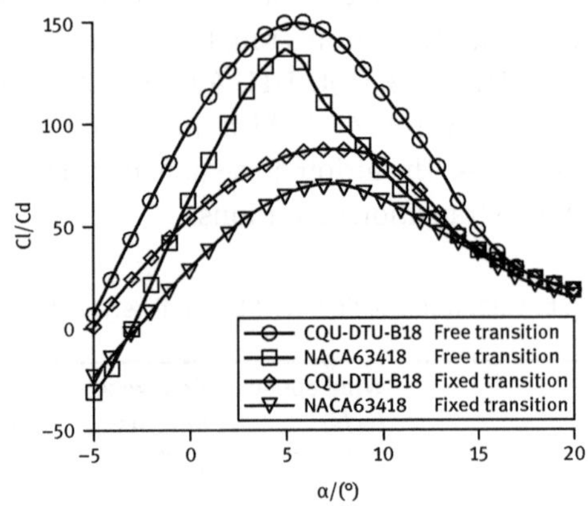

(b) Lift to drag ratio

Fig. 4.69: Lift and lift-to-drag ratio of CQU-DTU-B18 and NACA 63418 at $Re = 2.0 \times 10^6$.

Tab. 4.23: Aerodynamic performance of CQU-DTU-B18 and NACA 63418 at $Re = 2.0 \times 10^6$.

	Maximum lift coefficient	Design AOA	Design lift coefficient	Maximum lift-to-drag ratio	Noise	Power coefficient
NACA 64318	1.55(1.55)	7	1.01	136.2(70.1)	97.9	0.53
CQU-DTU-B18	1.92(1.81)	7	1.38	149.9(86.9)	94.8	0.541

Note: the value under fixed transition conditions is in parentheses.

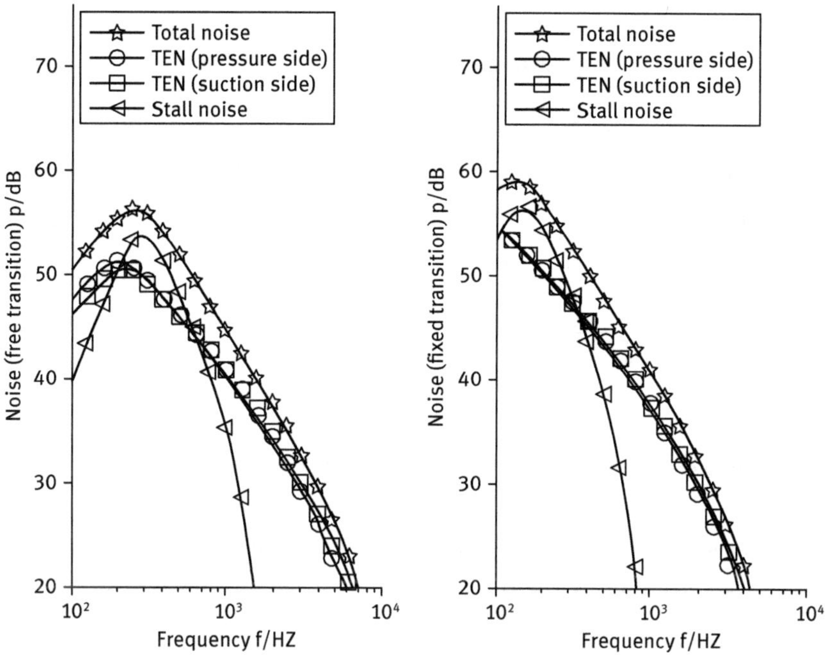

Fig. 4.70: Noise of NACA 63418 at $Re = 2.0 \times 10^6$; left: free transition condition; right: fixed transition condition; TEN: trailing edge noise.

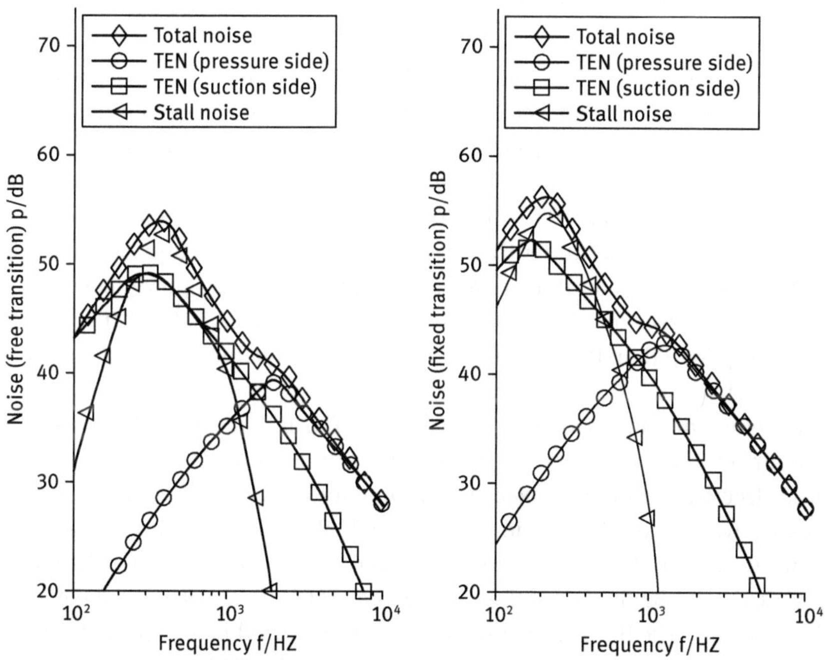

Fig. 4.71: Noise of CQU-DTU-B18 at $Re = 2.0 \times 10^6$; left: free transition condition; right: fixed transition condition; TEN: trailing edge noise.

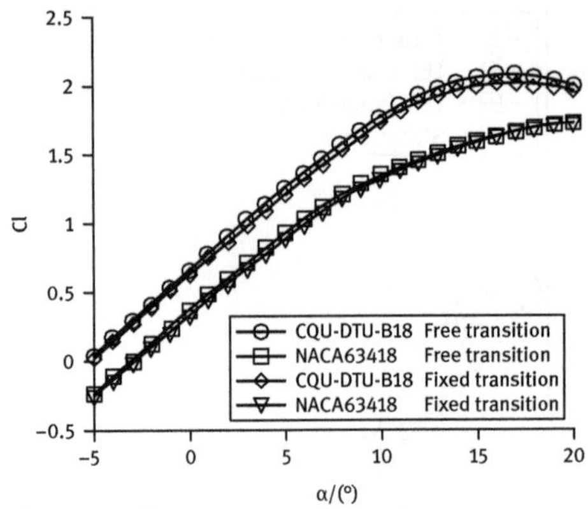

(a) Lift coefficient

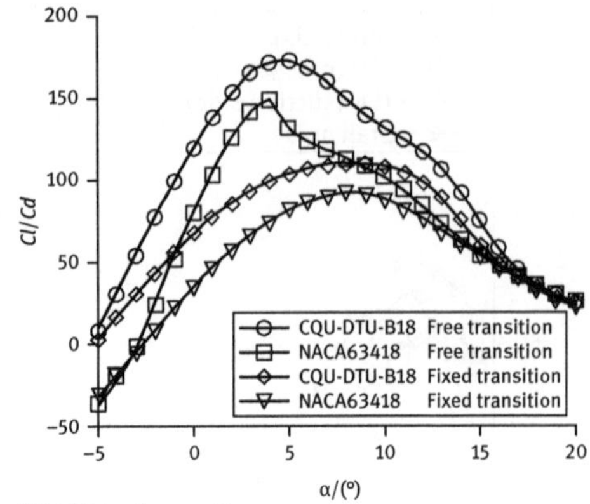

(b) Lift to drag ratio

Fig. 4.72: Lift and lift-to-drag ratio of CQU-DTU-B18 and NACA 63418 at $Re = 6.0 \times 10^6$.

Tab. 4.24: Aerodynamic performance of CQU-DTU-B18 and NACA 63418 at $Re = 6.0 \times 10^6$.

	Maximum lift coefficient	Design AOA	Design lift coefficient	Maximum lift-to-drag ratio	Noise	Power coefficient
NACA 64318	1.72(1.71)	7	1.20	148(93)	98.8	0.543
CQU-DTU-B18	2.15(2.07)	7	1.46	171(107)	95.4	0.553

Note: the value under fixed transition conditions is in parentheses.

The lift coefficient and lift-to-drag ratio of CQU-DTU-B18 and NACA 63418 at Reynolds number of $Re = 6.0 \times 10^6$, under free transition and fixed transition conditions, are shown in Fig. 4.72. The main aerodynamic characteristics, noise and weighted power coefficient (within the AOA range of 5–14°) are listed in Tab. 4.24. The results are similar to those obtained at $Re = 2.0 \times 10^6$. The maximum lift coefficient and maximum lift-to-drag ratio are 2.15 and 171 (under the free transition condition) and 2.07 and 107 (under the fixed transition condition), respectively. The weighted power coefficient of CQU-DTU-B18 increases by 1.55 % compared with that of NACA 63418. The noise of CQU-DTU-B18 is quieter than that of NACA 63418, with a reduction of 3.1 dB (shown in Fig. 4.73 and 4.74).

The lift coefficient and lift-to-drag ratio of CQU-DTU-B21 and the other two commonly used airfoils were compared at the Reynolds number of $Re = 2.0 \times 10^6$. The results under free transition and fixed transition conditions are shown in Fig. 4.75 and 4.76 respectively. The lift coefficient of CQU-DTU-B21 is higher than those of the other two airfoils both under free transition and fixed transition conditions. The lift-to-drag ratio of CQU-DTU-B21 is almost the same as that of DU 93-W-210 when the AOA is smaller than 6.0° and the lift-to-drag ratio of the former airfoil is higher than that of the latter airfoil, implying a smaller drag coefficient of CQU-DTU-B21.

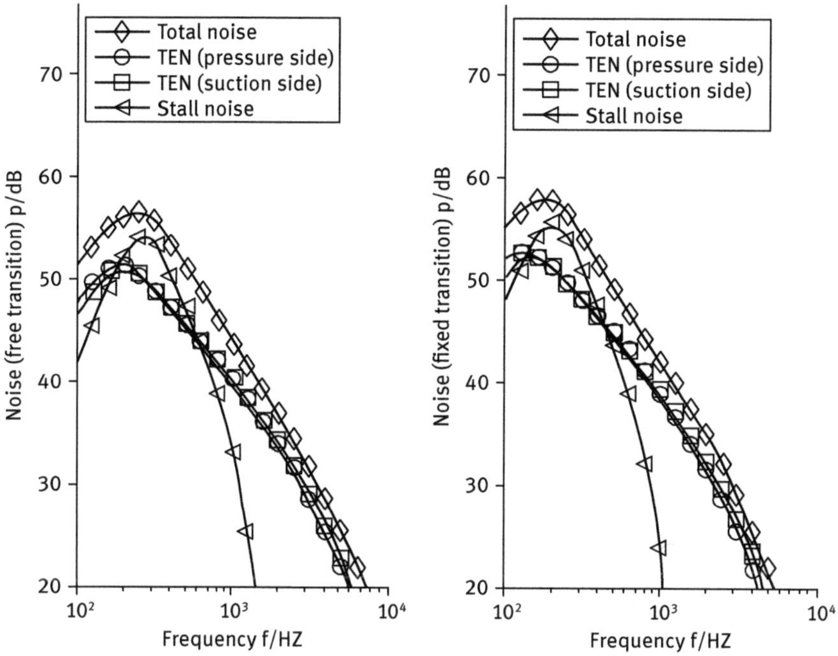

Fig. 4.73: Noise of NACA 63418 at $Re = 6.0 \times 10^6$; left: free transition condition; right: fixed transition condition; TEN: trailing edge noise.

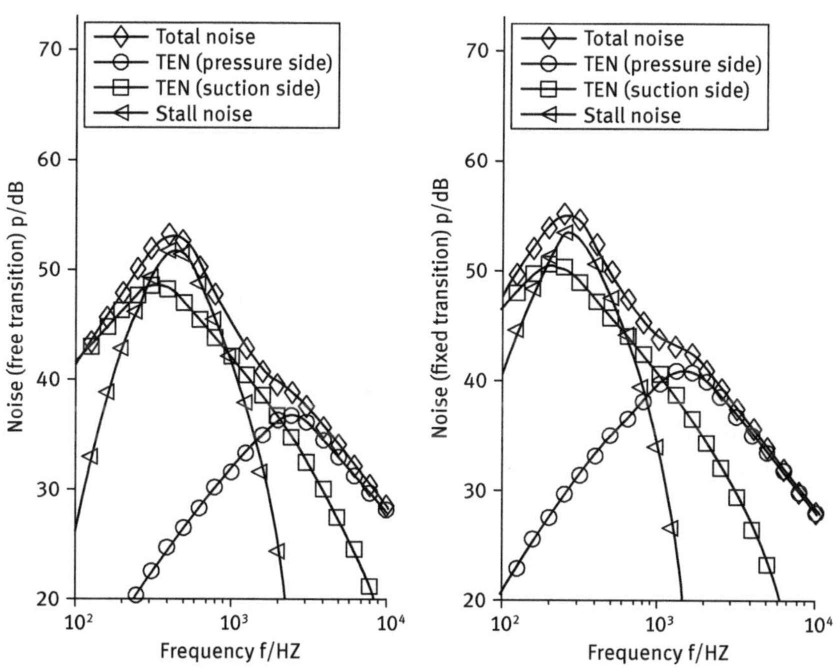

Fig. 4.74: Noise of CQU-DTU-B18 at $Re = 6.0 \times 10^6$; left: free transition condition; right: fixed transition condition; TEN: trailing edge noise.

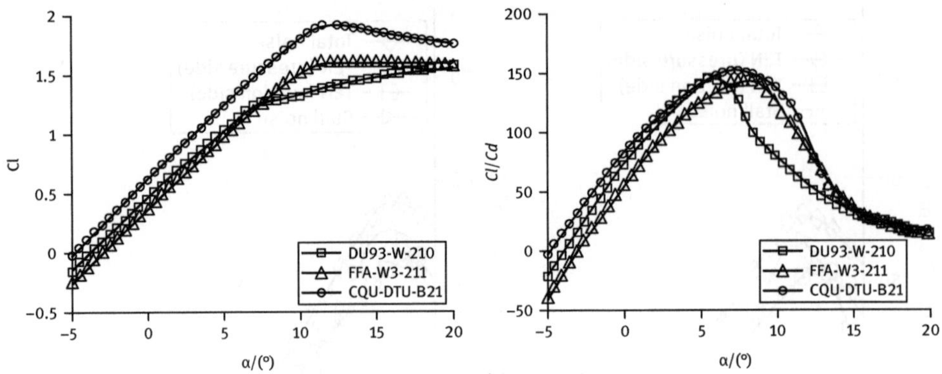

(a) Lift coefficient (b) Lift to drag ratio

Fig. 4.75: Lift and lift-to-drag ratio of three airfoils at $Re = 2.0 \times 10^6$ (free transition).

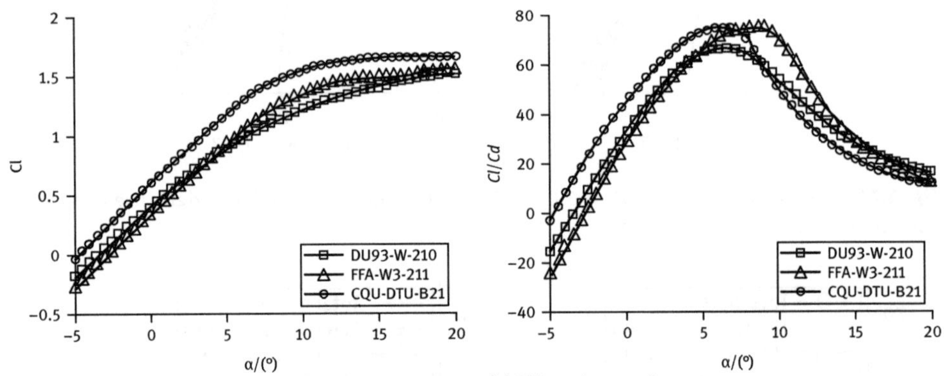

(a) Lift coefficient (b) Lift to drag ratio

Fig. 4.76: Lift and lift-to-drag ratio of three airfoils at $Re = 2.0 \times 10^6$ (fixed transition).

Compared with FFA-W3-211, the lift coefficient of CQU-DTU-B21 is higher in almost the whole AOA range under the free transition condition. Under the fixed transition condition, the lift-to-drag ratio of CQU-DTU-B21 is higher than that of FFA-W3-211 (when AOA is smaller than 7.5°) and smaller than that of FFA-W3-211 (when AOA is larger than 7.5°).

The comparison of aerodynamic performance between the three airfoils at the Reynolds number of $Re = 6.0 \times 10^6$ is shown in Fig. 4.77 and 4.78. The results are similar to those at $Re = 2.0 \times 10^6$.

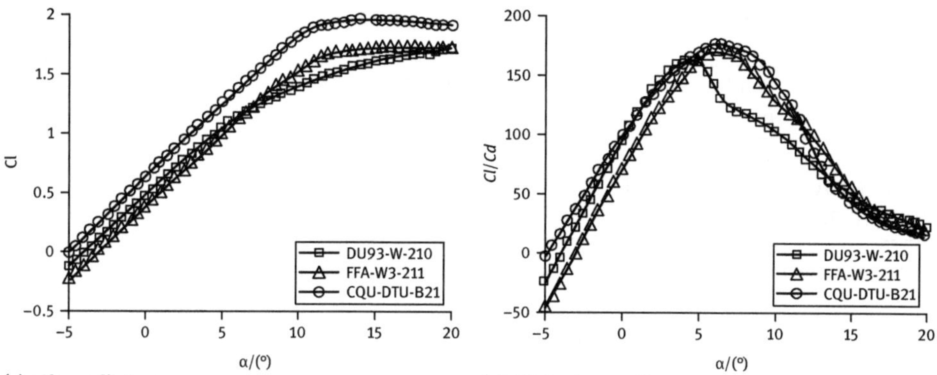

(a) Lift coefficient (b) Lift to drag ratio

Fig. 4.77: Lift and lift-to-drag ratio of three airfoils at $Re = 6.0 \times 10^6$ (free transition).

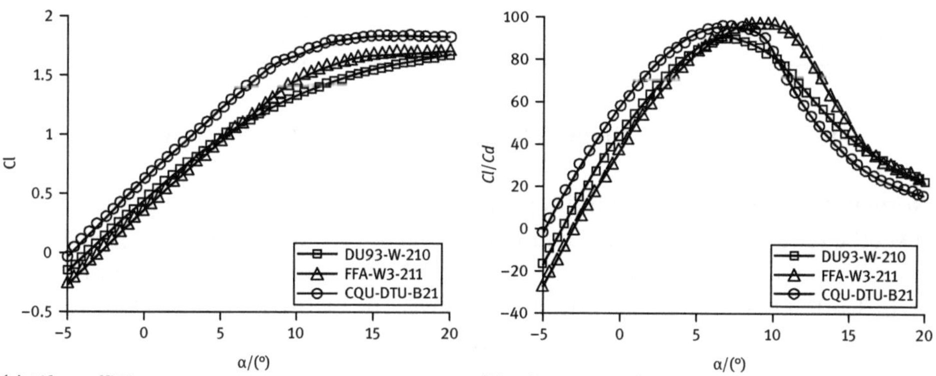

(a) Lift coefficient (b) Lift to drag ratio

Fig. 4.78: Lift and lift-to-drag ratio of three airfoils at $Re = 6.0 \times 10^6$ (fixed transition).

The noise characteristics of the three airfoils at Reynolds number of $Re = 2.0 \times 10^6$ and $Re = 6.0 \times 10^6$ are listed in Tab. 4.25 and 4.26 respectively. It can be seen that the noise of CQU-DTU-B21 is the almost same as that of FFA-W3-211. And the noise performance of these two airfoils is better than that of DU 93-W-210 at $Re = 2.0 \times 10^6$ and $Re = 6.0 \times 10^6$ (both under free and fixed transition conditions). Through comparing the results under two Reynolds numbers, it is found that the influence of the chord is not obvious.

In conclusion, the CQU-DTU-B21 airfoil has good aerodynamic performance (high lift coefficient and lift-to-drag ratio) and the emitted noise is similar or lower than other dedicated wind turbine airfoils.

Tab. 4.25: Noise of the three airfoils at $Re = 2.0 \times 10^6$ under free and fixed transition conditions.

	Wind speed (m/s)	Spanwise length (m)	Chord (m)	AOA (°)	Noise (dB)
FFA-W3-211	70	1	0.3	6	94.4(97.6)
DU 93-W-210	70	1	0.3	6	95.9(99.7)
CQU-DTU-B21	70	1	0.3	6	94.1(98.2)

Note: the value under fixed transition conditions is in parentheses.

Tab. 4.26: Noise of the three airfoils at $Re = 6.0 \times 10^6$ under free and fixed transition conditions.

	Wind speed (m/s)	Spanwise length (m)	Chord (m)	AOA (°)	Noise (dB)
FFA-W3-211	70	1	1	6	93.3(96.4)
DU 93-W-210	70	1	1	6	96.0(98.3)
CQU-DTU-B21	70	1	1	6	93.0(96.7)

4.7.4 Influence of airfoil trailing edge on the performance of the airfoil

Normally, the original airfoil trailing edge shape and trailing edge thickness cannot meet the requirements of blade manufacturing. In fact, the sharp trailing edge does not exist in real manufacturing. The design method of blunt trailing edge for the CQU-DTU-B airfoil has been briefly introduced above. The CQU-DTU-B21 was selected here to further study the influence of a blunt trailing edge. The aerodynamic performance of three airfoils with trailing edge thickness of 0, 0.5 %c and 1 %c (c is the chord) were compared. In Fig. 4.79, TE-I, II and III respectively represent the airfoil with a sharp trailing edge, with a blunt trailing edge thickness of 0.5 %c, and with a blunt trailing edge thickness of 1 %c. TE-II was obtained through clockwise rotating the lower airfoil surface by 0.275°. TE-III was obtained by clockwise rotating the lower airfoil surface by 0.572°. The gap at the trailing edge was closed by a circle with the same diameter as the trailing edge thickness.

The airfoil aerodynamic performance of the three airfoils was calculated using the software EllipSys2D [152, 153]. Combining the Reynolds-averaged Naiver–Stokes equations and the k-ωSST turbulent model, the lift coefficient and lift-to-drag ratio of airfoils at Reynolds number of $Re = 3 \times 10^6$ were obtained. The same kind of mesh configurations were adopted for the three cases. The transition model of e^N criterion (here N is 9) was adopted. The CFD results are shown in Fig. 4.80 and 4.81.

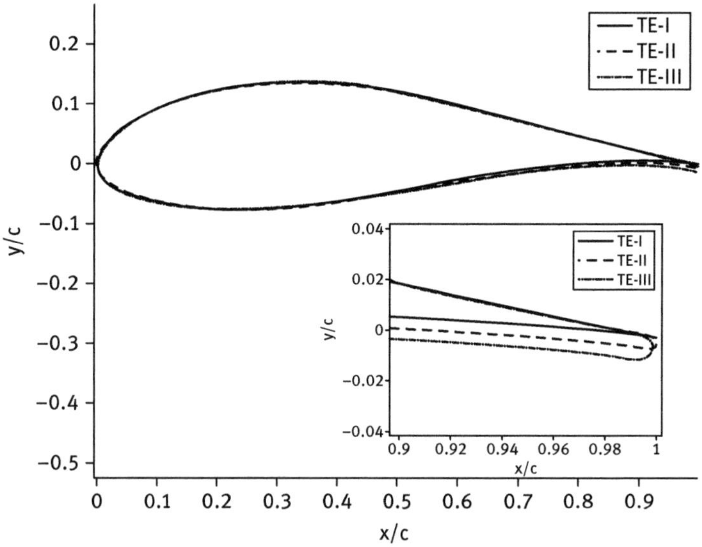

Fig. 4.79: Trailing edge design of CQU-DTU-B21 airfoil.

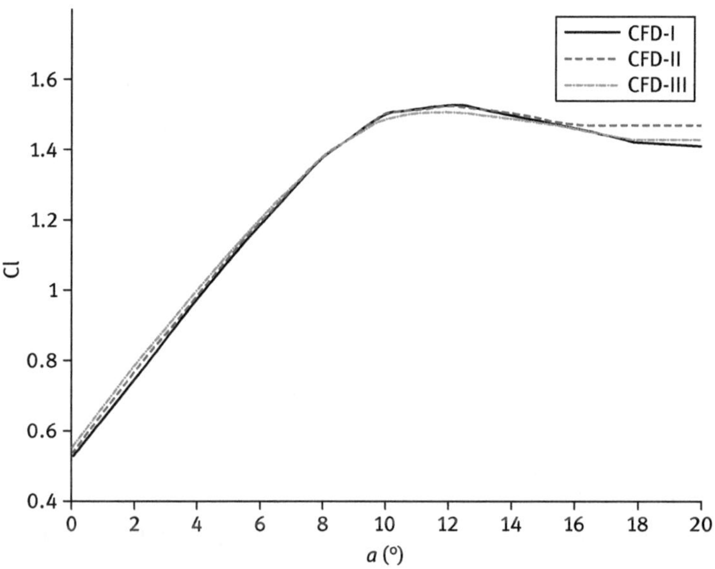

Fig. 4.80: Lift coefficient of airfoils with three trailing edge thicknesses.

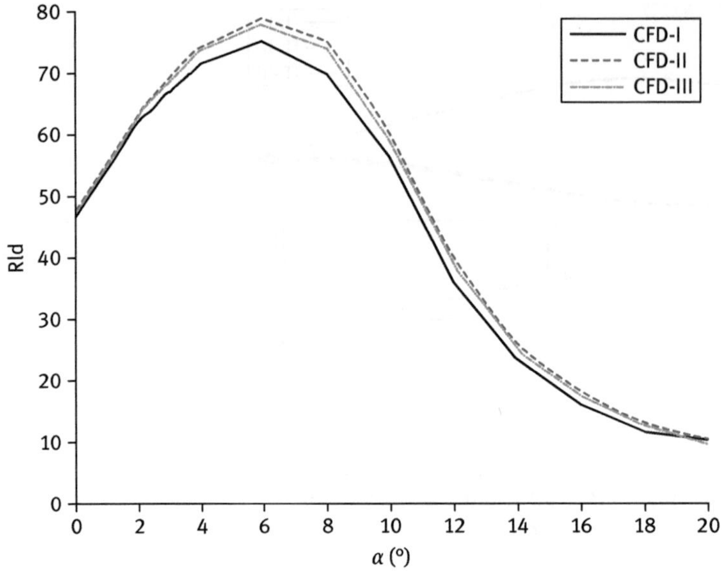

Fig. 4.81: Lift-to-drag ratio of airfoils with three trailing edge thicknesses.

As can be seen from Fig. 4.80, the lift coefficients of airfoils with three kinds of trailing edge thickness are similar in the linear region of the curve. And TE-III (with the maximum trailing edge thickness) stalls at relative smaller AOA compared to the other two airfoils. The variations in the lift-to-drag ratios of the three airfoils are shown in Fig. 4.81. It can be seen that the TE-II has the maximum lift-to-drag ratio over the whole AOA range. As can be seen from the variation in lift coefficient and lift-to-drag ratio, the thickness of the airfoil trailing edge has little influence on the lift coefficient and lift-to-drag ratio. For the CQU-DTU-B21 airfoil with relative thickness of 21 %, the trailing edge thickness of 0.5 $\%c$ is appropriate. A round blunt trailing edge can produce a higher lift coefficient compared with the sharp trailing edge. Thus the design method of blunt trailing edge adopted in the optimization of the CQU-DTU-B airfoil series has been validated.

4.8 Improved design of airfoils using smooth curvature technique

We found that the geometric scale factor a was not absolutely equal to 0.25 but close to it in the general integral equation. The geometric scale factor a was determined by experience in previous research, where its value was dependent on the airfoil, that is to say, for different airfoils, there will be different values for geometric scale factor a. Consequently, it takes a long time and requires skillful experience to correct the geometric scale factor a when fitting different airfoils.

Therefore, in this section a new method is presented to efficiently correct the geometric scale factor a based on curvature smooth continuity theory. By using this approach, the geometric scale factors a of twelve typical airfoils are calculated accurately. The curvature and curvature variation of the profiles for the integration shape function have better smooth continuity compared with the original airfoil. Then, the relationship between the curvature and the flow condition for the airfoil surface is discussed in detail. It is shown that the curvature plays a very important role in the pressure distribution of the airfoil surface. Finally, based on the proposed analysis, the DU 93-W-210 airfoil is improved and optimized using GA method, leading to a new airfoil. The new airfoil exhibits higher lift-to-drag ratio, higher lift coefficient and lower roughness sensitivity for both smooth surface flow and rough surface flow, compared with the original DU 93-W-210 airfoil.

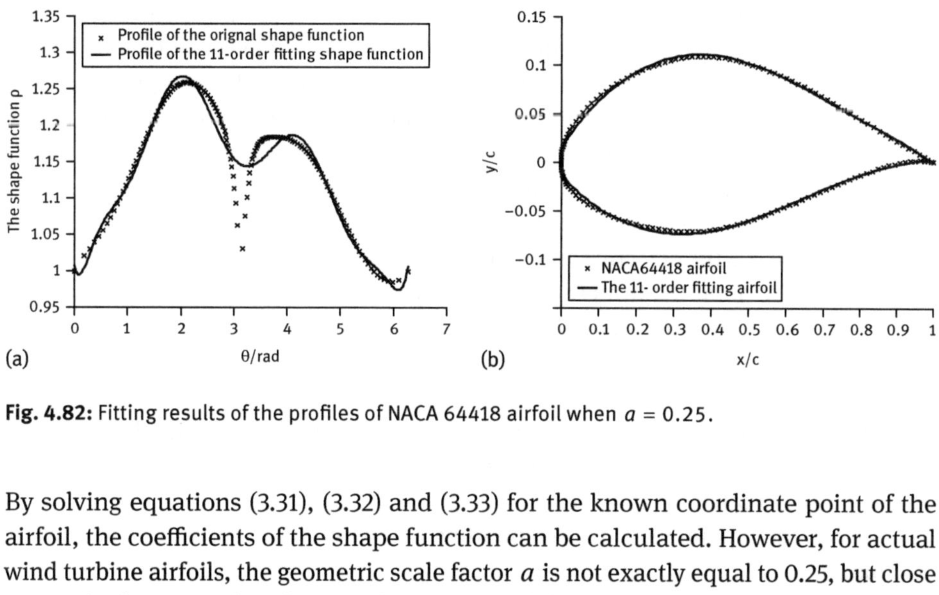

Fig. 4.82: Fitting results of the profiles of NACA 64418 airfoil when $a = 0.25$.

By solving equations (3.31), (3.32) and (3.33) for the known coordinate point of the airfoil, the coefficients of the shape function can be calculated. However, for actual wind turbine airfoils, the geometric scale factor a is not exactly equal to 0.25, but close to 0.25 in the general and integral equations. If the geometric scale factor a is 0.25, the fitting results of the profiles for both shape function and airfoil do not correspond well with the original airfoil. Fig. 4.82 shows the fitting results of for the NACA 64418 airfoil when the polynomial order of shape function is eleven. Since the shape function profile is not smooth and continuous, the fitting results are very poor, especially at the locations where the argument θ is zero, π and 2π correspond to the leading edge and trailing edge of the airfoil. So it is essential to correct the geometric scale factor a. Hence an improvement must be made to correct the geometric scale factor a based on curvature smooth continuity.

4.8.1 Smooth continuity of the profile for airfoil shape function

A smooth airfoil shape is important for the optimization results. Most airfoils meet the requirement of curvature smooth continuity. Since the curvature smooth continuity of the profile for an airfoil is required, the profile of the airfoil shape function must also meet that demand. Usually the curvature is expressed as the reciprocal of the curvature radius, as shown in equation (4.120). Equation (4.121) is the curvature variation:

$$C = 1/R = \frac{y''}{[1 + y'^2]^{3/2}}, \tag{4.120}$$

$$C' = \frac{y'''[1 + y'^2] - 3y'y''^2}{[1 + y'^2]^{5/2}}. \tag{4.121}$$

The geometric scale factor a is included in equations (4.120) and (4.121). In order to calculate the factor a, an iterative program of curvature for the profile of the airfoil shape function was written. The smaller the curvature is, the better the smooth continuity it exhibits. This is the main reason why the curvature is introduced to compute the factor a. In other words, if the curvature becomes a minimum after the iterative calculation for the profile of the airfoil shape function, the profile exhibits good smooth continuity, meanwhile the factor a is determined. Based on this principle and assuming a range for the geometric scale factor a of $[0.24, 0.26]$, which is close to 0.25, the minimum curvature (which is the objective function in this case) of the profile for the airfoil shape function can be obtained from the iterative solution and then the corrected factor a can be found. The specific steps are as follows:

(1) the parameter a lies within the range $[0.24, 0.26]$ with a size step of 0.000 01;
(2) choose a type of airfoil;
(3) solve equations (3.31)–(3.33), where both ρ and θ can be determined;
(4) according to equations (4.121) and (4.122), the curvature and curvature variation can be calculated for each iteration step;
(5) the corrected parameter a corresponding to the minimum curvature will then be found.

Now twelve typical wind turbine airfoils are fitted with an 11th-order function. The number of geometry points of the profiles for the twelve shape functions has the same value of 201. The number of geometry points of the profiles for the twelve airfoils is 160. Even though a smoother curvature can be obtained by using more geometry points, it was proven that the number of geometry points is sufficient to calculate the aerodynamics of wind turbine airfoils. The corrected geometric scale factors a are shown in Tab. 4.27. It turns out that the corrected geometric scale factors a are not 0.25, but less than 0.25 in the integral equation.

In order to compare the results of the profiles for airfoil shape functions between the fitted airfoils and the original airfoils, three airfoils (NACA 64418, S809 and DU 93-W-210 airfoils) were selected for investigation.

Tab. 4.27: Results of corrected geometric scale factor *a* for different airfoils.

Airfoil name	Parameter *a*	Airfoil name	Parameter *a*
NACA 63415	0.24817	S814	0.24726
NACA 63418	0.24751	FFA-W3-211	0.24756
NACA 63421	0.24693	FFA-W3-241	0.24582
NACA 64418	0.24737	FX 66-S196-V1	0.24875
NACA 64421	0.24694	DU 91-W2-250	0.24592
S809	0.24882	DU 93-W-210	0.24410

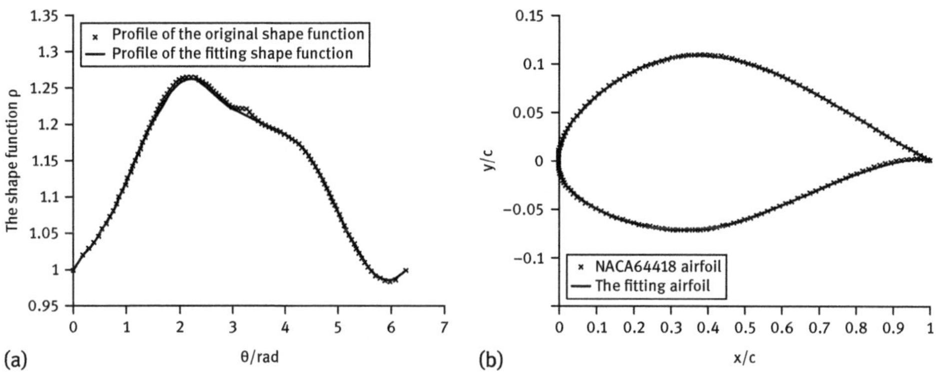

Fig. 4.83: Comparison of the fitting results and NACA 64418 airfoil results.

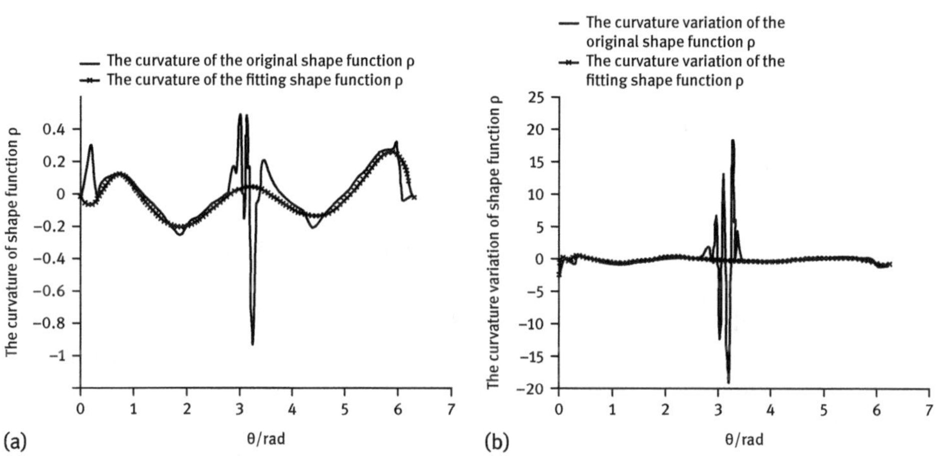

Fig. 4.84: Comparison of curvature and curvature variation for the profiles of the NACA 64418 airfoil shape function and fitted airfoil shape function.

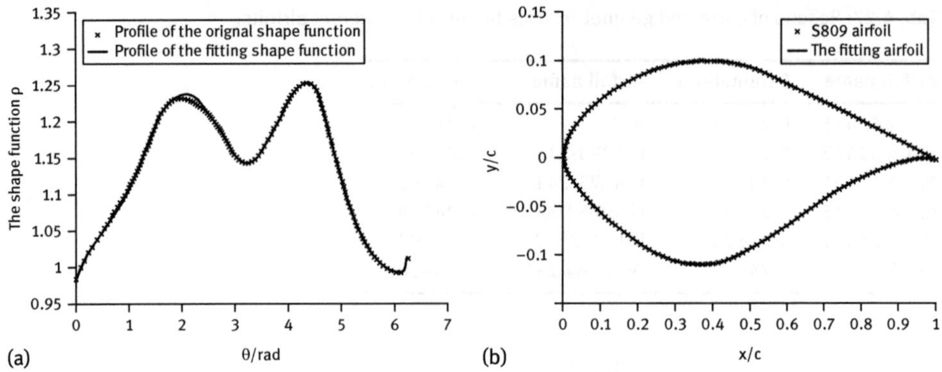

Fig. 4.85: Comparison of the fitting results and S809 airfoil results.

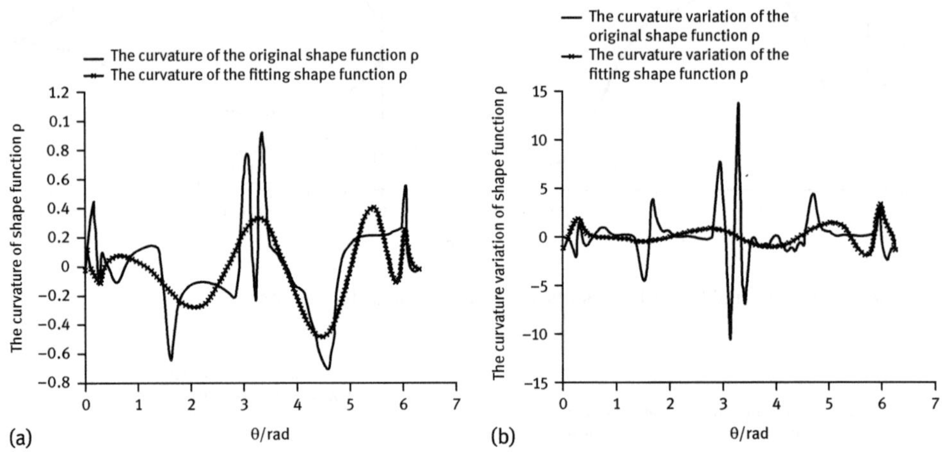

Fig. 4.86: Comparison of curvature and curvature variation for profiles of S809 airfoil shape function and fitted airfoil shape function.

Comparing Fig. 4.82 and 4.83, it can be seen that the profile of the fitting airfoil shape function tends to be more smooth and does not display sharp points after correcting the geometric scale factor a, especially when the argument θ is π. Fig. 4.85 and 4.87 show the fitting results for S809 and DU 93-W-210 airfoil respectively. The results are similar to those in Fig. 4.83. It is beneficial to accurately determine the coefficients of the shape function.

Investigations of the curvatures of the profiles of shape functions for three typical airfoils (NACA 64418, S809 and DU 93-W-210 airfoils) were also conducted in detail. The curvature distribution and curvature variation of the profiles for the three airfoil shape functions are plotted in Fig. 4.84, 4.86 and 4.88. Fig. 4.84 shows the comparison of curvature distribution and curvature variation of the profiles of the fitted shape function and NACA 63418 airfoil shape function. The curvature of the profiles for the

NACA 63418 airfoil shape function changes abruptly, and the location of the leading edge where the argument θ equals π has a very sharp point. This means that the original airfoil exhibits poor smooth continuity. Based on the integral theory of airfoil profiles, different kinds of airfoil can be well expressed. Moreover, compared with the profile of the original airfoil shape function, the curvature distribution and curvature variation exhibit a smooth continuity, particularly at the location of the leading edge where the sharp points become smoother. Fig. 4.86 and 4.88 are also similar to Fig. 4.84, but there is a small difference at the location of the trailing edge where the value of argument θ is zero and 2π. At these particular points, the curvature distribution and curvature variation of the profile of the fitted shape function and the original shape function both have sharp points. The main reason is that the camber of the airfoil at the trailing edge is greater (S-shaped tail) which could cause the larger curvature change.

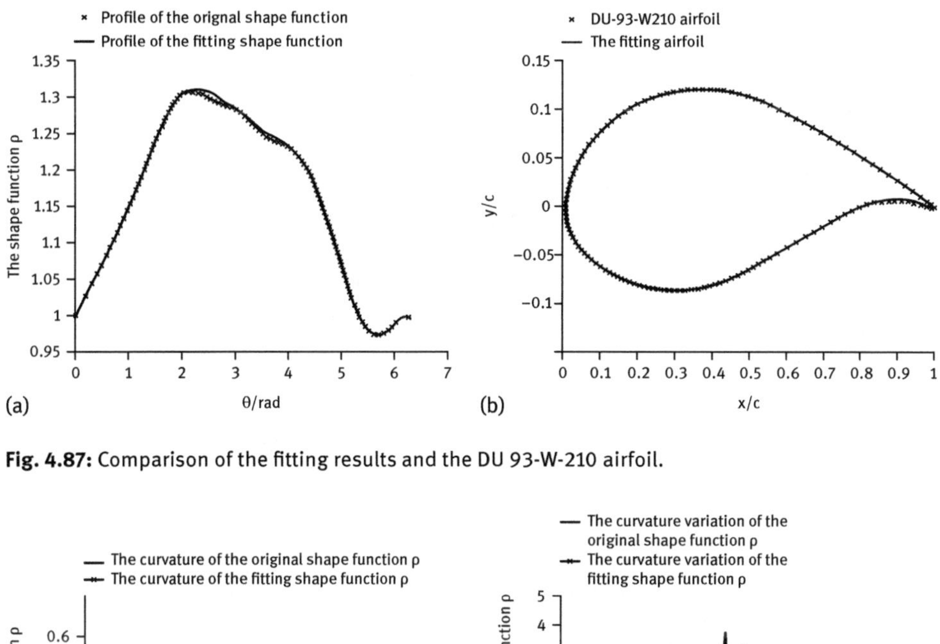

Fig. 4.87: Comparison of the fitting results and the DU 93-W-210 airfoil.

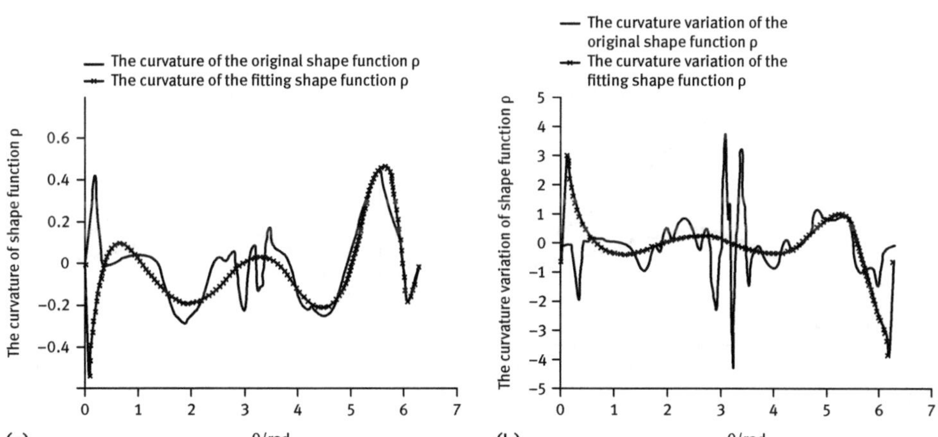

Fig. 4.88: Comparison of curvature and curvature variation for profiles of DU 93-W-210 airfoil shape function and fitted airfoil shape function.

4.8.2 Curvature of profile for airfoil shape function

To illustrate the advantages and the importance of curvature for the profiles of airfoils, the curvature for the profiles of typical airfoils and the fitted airfoils are studied in detail. The comparison of curvature and curvature variation for the profiles of the NACA 64418 airfoil and its fitted airfoil are shown in Fig. 4.89. The curvature for the profiles of the fitted airfoil is smoother than the original NACA 64418 airfoil, especially near the location of $x/c = 0.4$ where the curvature of the profile for NACA 64418 airfoil exhibits no smooth continuity. The pressure coefficients distribution of the NACA 64418 airfoil and its fitted airfoil ($Re = 3.0 \times 10^6$, $Ma = 0.15$, $\alpha = 6°$) are plotted in Fig. 4.90 to study the relationship between curvature and the pressure distribution of the airfoil surface.

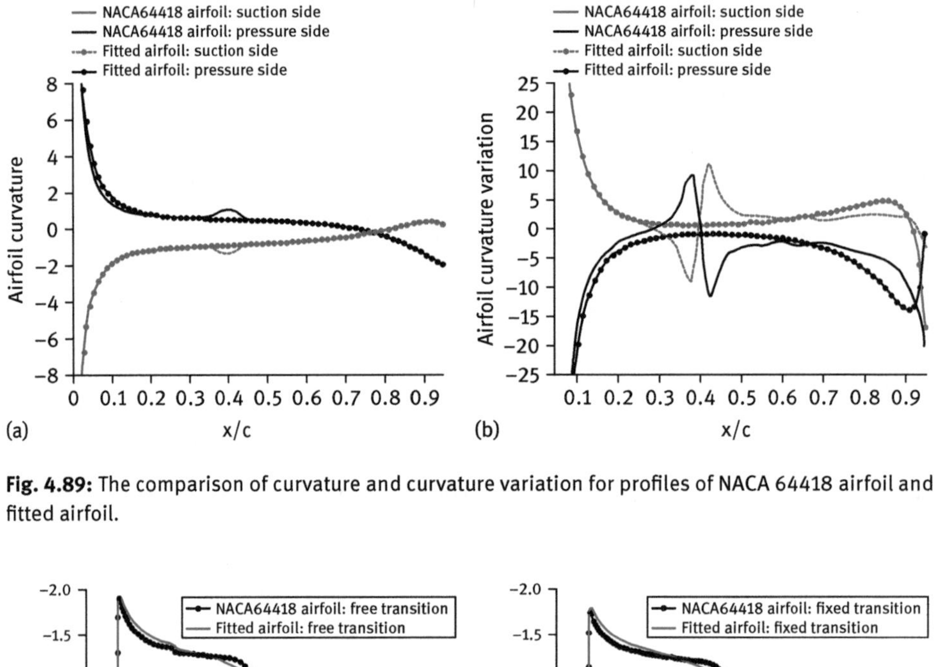

Fig. 4.89: The comparison of curvature and curvature variation for profiles of NACA 64418 airfoil and fitted airfoil.

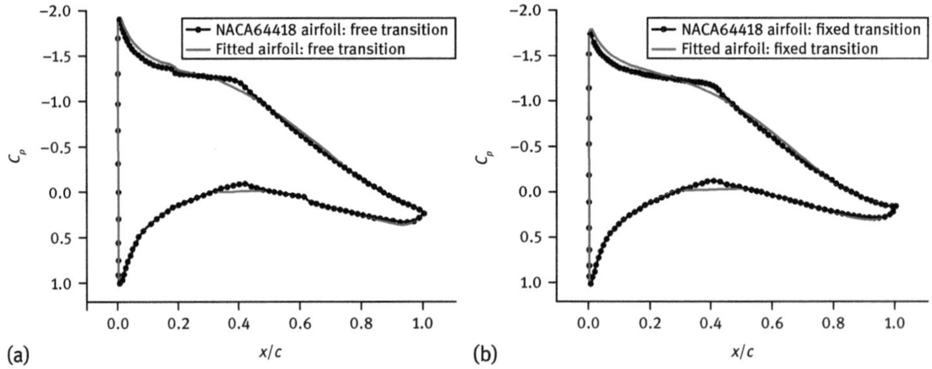

Fig. 4.90: Pressure coefficients distribution of NACA 64418 and fitted airfoil ($Re = 3.0 \times 10^6$, $Ma = 0.15$, $\alpha = 6°$).

As shown in Fig. 4.90, both the NACA 41-418 airfoil and its fitted airfoil exhibit smooth continuity in the free transition and fixed transition cases, except at the position of $x/c = 0.4$ for the NACA 64418 airfoil. It means that there is unexpected flow acceleration or deceleration for the NACA 64418 airfoil surface at this location (about $x/c = 0.4$) due to sudden changes of surface curvature (as shown in Fig. 4.89 at the same position). In addition, Fig. 4.91 shows the comparison of curvature and curvature variation for the profiles of the DU 93-W-210 airfoil and its fitted airfoil. The curvature of the fitted airfoil is as smooth as the curvature of the DU 93-W-210 airfoil. Therefore, the pressure distribution of the two airfoils displays smooth continuity (as shown in Fig. 4.92). From

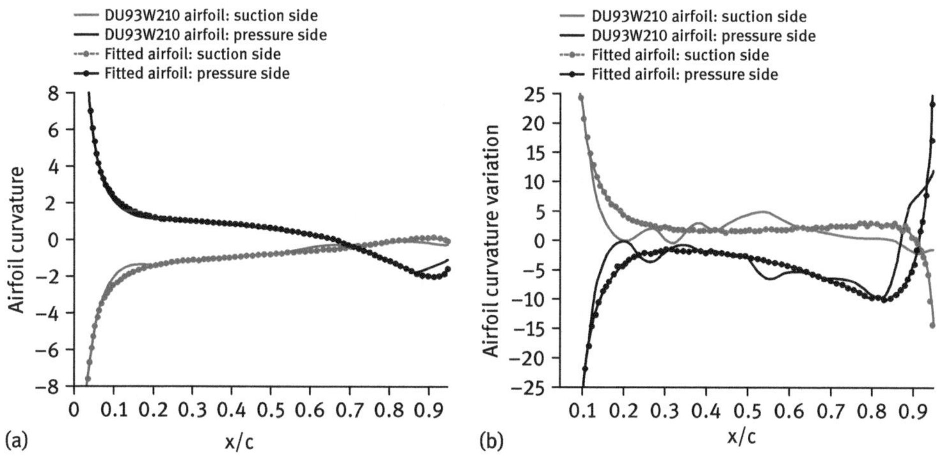

Fig. 4.91: Comparison of curvature and curvature variation for profiles of DU 93-W-210 airfoil and fitted airfoil.

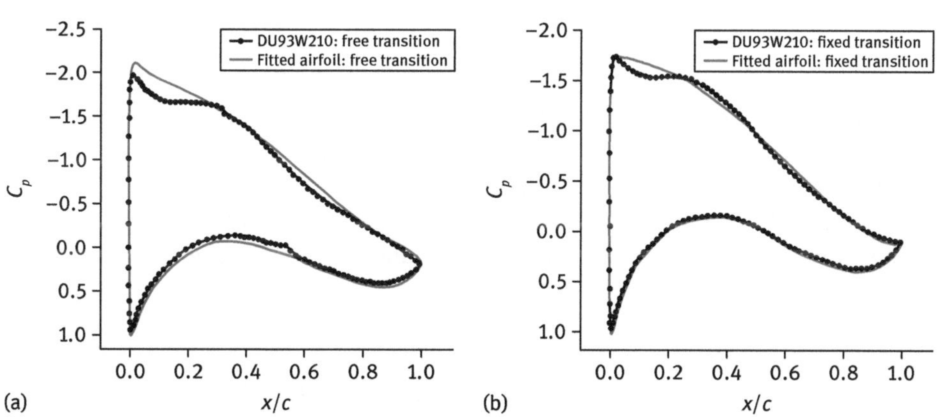

Fig. 4.92: Pressure coefficients distribution of DU 93-W-210 and fitted airfoil ($Re = 3.0 \times 10^6$, $Ma = 0.15$, $\alpha = 6°$).

the above comparison analysis, it can be seen that the curvature of airfoils plays a very important role under the flow conditions for the airfoil's surface. The profiles of the fitted airfoils obtained by the method presented here show curvature smooth continuity. As a result, the pressure distribution of the fitted airfoils also displays smooth continuity.

4.8.3 Improvement and optimization of the airfoil

There are many factors affecting airfoil performance, including structural, aerodynamic and other interdisciplinary factors. Based on Blade Element Momentum theory (BEM), the energy coefficient of blades is related directly to the lift-to-drag ratio of the airfoil. In addition, roughness sensitivity in the airfoil's leading edge region is also an important factor and cannot be ignored. Roughness in the airfoil leading edge region is formed by accumulation of dust, dirt and bugs. Blade erosion and manufacturing imperfections may also have an influence on roughness. Meanwhile, a moderately high maximum lift coefficient is necessary to restrict the blade chord and to reduce the blade area so as to decrease the aerodynamic load. Based on the analysis described in the following sections, the DU 93-W-210 wind turbine airfoil is improved and optimized. Our aim is to make a new airfoil, exhibiting a higher lift-to-drag ratio, lift coefficient, and lower roughness sensitivity than the original one.

4.8.3.1 Objective function

In order to maximize the energy efficiency of the wind turbine at a given condition of corresponding Reynolds number and Mach number, the objective function is proposed to maximize the lift-to-drag ratio of both smooth and rough surface working conditions at a certain angle of attack:

$$f(x) = \max(\mu_1 \cdot c_l/c_d + \mu_2 \cdot c_l'/c_d'), \tag{4.122}$$

where μ_1, μ_2 are the weighting factors, $\mu_1, \mu_2 \in [0, 1]$, $\mu_1 + \mu_2 = 1$; c_l/c_d, c_l'/c_d' are the lift-to-drag ratio for smooth surface flow and rough surface flow, respectively. The free transition model is used to simulate the smooth surface flow, while the fixed transition model is used to simulate the rough surface flow (the transition points for the suction and pressure sides are fixed at the chord of 1 % and 10 %, respectively) [4, 5].

4.8.3.2 Design variables

Based on the corrected geometric scale factor a in the integral equation, the control coefficients of airfoil shape function ρ are to be determined. Thus, the second to tenth coefficients of airfoil shape function ρ are chosen as the design variables:

$$X = (C_1, C_2, C_3, C_4, C_5, C_6, C_7, C_8, C_9, C_{10})^T. \tag{4.123}$$

4.8.3.3 Design constraints

Based on the 11-order fitting control coefficients of the DU 93-W-210 airfoil shape function, the upper and lower limits are imposed on the design variables:

$$X_{\min} \leq X \leq X_{\max}. \tag{4.124}$$

To make sure that a relatively high lift coefficient for the optimized airfoil can be achieved, the lift coefficient is restricted to be higher than 0.8 at flow conditions $Re = 3.0 \times 10^6$, $Ma = 0.15$, $\alpha = 6°$:

$$c_l \geq 1.0. \tag{4.125}$$

The airfoil leading edge radius should not be too small in the optimization process to avoid a sharp leading edge. If the leading edge radius is small, it may not convergence when RFOIL is used to calculate the aerodynamic results. In addition, the larger the airfoil leading edge radius is, the higher the maximum lift coefficient will be. This restriction can be satisfied by controlling the thickness from the upper point to the lower point at the 10 % chord location:

$$t|_{x=0.1} \geq 0.08. \tag{4.126}$$

Considering that the aim of the optimization is to improve the performance of the DU 93-W-210 airfoil which has a maximum thickness to chord ratio of about 0.21, it is essential to restrict the maximum thickness of the improved airfoil:

$$\frac{th}{c} = t \in [0.208, 0.21]. \tag{4.127}$$

Geometric compatibility is important for airfoil families of a wind turbine blade. Since the maximum thickness of the DU 93-W-210 airfoil is located at 33.6 % chord length, the location of maximum thickness is set between 25 % and 35 % chord length:

$$0.24 \leq L_{\max} \leq 0.36. \tag{4.128}$$

4.8.4 Optimization results

The multiobjective GA optimization method is used to optimize the original airfoil. RFOIL software is used to calculate the aerodynamic performance of the airfoil. Some GA parameters are: maximum number of generations is 300, crossover probability and mutation probability are taken as 0.01 and 0.7, and population size has a value of 50. In order to keep high C_L, and insensitive to roughness, the weight coefficients of free transition and fixed transition are taken as $\lambda_1 \in [0.25, 0.5]$ and $\lambda_2 = 1 - \lambda_1$ respectively. Then the DU 93-W-210 airfoil is improved and optimized by using the GA method based on the integral theory of airfoil profiles. After optimizing the airfoil, it is found that the new airfoil has better aerodynamic performance when the weight

Tab. 4.28: Coefficients of the airfoil shape functions.

Airfoil name	Coefficients				
	C_1	C_2	C_3	C_4	C_5
DU 93-W-210	0.174971	−0.085675	−0.039083	0.277646	−0.274520
CQU210	−0.046712	0.987714	−1.894472	1.4574131	−2.744097
	C_6	C_7	C_8	C_9	C_{10}
DU 93-W-210	0.122929	−0.0297182	0.003998	−2.80288e−4	7.92848e−6
CQU210	0.177919	−0.051443	0.007869	−6.26506e−4	2.04977e−5

coefficients λ_1 is set to 0.40 and λ_2 to 0.60. The new airfoil is named CQU210. Tab. 4.28 shows the control coefficients of the shape functions for the DU 93-W-210 airfoil and the CQU210 airfoil.

Fig. 4.93 shows the CQU210 airfoil and the DU 93-W-210 airfoil profiles. Tab. 4.29 shows the comparison of geometric characteristics for the two airfoils. The maximum thickness of the CQU210 airfoil is $th/c = 0.20994$ at the location of 0.314 chord length from the leading edge. The maximum camber of the CQU210 airfoil is $cam/c = 0.02158$ at $x/c = 0.797$. Compared with the DU 93-W-210 airfoil, the maximum thickness of the new airfoil is a bit thicker.

To demonstrate the performance of the optimized airfoil, a comparison is made between the new airfoil and the DU 93-W-210 airfoil. Fig. 4.94 shows the aerodynamic performance of the new airfoil and the DU 93-W-210 airfoil in the same working conditions ($Re = 6.0 \times 10^6$, $Ma = 0.15$). The maximum lift-to-drag ratio of the new airfoil is

Tab. 4.29: Comparison of geometric characteristic of airfoils.

Airfoil name	Max. thickness	at $x =$	Max. camber	at $x =$
DU 93-W-210	0.20689	0.336	0.02735	0.715
CQU210	0.20994	0.314	0.02158	0.797

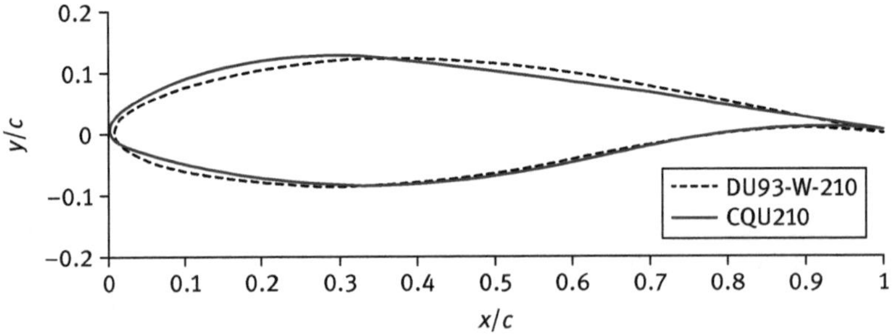

Fig. 4.93: CQU210 and DU 93-W-210 airfoil profiles.

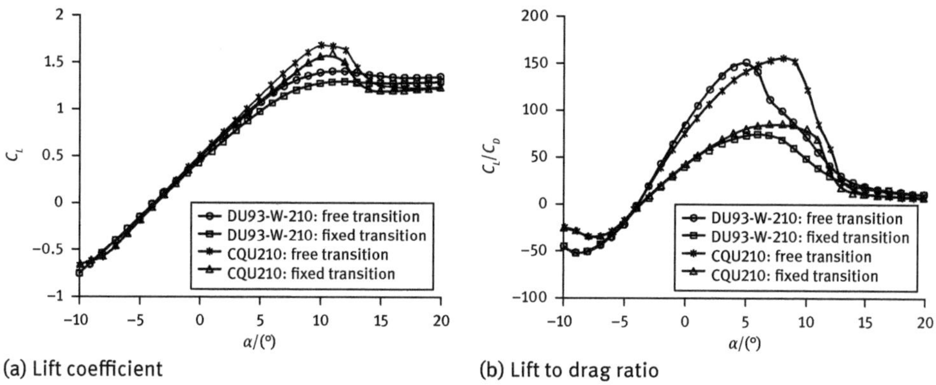

Fig. 4.94: Aerodynamic performance of the CQU210 airfoil and the DU 93-W-210 airfoil.

up to 156.113 in the working conditions of smooth surface flow, at the angle of attack $\alpha = 7°$. The new airfoil is insensitive to leading edge roughness and exhibits good stall characteristics. Within the main angle of attack, the new airfoil shows higher lift and lift-to-drag ratio than the DU 93-W-210 airfoil in the working condition of fixed transition. Conversely, in the condition of free transition, the lift-to-drag ratio of the new airfoil is a little lower than the DU 93-W-210 airfoil at the angle of attacks approximately from $\alpha = 0°$ to 6°. This can be explained by the drag coefficients of the new airfoil being greater than those of the DU 93-W-210 airfoil within those angles of attack. Nevertheless, the maximum lift coefficients of the new airfoil are still higher than those of the UU93-W-210 airfoil before stall. When the angle of attack is greater than 12°, the airfoil is in the state of stall. In this situation, the aerodynamic performance of the airfoil can hardly be predicted accurately; the drag coefficient is dramatically increased. This is the reason why the lift-to-drag ratio for the CQU210 airfoil is lower than the DU 93-W-210 airfoil at the angle of attack greater than 13°.

Tab. 4.30 shows the main aerodynamic parameters of the DU 93-W-210 airfoil and the CQU210 airfoil. Compared with the DU 93-W-210 airfoil, the maximum C_L/C_D of the CQU210 airfoil increased by 3.4 % and 14.7 % in the working conditions of free transition and fixed transition respectively. The maximum C_L of the CQU210 airfoil also increased by 18.9 % and 22.8 % in the working conditions of free transition and fixed transition, respectively.

Tab. 4.30: Aerodynamic performance of airfoils.

Airfoil name	Free transition		Fixed transition	
	$c_{l,max}$	L/D_{max}	$c_{l,max}$	L/D_{max}
DU 93-W-210	1.409	150.910	1.289	74.410
CQU210	1.676	156.113	1.583	85.326

Consequently, the new improved CQU210 airfoil is found to have much better on-design and off-design operational condition characteristics, and it shows higher lift coefficient and larger lift-to-drag ratio in fixed transition conditions in the main angles of attack of working range. The results show that the new method is feasible for optimizing wind turbine airfoils.

4.9 Design of wind turbine airfoils with high performance

Equations (3.24), (3.25) and (3.26) were adopted to express the airfoil shape in the optimization of the relatively thinner airfoils. Various airfoils could be obtained by controlling the coefficient of the function which has good integration and versatility properties. However, it was found that the airfoil trailing edge generated by this method was relatively smooth. Especially for the medium-thickness and high-thickness airfoil, the curves of the trailing edge (on both pressure and suction side) were close to a straight line with small camber. Due to the increase in the number of control variables, it was hard to determine the range of the variables, which made the optimization very difficult. For the optimization of thinner airfoils, the number of control variables was increased from 6 to 12 and the improved particle swarm optimization algorithm was used.

4.9.1 Objective function

Under smooth and rough conditions, with Reynolds number $Re = 6.0 \times 10^6$ and Mach number $Ma = 0.15$, the maximum lift-to-drag ratio was taken as objective function at AOA of 6°.

$$f(x) = \max(\mu_1 \cdot c_l/c_d + \mu_2 \cdot c_l'/c_d'), \tag{4.129}$$

where μ_1, μ_2 are the weight factors under the smooth and rough conditions respectively ($\mu_1, \mu_2 \in [0, 1]$ and $\mu_1 + \mu_2 = 1$); the linear weighting factor method was used ($\mu_1 = [0.10{:}0.05{:}0.90]$); c_l/c_d and c_l'/c_d' are the lift-to-drag ratios of the airfoil under smooth and rough conditions respectively. The free transition conditions were used to simulate smooth conditions; the fixed transition conditions were used to simulate rough conditions. The transition point was fixed at the chordwise location of 1 % on suction surface and 10 % chord on pressure surface to simulate the worst airfoil roughness conditions [4, 5].

4.9.2 Design variables

The shape control could be better if the number of design variables was as high as possible. However, the increasing the number of variables makes it more difficult to define the ranges of variables. Therefore, the first to twelfth coefficients in the integrated expression function $\varphi(\theta)$ were chosen as design variables to control the airfoil shape (especially the shape of the airfoil trailing edge). And the design variables were as follows:

$$X = (a_1, b_1, a_2, b_2, a_3, b_3, a_4, b_4, a_5, b_5, a_6, b_6)^{\mathrm{T}}. \tag{4.130}$$

4.9.3 Design constraints

In the optimization of wind turbine airfoils, the shape function would no longer have the airfoil shape feature if the control variables exceeded a certain range. And the boundary conditions of variables were set as follows:

$$X_{\min} \le X \le X_{\max}. \tag{4.131}$$

The range of design variables are shown in Tab. 4.31.

Tab. 4.31: Constraint boundaries of optimization variables.

	a_1	b_1	a_2	b_2	a_3	b_3	a_4	b_4	a_5	b_5	a_6	b_6
Maximum	0.4	0.2	0.2	0.2	0.1	0.1	0.1	0.1	0.1	0.1	0.1	0.1
Minimum	0.0	−0.1	−0.1	−0.1	−0.1	−0.1	−0.1	−0.1	−0.08	−0.08	−0.08	−0.08

Airfoils with maximum relative thickness of 18 % are commonly used in the main power-generating zone of the wind turbine blade, which requires higher lift-to-drag ratio and lower roughness sensitivity. Therefore, the improved particle swarm algorithm was used in the design of three airfoils with maximum relative thickness of 15 %, 18 %, 21 %. The thickness constraints were set as follows:

$$\frac{th}{c} = t \in [0.15, 0.21]. \tag{4.132}$$

Besides the thickness constraints of airfoils, the chordwise position of the maximum relative thickness was also constrained:

$$0.24 \le L_{\max} \le 0.35. \tag{4.133}$$

Normally, the radius of the leading edge of a wind turbine airfoil should not be too small. This requirement is satisfied through controlling the distance between the point (10 % chordwise) on the pressure side and the point (10 % chordwise) on the suction side:

$$t|_{x/c=0.1} \ge 0.08. \tag{4.134}$$

4.9.4 Optimization results and analysis of thin airfoil series

The improved particle swarm algorithm [154–157] was coupled with the software RFOIL [5] in optimization. The basic parameters of the algorithm are as follows: the inertia weight is 0.85; the learning factors C_1, C_2 are 0.5; the variable dimension is 12; the population size is 30 and the maximum iteration number is 400. The optimization flow chart is shown in Fig. 4.95. Three kinds of airfoils with good performance were obtained after optimization, which were called CQU-A150, CQU-A180 and CQU-A210. In the names of this airfoil series, CQU represents Chongqing University, A represents high performance airfoils, and the number represents 100 times the maximum relative thickness of the airfoil. The geometric characteristics and aerodynamic performance of the airfoils with different thickness are compared with traditional airfoils as follows.

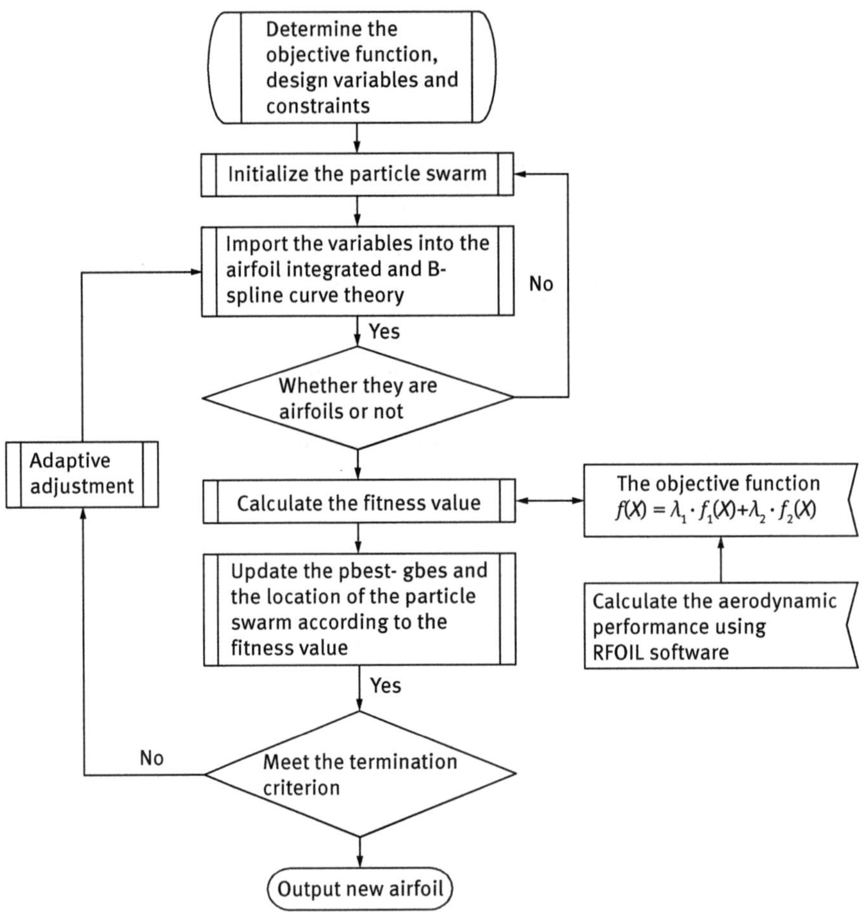

Fig. 4.95: Flow chart of improved particle swarm optimization.

4.9.4.1 CQU-A150 airfoil

Fig. 4.96 presents the shapes of CQU-A150 and NACA 64415. Due to the increased number of control variables, the shape of the airfoil can be better controlled. Based on the Tab. 4.32, the maximum relative thickness of the new airfoil has decreased compared with the NACA 64415 airfoil, with chordwise location of the maximum relative thickness closer to the airfoil leading edge. This will help to improve the maximum lift coefficient of the airfoil. And the maximum camber is larger than that of NACA 64415, which is conducive to improving the maximum lift-to-drag ratio of the airfoil.

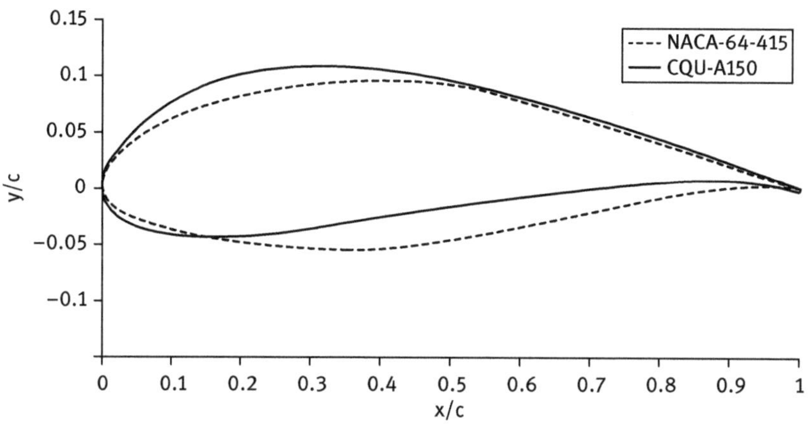

Fig. 4.96: Shapes of airfoils.

Tab. 4.32: Comparison of geometric characteristics of airfoils.

Airfoil name	Maximum relative thickness	Chordwise location of maximum relative thickness	Maximum camber	Chordwise location of maximum camber
NACA 64415	0.14998	0.374	0.02047	0.531
CQU-A150	0.14717	0.248	0.04095	0.435

The aerodynamic performance of CQU-A150 and NACA 64415 (with same relative thickness) under the same conditions ($Re = 6.0 \times 10^6$, $Ma = 0.15$) are shown in Fig. 4.97. In the main AOA range, the lift coefficient and lift-to-drag ratio of the new airfoil are higher both under rough condition or smooth condition compared with those of NACA 64415. The aerodynamic performance of the new airfoil has been greatly improved.

The noise performance of CQU-A150 and NACA 64415 under free and fixed transition conditions are shown in Fig. 4.98. The noise of the new airfoil is lower than that of NACA 64415 before stall.

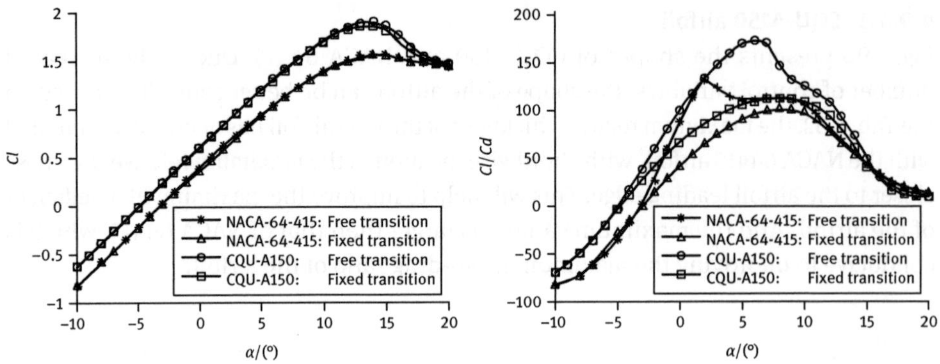

(a) Lift coefficient (b) Lift to drag ratio

Fig. 4.97: Aerodynamic characteristics of CQU-A150 and NACA 64415.

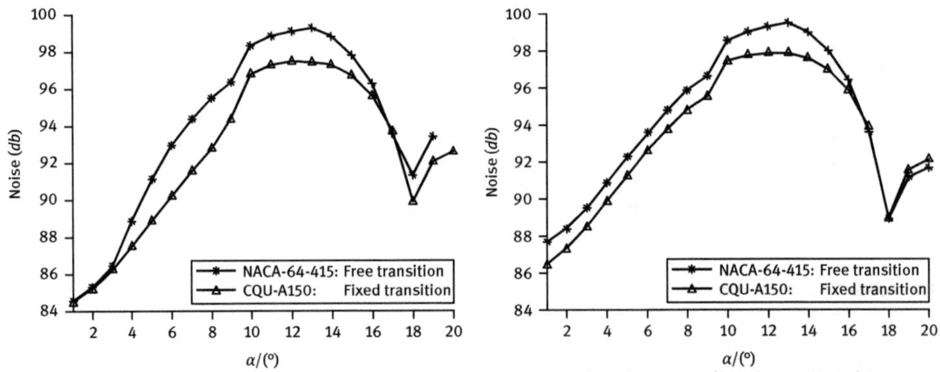

(a) Comparison of noise (free transition) (b) Comparison of noise (fixed transition)

Fig. 4.98: Noise characteristics of CQU-A150 and NACA 64415.

The key aerodynamic parameters are listed in Tab. 4.33. Under the free transition condition, the maximum lift coefficient of the new airfoil is 1.889 (at AOA of 14°) which is an improvement of 21.448 %; the maximum lift-to-drag ratio is 173.373 (at AOA of 6°) which is an improvement of 30.655 %. Under fixed transition, the maximum lift coefficient is 1.849 (at AOA of 14°), which is an improvement of 20.403 %; the maximum lift-to-drag ratio is 112.287 (at AOA of 9°), which is an improvement of 10.204 %. In addition, the new airfoil has lower roughness sensitivity and lower noise. Compared with the commonly used wind turbine airfoil, the performance has been significantly improved.

Tab. 4.33: Comparison of aerodynamic parameters of airfoils.

Airfoil name	Free transition		Fixed transition $X_{tr,s} = 0.01; X_{tr,p} = 0.1$		Roughness sensitivity	Noise (dB)
	$c_{l,max}$	L/D_{max}	$c_{l,max}$	L/D_{max}		
NACA 64415	1.5548(15°)	132.6957(2°)	1.5360(15°)	100.9748(8°)	1.2 %	95.8
CQU-A150	1.8889(14°)	173.3733(6°)	1.8494(14°)	112.2868(9°)	2.1 %	95.5
	21.488 %	30.655 %	20.403 %	11.203 %		

Note: The data in parentheses represent the AOA with the maximum lift coefficient and maximum lift-to-drag ratio; the noise is calculated at the AOA with the maximum lift-to-drag ratio and under the fixed transition condition.

4.9.4.2 CQU-A180 airfoil

The aerodynamic performance of the CQU-A180 airfoil was compared with that of the selected DF180 airfoil with equal thickness (used on 2.5 MW wind turbine of Dongfang Steam Turbine Company). Fig. 4.99 presents the shapes of CQU-A180 and DF180. As shown in Fig. 4.99, due to the increased number of control variables, the shape of the airfoil can be better controlled. The shape of the airfoil trailing edge has larger curvature and shows an “s”-style, which is conducive to improving the aerodynamic performance of the airfoil. Based on Tab. 4.34, the chordwise position of the maximum relative thickness is closer to the airfoil leading edge compared with DF180, which is beneficial to improving the maximum lift coefficient. And the maximum camber is larger, which is conducive to improving the maximum lift-to-drag ratio of the airfoil.

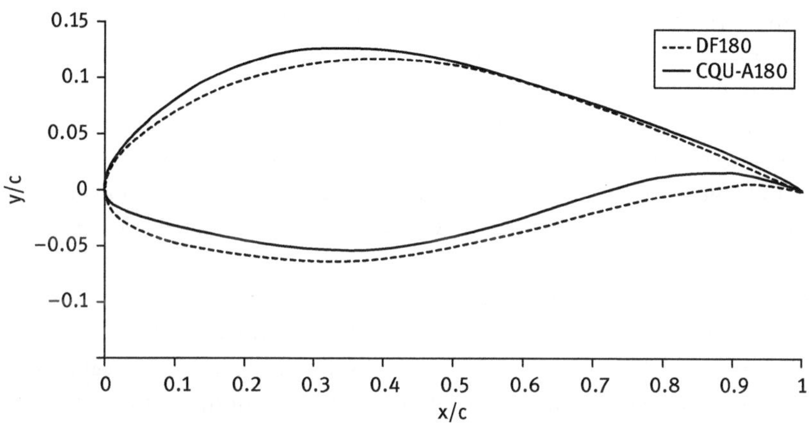

Fig. 4.99: Shapes of airfoils.

Tab. 4.34: Comparison of geometric characteristics of airfoils.

Airfoil name	Maximum relative thickness	Chordwise location of maximum relative thickness	Maximum camber	Chordwise location of maximum camber
DF180	0.17988	0.360	0.03029	0.550
CQU-A180	0.18064	0.344	0.03602	0.644

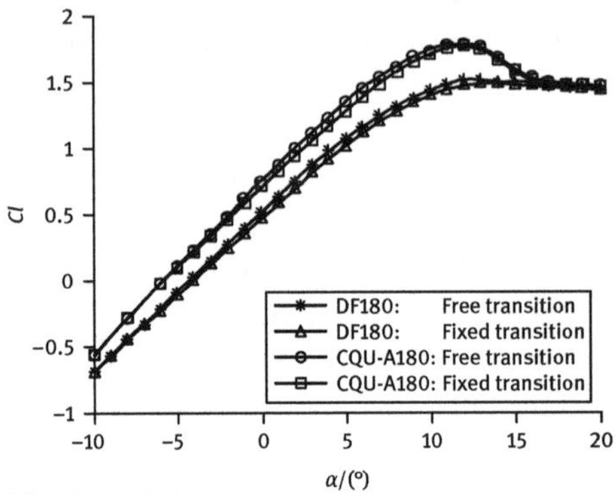

(a) Lift coefficient

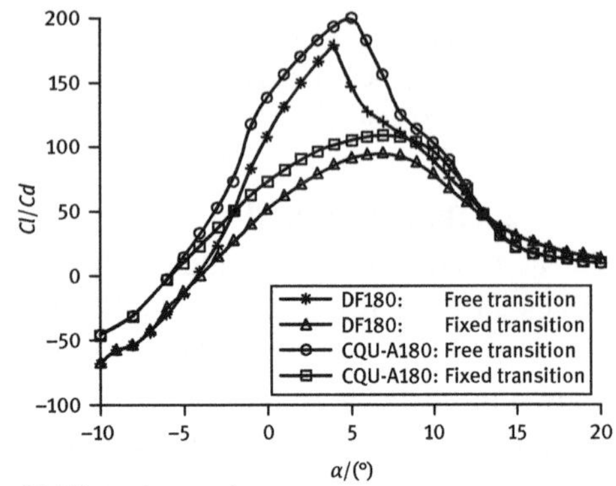

(b) Lift to drag ratio

Fig. 4.100: Aerodynamic characteristics of CQU-A180 and DF180.

Fig. 4.100 presents the aerodynamic performance of airfoils CQU-A180 and DF180 under the same conditions ($Re = 6.0 \times 10^6$, $Ma = 0.15$). In the main range of the AOA, compared with DF-180 airfoil, the lift coefficient and lift-to-drag ratio of the new airfoil are higher under rough and smooth conditions. And the aerodynamic performance has been obviously improved.

The variation of noise with respect to AOA under the free transition condition and fixed transition are shown in Fig. 4.101. The noise of the new airfoil is lower than that of the DF180 airfoil before the stall. After the stall, it's very difficult to accurately predict the aerodynamic performance of the airfoil, which will cause a large error in the noise calculation.

Tab. 4.35 presents the key aerodynamic parameters obtained. Under free transition condition, the maximum lift coefficient of the new airfoil is 1.780 (at AOA of 12°), which is an improvement of 18.277 %; the maximum lift-to-drag ratio is 201.358 (at AOA of 5°), which is an improvement of 12.545 %. Under fixed transition, the maximum lift coefficient is 1.758 (at AOA of 12°), which is an improvement of 19.435 %; the maximum lift-to-drag ratio is 107.968 (at AOA of 7°), which is an improvement of 14.614 %. What is more, the new airfoil has lower roughness sensitivity and lower noise. Compared with the commonly used wind turbine airfoils, performance has been significantly improved.

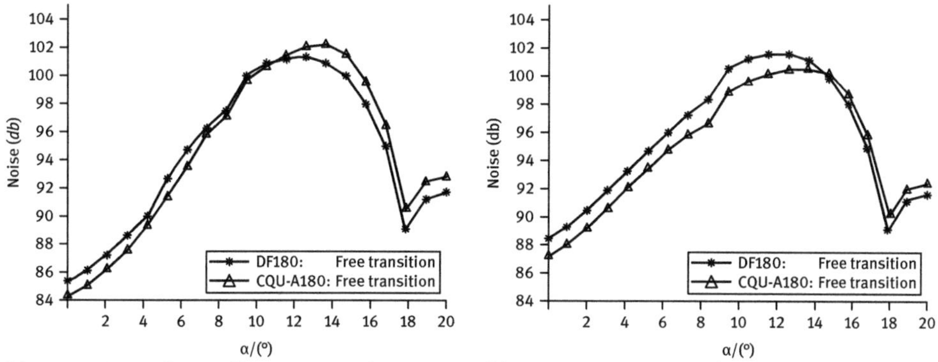

(a) Comparison of noise (free transition) (b) Comparison of noise (fixed transition)

Fig. 4.101: Noise characteristics of CQU-A180 and DF180.

Tab. 4.35: Comparison of aerodynamic parameters of airfoils.

Airfoil name	**Free transition**		**Fixed transition** $X_{tr,s} = 0.01; X_{tr,p} = 0.1$		**Roughness sensitivity**	**Noise (dB)**
	$c_{l,max}$	L/D_{max}	$c_{l,max}$	L/D_{max}		
DF180	1.5046(13°)	178.9134(4°)	1.4721(13°)	94.2013(7°)	2.2 %	96.1
CQU-A180	1.7796(12°)	201.3575(5°)	1.7582(12°)	107.9677(7°)	1.2 %	94.8
	18.277 %	12.545 %	19.435 %	14.614 %		

Note: The data in parentheses represent the AOA with the maximum lift coefficient and maximum lift-to-drag ratio; the noise is calculated at the AOA with the maximum lift-to-drag ratio and under fixed transition condition.

4.9.4.3 CQU-A210 airfoil

The CQU-A210 airfoil was compared with the selected DF210 airfoil with the same relative thickness (used on 2.5 MW wind turbine of Dongfang Steam Turbine Company). Fig. 4.102 presents the shapes of CQU-A210 and DF210. Due to the increased number of control variables, the shape of airfoil (especially the trailing edge) can be better controlled. The shape of the airfoil trailing edge has larger curvature and shows an "s"-style, which is conducive to improving the aerodynamic performance of the airfoil. It is known from Tab. 4.36 that the chordwise position of the maximum relative thickness is closer to the airfoil leading edge compared with DF210, which is beneficial to improving the maximum lift coefficient. And the maximum camber is larger, which is conducive to improving the maximum lift-to-drag ratio of the airfoil.

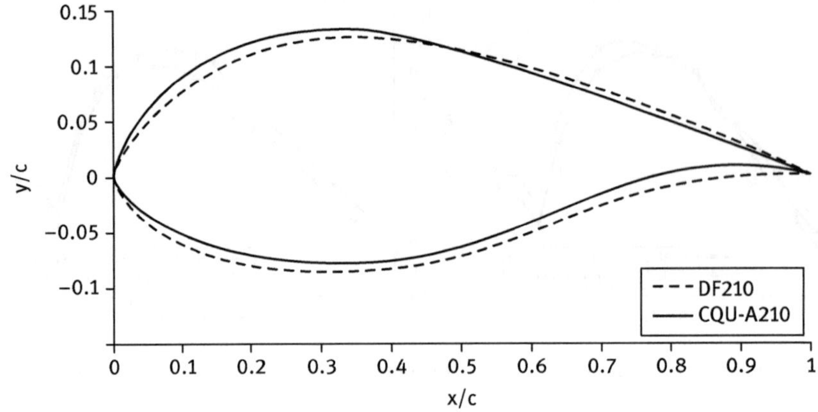

Fig. 4.102: Shapes of airfoils.

Tab. 4.36: Comparison of geometric characteristics of airfoils.

Airfoil name	Maximum relative thickness	Chordwise location of maximum relative thickness	Maximum camber	Chordwise location of maximum camber
DF210	0.21036	0.360	0.02531	0.700
CQU-A210	0.21019	0.320	0.02741	0.731

Fig. 4.103 presents the aerodynamic performance contrast of CQU-A210 and DF210 under the same conditions ($Re = 6.0 \times 10^6$, $Ma = 0.15$). As is shown in the figure, in the main range of the AOA, the lift coefficient and lift-to-drag ratio of the new airfoil are higher under rough or smooth conditions compared with DF-210. And the aerodynamic performance has been largely improved.

The change in noise with respect to AOA under the free transition condition and fixed transition is shown in Fig. 4.104. As is shown in the figure, the noise of the new airfoil is lower than that of the DF210 airfoil before the stall with reducing impacts on the surrounding environment.

Tab. 4.37 presents the key aerodynamic parameters of the two airfoils. Under free transition condition, the maximum lift coefficient of the new airfoil is 1.731 (at AOA of 11°), which is an improvement of 11.823 %; the maximum lift-to-drag ratio is 187.384 (at AOA of 6°), which is an improvement of 10.489 %. Under fixed transition, the maximum lift coefficient is 1.660 (at AOA of 12°), which is an improvement of 12.976 %; the maximum lift-to-drag ratio is 99.231 (at AOA of 7°), which is an improvement of 8.032 %. In addition, the new airfoil has lower roughness sensitivity and lower noise. Compared with the DF210 airfoil, performance has been significantly improved.

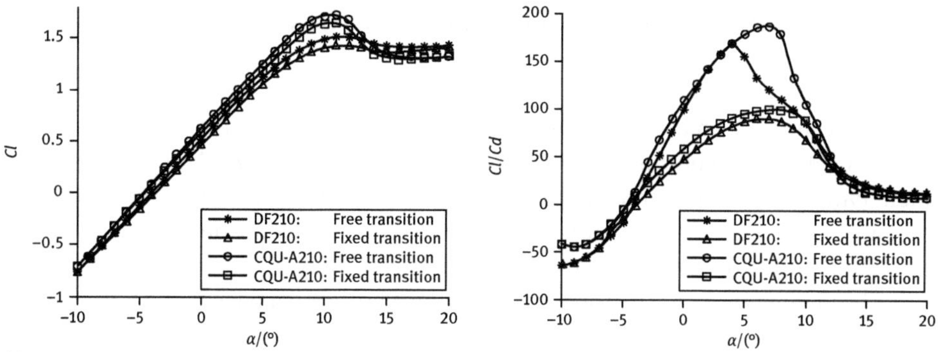

(a) Lift coefficient (b) Lift to drag ratio

Fig. 4.103: Aerodynamic characteristics of CQU-A210 and DF210.

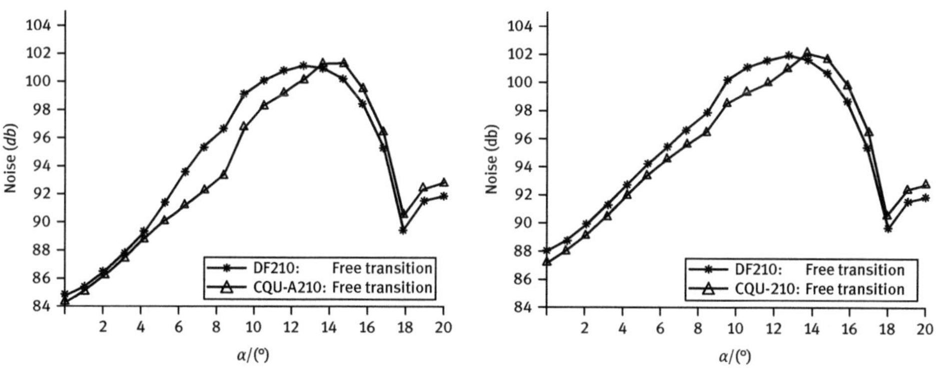

(a) Comparison of noise (free transition) (b) Comparison of noise (fixed transition)

Fig. 4.104: Noise characteristics of CQU-A210 and DF210.

Tab. 4.37: Comparison of aerodynamic parameters of airfoils.

Airfoil name	Free transition		Fixed transition $X_{tr,s} = 0.01; X_{tr,p} = 0.1$		Roughness sensitivity	Noise (dB)
	$c_{l,max}$	L/D_{max}	$c_{l,max}$	L/D_{max}		
DF210	1.5478(12°)	169.5953(4°)	1.4689(12°)	91.853(7°)	5.1 %	95.5
CQU-A210	1.7308(11°)	187.3842(6°)	1.6595(12°)	99.2308(7°)	4.1 %	94.6
	11.823 %	10.489 %	12.976 %	8.032 %		

Note: The data in parentheses represent the AOA with the maximum lift coefficient and maximum lift-to-drag ratio; the noise is calculated at the AOA with the maximum lift-to-drag ratio and under fixed transition condition.

4.9.5 A new direct design method for medium thickness wind turbine airfoils

However, it is hard to adjust the coefficients of the integral function for the medium thickness airfoil. B-spline curve has the advantage of local adjustment, which makes it easier to effectively control the airfoil profiles at the trailing edge. And some improvement can be obtained just by increasing the number of control variables. Therefore, a new direct design method for the medium thickness wind turbine airfoil based on airfoil integral expression and B-spline curve is presented in this section.

Airfoil integral theory method is used at the whole upper edge of the airfoil and at the location from the airfoil leading edge to $0.4c$ (c is the airfoil chord) of the airfoil lower edge, and the B-spline curve is introduced to control the points at the location from $0.4c$ to the airfoil trailing edge at the lower surface, which is shown in Fig. 4.105.

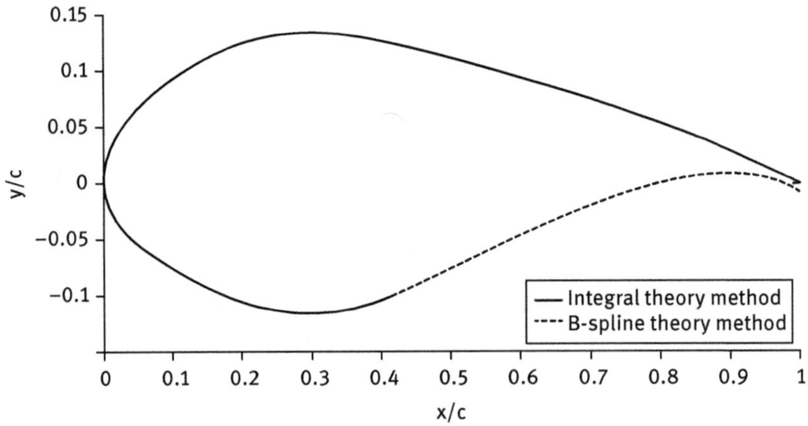

Fig. 4.105: Combined airfoil integral and B-spline theory method.

The coordinates of an airfoil shape using integral theory can be expressed by equations (3.24)–(3.26). The coordinates of the remaining part of the airfoil can be expressed by B-spline curve as follows:

$$p(u) = \sum_{i=0}^{n} d_i N_i, k(u), \tag{4.135}$$

where d_i, $i = 0, 1, 2, \dots, n$ are control points; N_i, $k(u)$, $i = 0, 1, 2, \dots, n$ is a k times based function, u is the node vector.

The cubic B-spline curve in matrix form represents the coordinate at the position from $0.4c$ to the trailing edge of the airfoil lower surface:

$$P_{0,3}(t) = \frac{1}{6}\begin{bmatrix}1 & t & t^2 & t^3\end{bmatrix}\begin{bmatrix}1 & 4 & 1 & 0\\ -3 & 0 & 3 & 0\\ 3 & -6 & 3 & 0\\ -1 & 3 & -3 & 1\end{bmatrix}\begin{bmatrix}P_0\\ P_1\\ P_2\\ P_3\end{bmatrix}, \quad t \in [0, 1], \tag{4.136}$$

where P_0, P_1, P_2, P_3 are the four controlled points.

The new wind turbine airfoil design method is performed by applying equations (3.24)–(3.26), (4.137) and (4.138). In order to obtain smooth and continuous airfoil profiles at the position of $0.4c$ at the lower surface, the controlled points $P_{0,3}(t)$ are two given fixed points, where $P_{0,3}(0)$ is the point at the trailing edge of $0.4c$ at the airfoil lower surface, and $P_{0,3}(1)$ is the fixed point at $(1, 0)$. So knowing the two points, the other two points can be deducted from equation (4.138). In fact, the controlled parameter variables are the two points P_1 and P_2.

4.9.6 Optimal model of thick airfoil series

4.9.6.1 Design variables

In the combination of integrated expression and B-spline curve method, the first to eighth coefficients $\varphi(\theta)$ of the integrated shape expression and the control parameters P_1 and P_2 of the B-spline curve were taken as the design variables:

$$X = (a_1, b_1, a_2, b_2, a_3, b_3, a_4, b_4, P_1, P_2). \tag{4.137}$$

4.9.6.2 Design objects

To improve the energy efficiency of the wind turbine, which is related to the lift-to-drag ratio, the objective function is introduced to maximize the lift/drag of both smooth and rough surfaces in working conditions at a certain angle of attack and for a corresponding Reynolds number and Mach number ($Re = 3.0 \times 10^6$, $Ma = 0.15$, $\alpha = 6°$)

$$f(x) = \max(\mu_1 \cdot c_l/c_d + \mu_2 \cdot c_l'/c_d'), \tag{4.138}$$

where μ_1, μ_2 are the weighting factors at the condition of smooth surface and rough surface, respectively, $\mu_1, \mu_2 \in [0, 1]$, $\mu_1 + \mu_2 = 1$, and using a linear weighting factor while setting $\mu_1 = [0.10{:}0.05{:}0.90]$. c_l/c_d, c_l'/c_d' are the lift-to-drag ratio for the smooth surface flow and rough surface flow, respectively. The free transition model is used to simulate the smooth surface flow, while the fixed transition model is introduced to simulate the rough surface flow (the transition points for the suction and pressure sides are fixed to 1 % and 10 % respectively) [4, 5].

4.9.6.3 Design constraints

The control function $\varphi(\theta)$ in the airfoil integral theory is a high-order trigonometric polynomial. And it is known that the more coefficients there are, the higher the order is. This is favorable for controlling the airfoil profiles. As a result, an airfoil with good performance will be designed in the range of the design space. Nevertheless, it is more difficult to decide the range of the design variables when the number of control variables is increased. Thus, in order to control the airfoil profiles easily, especially for the trailing edge at the lower surface, the coefficients of the function $\varphi(\theta)$ from 1 to 12 and the parameters P_1 and P_2 of the B-spline curve were selected as optimization design variables.

Once the values of the design variables exceed a certain range of the control variables, the airfoil profiles would no longer have airfoil shape characteristics. So the variable boundary conditions are as follows:

$$X_{\min} \leq X \leq X_{\max}. \tag{4.139}$$

The task is to optimize the medium thickness airfoil, which is generally located at the middle of the blade spanwise location. So, not only high aerodynamic performance is needed, but also a certain structural strength. Therefore, it is essential to restrict the maximum relative thickness for the design airfoil:

$$\frac{th}{c} = t \in [0.25, 0.40]. \tag{4.140}$$

Geometric compatibility is important in airfoil families for a wind turbine blade. The location of maximum relative thickness is between 25 % and 35 % chord:

$$0.24 \leq L_{\max} \leq 0.35. \tag{4.141}$$

4.9.7 Optimization results

In this optimization, the improved PSO algorithm is used to optimize wind turbine airfoils. RFOIL software is chosen to calculate the aerodynamic performance of the airfoil. The basic parameters of the improved algorithm are the same as those in Chapter 3. The flow chart of the new airfoil optimization design that combines numerical optimization with the tools for aerodynamic calculations is shown in Fig. 4.95. Four new airfoils with four different thicknesses were designed based on the new design method, named as CQU-A250, CQU-A300, CQU-A350, CQU-A400 airfoils. In the following, the comparison of the aerodynamic performance of these four airfoils with those of representative airfoils is presented.

4.9.7.1 CQU-A250 airfoil

The CQU-A250 airfoil was compared with the selected DF250 airfoil with the same relative thickness (used on 2.5 MW wind turbine of Dongfang Steam Turbine Company). Fig. 4.106 presents the shapes of CQU-A250 and DF250. The maximum relative thickness of the new airfoil is $t/c = 0.24945$ with the chordwise position at $x/c = 0.301$, and the maximum relative camber is $cam/c = 0.02972$ with the chordwise position at $x/c = 0.750$ (which are shown in Tab. 4.38). In terms of its geometric characteristics, the airfoil has good structural characteristics and excellent compatibility with other wind turbine airfoils. Considering the manufacturing process and the structural strength of the blade, the airfoil has a blunt trailing edge.

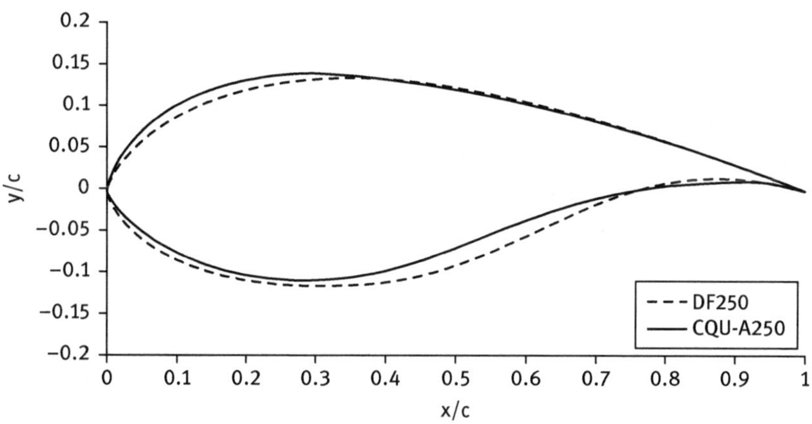

Fig. 4.106: Shapes of airfoils.

Tab. 4.38: Comparison of geometric characteristics of airfoils.

Airfoil name	Maximum relative thickness	Chordwise location of maximum relative thickness	Maximum camber	Chordwise location of maximum camber
DF250	0.25049	0.330	0.02671	0.800
CQU-A250	0.24945	0.301	0.02972	0.750

The comparison of aerodynamic characteristics between the newly-designed airfoil and the DF250 airfoil is shown in Fig. 4.107. The comparison was made under the same conditions ($Re = 6.0 \times 10^6$, $Ma = 0.15$). As is shown in the figure, the lift coefficient (smooth condition) of the CQU-A250 airfoil is 1.813 at the AOA of 13°; the maximum lift-to-drag ratio (smooth condition) is 168.558 at the AOA of 7°; the lift coefficient (rough condition) of the new airfoil is 1.688 at AOA of 12°; the maximum lift-to-drag ratio (rough condition) is 93.554 at the AOA of 8°.

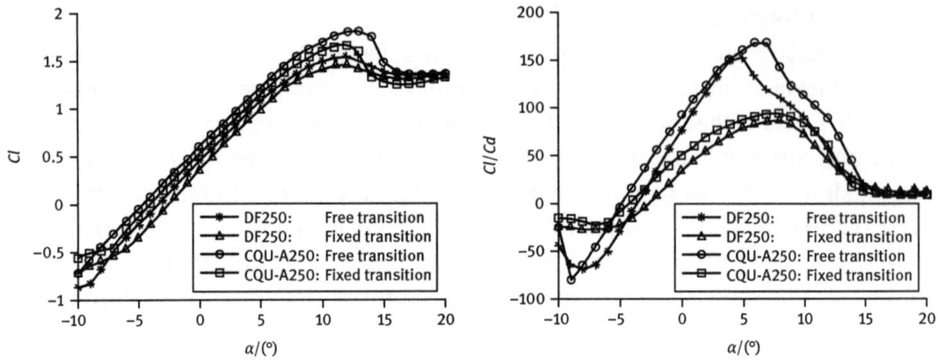

(a) Lift coefficient (b) Lift to drag ratio

Fig. 4.107: Aerodynamic characteristics of CQU-A250 and DF250.

Fig. 4.108 presents the variation of noise with respect to AOA under the free transition condition and fixed transition. The noise of the new airfoil is lower than that of the DF250 airfoil before the stall. The newly-designed airfoil has lower noise before stall, thus reducing the impact on the surrounding environment.

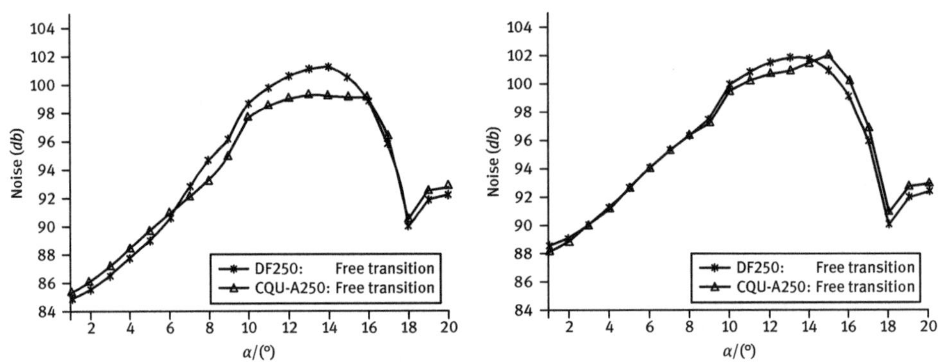

(a) Comparison of noise (free transition) (b) Comparison of noise (fixed transition)

Fig. 4.108: Noise characteristics of CQU-A250 and DF250.

Tab. 4.39 presents the comparison of aerodynamic characteristics of these two airfoils. Compared with the DF250 airfoil, the maximum lift coefficient (smooth condition) of CQU-A250 has increased by 17.939 % and the maximum lift-to-drag ratio (smooth condition) has increased by 10 % under the smooth condition. And the maximum lift coefficient (rough condition) of the CQU-A250 airfoil has been increased by 15.364 % and the maximum lift-to-drag ratio (smooth condition) has been increased by 8.474 %. Results show that the aerodynamic performance of the newly-designed airfoil has been greatly improved in the main AOA range under both smooth and rough conditions.

Tab. 4.39: Comparison of aerodynamic parameters of airfoils.

Airfoil name	Free transition		Fixed transition $X_{tr,s} = 0.01; X_{tr,p} = 0.1$		Roughness sensitivity	Noise (dB)
	$c_{l,max}$	L/D_{max}	$c_{l,max}$	L/D_{max}		
DF250	1.5369(12°)	153.2345(5°)	1.4632(12°)	86.2459(8°)	4.8 %	96.3
CQU-A250	1.8126(13°)	168.5582(7°)	1.6880(12°)	93.5541(8°)	6.9 %	96.3
	17.939 %	10.000 %	15.364 %	8.474 %		

Note: The data in parentheses represent the AOA with the maximum lift coefficient and maximum lift-to-drag ratio; the noise is calculated at the AOA with the maximum lift-to-drag ratio and under fixed transition condition.

4.9.7.2 CQU-A300 airfoil

For the wind turbine airfoil with maximum relative thickness of 30 %, which is generally arranged at 40 % spanwise location, high aerodynamic performance as well as structural strength are required. Fig. 4.109 presents the shapes of the newly-designed CQU-A300 airfoil and the well-known DU 97-W-300 airfoil. The comparison of geometric characteristics of the two airfoils is listed in Tab. 4.40. Compared with the DU 97-W-300 airfoil, the thickness of the suction surface of the newly-designed airfoil has been increased in order to get the maximum lift coefficient. Meanwhile, roughness sensitivity has increased as well.

The comparison of aerodynamic characteristics between the newly-designed airfoil and the DU 9-W-300 airfoil under the same conditions ($Re = 6.0 \times 10^6$, $Ma = 0.15$) is made and shown in Fig. 4.110. Tab. 4.41 presents the comparisons of aerodynamic characteristics of these two airfoils. Under smooth condition, the lift coefficient of the CQU-A300 airfoil is 1.813 at the AOA of 11°; the maximum lift-to-drag ratio is 146.103

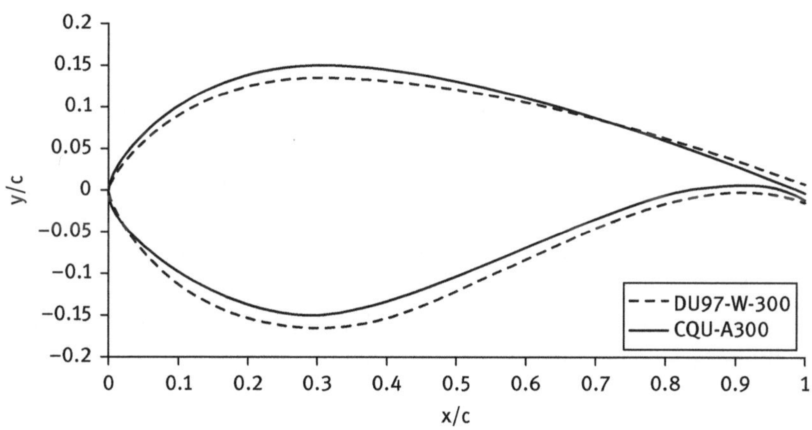

Fig. 4.109: Shapes of airfoils.

Tab. 4.40: Comparison of geometric characteristics of airfoils.

Airfoil name	Maximum relative thickness	Chordwise location of maximum relative thickness	Maximum camber	Chordwise location of maximum camber
DU 97-W-300	0.30000	0.292	0.02105	0.804
CQU-A300	0.29821	0.296	0.03132	0.784

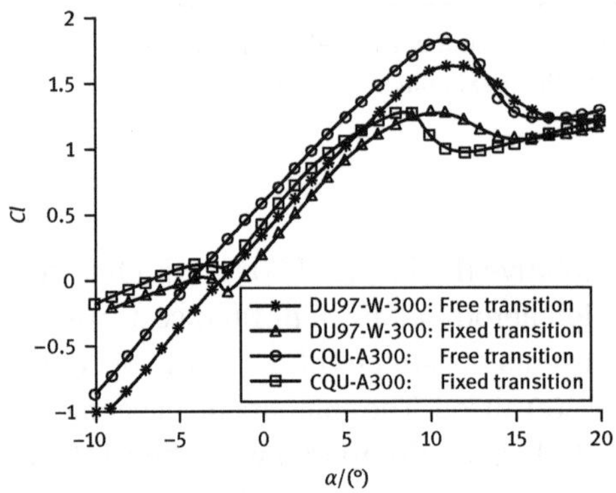

(a) Lift coefficient

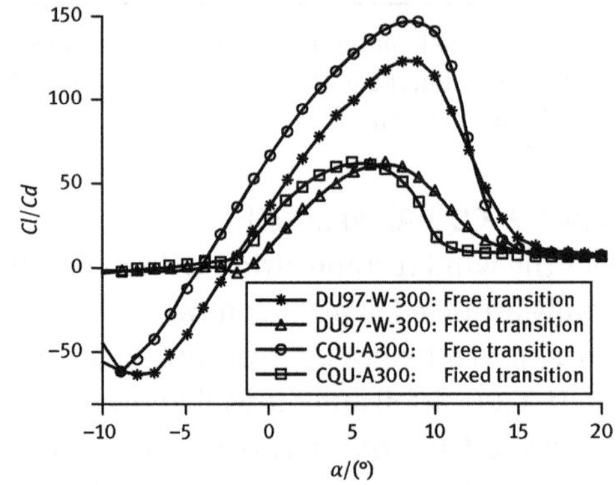

(b) Lift to drag ratio

Fig. 4.110: Aerodynamic characteristics of CQU-A300 and DU 97-W-300.

Tab. 4.41: Comparison of aerodynamic parameters of airfoils.

Airfoil name	Smooth condition		Rough condition	
	$c_{l,max}$	L/D_{max}	$c_{l,max}$	L/D_{max}
DU 97-W-300	1.6256(11°)	122.6380(9°)	1.2683(11°)	61.8889(7°)
CQU-A300	1.8046(11°)	146.1030(9°)	1.2579(9°)	61.3650(5°)

at the AOA of 9°. Compared with the DU 97-W-300 airfoil, the new airfoil has better aerodynamic performance under the smooth condition. Under the rough condition, the lift coefficient of the new airfoil is 1.2579 at AOA of 9° and the maximum lift-to-drag ratio is 61.365 at the AOA of 5°. Although the aerodynamic performance of the new airfoil is as good as that of the DU 97-W-300 airfoil under the rough condition, the transition will start at a smaller AOA for CQU-A300. The reason is that the increasing thickness of the suction surface will lead to earlier separation. What is more, CQU-A300 has better aerodynamic characteristics and structural compatibility.

4.9.7.3 CQU-A350 airfoil

For airfoils with larger relative thickness (the maximum relative thickness reaches 35 % or larger), which is generally arranged in the vicinity of the blade root, enough strength is required. Except the trailing edge, the upper and lower surfaces of airfoils are similar to some degree. With increasing thickness, the position of the maximum relative thickness moves toward the airfoil trailing edge along the chord. Fig. 4.111 presents the shape of the newly-designed CQU-A350 airfoil and the well-known DU 00-W2-350 airfoil. Tab. 4.42 presents the comparison of geometric characteristics of the airfoils. The biggest difference in geometric shape between the newly-designed airfoil and the DU 00-W2-350 airfoil lies in the upper and lower surface of the leading edge, with the leading edge thickness of the upper surface decreasing slightly and that of the lower surface increasing slightly.

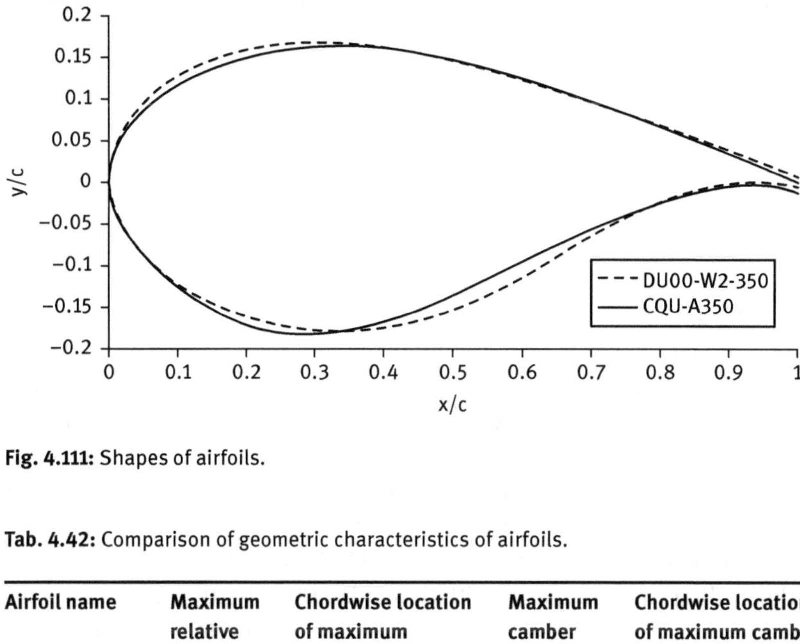

Fig. 4.111: Shapes of airfoils.

Tab. 4.42: Comparison of geometric characteristics of airfoils.

Airfoil name	Maximum relative thickness	Chordwise location of maximum relative thickness	Maximum camber	Chordwise location of maximum camber
DU 00-W2-350	0.34568	0.321	0.01944	0.831
CQU-A350	0.34889	0.319	0.02732	0.801

Research shows that the results of RFOIL software predictions and wind tunnel tests will be quite different for airfoils with larger relative thickness, but the overall trend of variation of lift coefficient is consistent [15]. Therefore, RFOIL was still used to get the aerodynamic performance of airfoils with larger relative thickness.

The comparison of aerodynamic performance between CQU-A350 and DU 97-W2-350 under the same conditions ($Re = 6.0 \times 10^6$, $Ma = 0.15$) is shown in Fig. 4.112. The lift coefficient and lift-to-drag ratio of CQU-A350 are higher than those of DU 00-W2-350 in the main AOA range (under smooth condition). According to the aerodynamic and structural characteristics of other airfoils with larger thickness, CQU-A350 has better aerodynamic characteristics and structural compatibility.

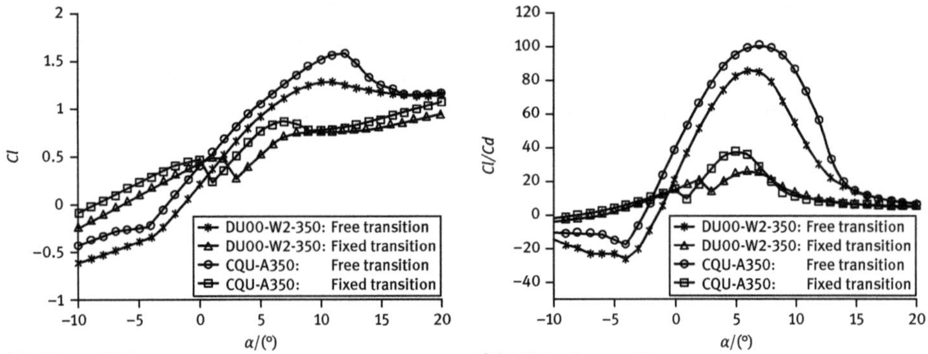

(a) Lift coefficient (b) Lift to drag ratio

Fig. 4.112: Aerodynamic characteristics of CQU-A350 and DU 00-W2-350.

4.9.7.4 CQU-A400 airfoil

For airfoils with larger relative thickness (the maximum relative thickness reaches 40 % or larger), which is generally arranged near the blade root, enough strength is required. With the increase in the relative thickness, the position of the maximum relative thickness moves toward the airfoil trailing edge. The shape of the CQU-A350 airfoil and the well-known DU 00-W2-350 airfoil are shown in Fig. 4.113. Tab. 4.43 presents the comparison of geometric characteristics of the two airfoils. As has been shown in Fig. 4.113, the chordwise position of the maximum relative thickness of the CQU-A400 is closer to the trailing edge compared with that of DU 00-W2-401, which makes the gravity center move toward the trailing edge and ensures the uniform distribution of mass in the blade root zone.

Tab. 4.43: Comparison of geometric characteristics of airfoils.

Airfoil name	Maximum relative thickness	Chordwise location of maximum relative thickness	Maximum camber	Chordwise location of maximum camber
DU 00-W2-401	0.39555	0.315	0.01859	0.827
CQU-A400	0.40401	0.327	0.02620	0.797

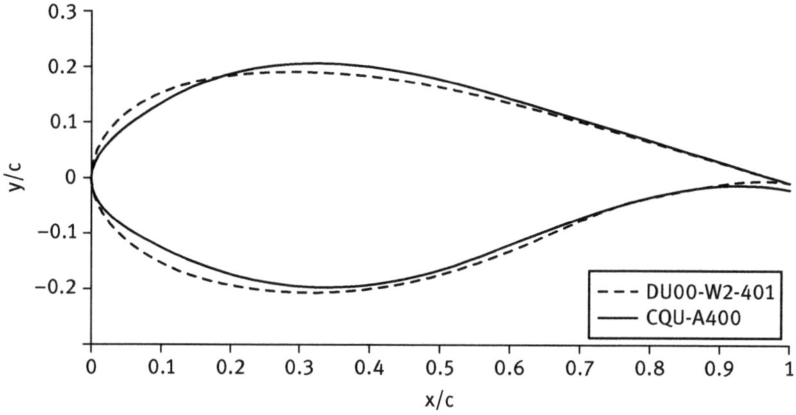

Fig. 4.113: Shapes of airfoils.

The comparison of aerodynamic performance of CQU-A400 and DU 00-W2-401 under the same conditions ($Re = 6.0 \times 10^6$, $Ma = 0.15$) is depicted in Fig. 4.114. The lift coefficient and lift-to-drag ratio of the CQU-A400 airfoil are higher than those of the DU 00-W2-401 airfoil under smooth condition. It can be seen that the CQU-A400 airfoil has better structural compatibility and larger relative thickness, which meets the structural requirements.

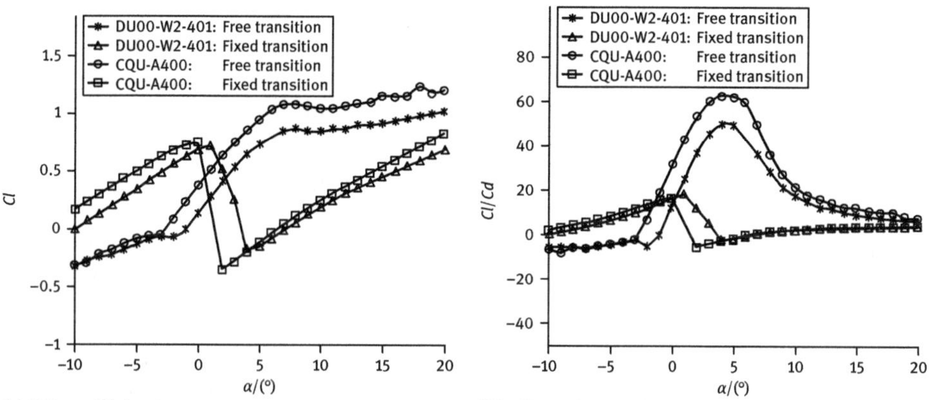

(a) Lift coefficient (b) Lift to drag ratio

Fig. 4.114: Aerodynamic characteristics of CQU-A400 and DU 00-W2-401.

4.10 Chapter conclusions

Based on the integrated expression of airfoils (combining Joukowsky transformation theory and series theory), optimization models for wind turbine airfoils were established. With the thorough consideration of multidisciplinary constraints (aerodynamic, structural, acoustic, etc.), several optimization models for wind turbine airfoils were proposed: the optimization model with the single objective of maximum lift-to-drag ratio; the multiple-objective optimization model with the objectives of maximum lift-to-drag ratio under rough and smooth conditions; and the optimization model with the objective of high RLD to SPL ratio. The aerodynamic characteristics were obtained through combining the integrated expression and XFOIL or RFOIL. CQU-DTU-A series airfoils, WT series airfoils, CQU-DTU-B series airfoils and CQU-A series airfoils were obtained after optimizations. The excellent aerodynamic performance of the optimized airfoils was verified through comparison with the aerodynamic performance of commonly used airfoils in the wind energy industry.

Secondly, after presenting many different airfoils with the integrated expression theory and thorough analysis, it was found that the geometric scale factor a is not absolutely equal to 0.25 but close to it in the integrated expression. Based on curvature smooth continuity theory, a new method was presented to correct the geometric scale factor a in the integrated expression for various kinds of airfoils. As a result, the curvature smooth continuity of the fitting profile had been greatly improved, compared with that of the original profile. Based on this new method, the DU 93-W-210 airfoil was improved with the corrected geometric scale factor a and the genetic algorithm (GA) by controlling the coefficients of the shape function, leading to a new airfoil. Comparatively, the aerodynamic performance of the new airfoil, such as maximum lift coefficient, maximum lift-to-drag ratio, roughness insensitivity and so forth, are better than those of the original DU 93-W-210 airfoil. The achieved results show that this novel method is feasible to optimize wind turbine airfoils.

The integrated expression of airfoil profiles based on Joukowsky conformal transform theory could be used to optimize the profiles for the thin thickness airfoil. However, it is hard to adjust the coefficient to the integral function for the medium thickness airfoil. B-spline curve has the advantage of local adjustment, which makes it easier to effectively control the airfoil profiles at the trailing edge. Therefore, a new direct design method for the medium thickness wind turbine airfoil based on airfoil integral expression and B-spline curve was proposed. Adopting the improved particle swarm algorithm, the optimization model was built to get airfoils with medium and large thickness. Four new airfoils (CQU-A250, CQU-A300, CQU-A350 and CQU-A400) were obtained. Compared with the commonly used wind turbine airfoils, the CQU-A airfoil series exhibits good aerodynamic and structural performance and satisfies the requirements for wind turbine airfoils, thus validating the feasibility and superiority of this method.

The design and optimization of airfoil series for wind turbine blades has been described in this chapter. Different airfoil series were designed considering different methods and theories. The aerodynamic performance and structural characteristics of the new airfoils were verified, which broadened the methods for design of and research on wind turbine airfoils. Furthermore, it also provides good foundations for the design of wind turbine blades.

5 Experiments on the wind turbine airfoil and data analysis

5.1 Introduction

The airfoil, as a basic element of blade, has a direct impact on the aerodynamic characteristics of the blade. To verify the general design theory and methodology for the wind turbine blade airfoil that has been proposed in this book, the high performance wind turbine airfoil WT180 (relative thickness is 18 %) introduced in Chapter 4 was chosen for wind tunnel experiments. The design theory has been verified through comparing the results with those from theoretical computations.

The experiment was carried out in the NF-3 wind tunnel at the airfoil research center of Northwestern Polytechnical University. The experimental Reynolds numbers are 2.0×10^6, 3.0×10^6, 4.0×10^6. The angle of attack (AOA) ranges from –10° to 25°. The tests were conducted under free and fixed transition conditions, with various Reynolds numbers and angles of attack, in order to study the airfoil's aerodynamic characteristics.

5.2 Design and manufacture of the airfoil model

Limited by the size of the experimental section in the wind tunnel, the model was designed to have a chord of 800 mm and span of 1595 mm [159]. The steel inner structure, which is drawn in Fig. 5.1, is equipped inside the model so that the model can bear strong winds. The photo of the real steel inner structure is shown in Fig. 5.2.

The exterior surface was made of red pine from northeastern China since it has low air-dried density and shrinkage rate [160]. As shown in Fig. 5.1, the hollow cylinder, through which the piezometric tubes pass, is located at the axis. There are holes in the steel holder to facilitate assembly of the model. At the initial stage of model manufacture, the battens are stuck onto the periphery of the steel holder to form the outer workblank of the airfoil model (as shown in Fig. 5.3). Then the outer batten model was planed gradually to approach the target airfoil profile. Six sample plates along the span were selected for examination and polished to a high quality surface (normally the process is repeated at least seven times). The sample plates, shown in Fig. 5.4, are manufactured by the line cutting method which strictly follows the airfoil coordinates. After the process, the curvature should be examined. To meet the smooth surface requirement for the airfoil, the model was spray painted. The final experimental airfoil model is shown in Fig. 5.5.

DOI 10.1515/9783110344387-006

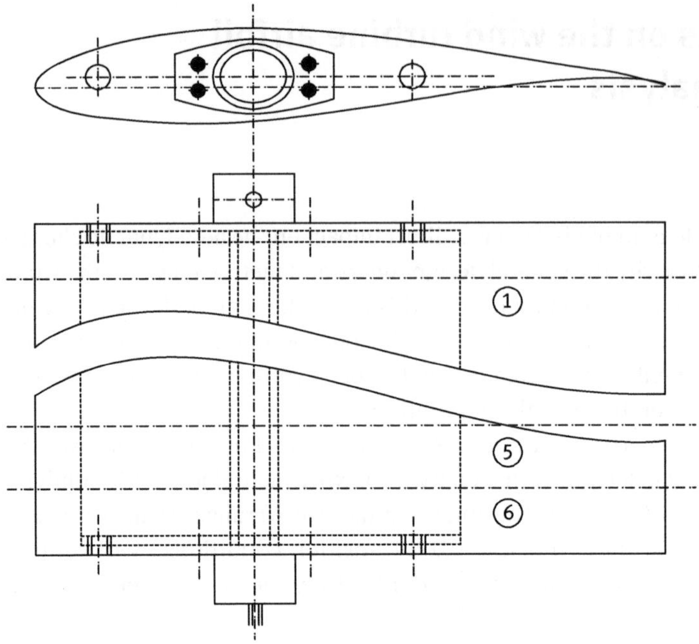

Fig. 5.1: Schematic diagram of the airfoil model.

Fig. 5.2: Photo of steel inner structure of the airfoil model.

Fig. 5.3: Workblank of the airfoil model.

Fig. 5.4: Sample plate of the airfoil and finishing.

To measure the static and total pressure, pitot-tubes were installed at different locations in the test section, shown in Fig. 5.6. The airfoil is equipped with 92 pressure taps (46 on the suction side, 46 on the pressure side, 1 on the leading edge and 1 on the trailing edge). The pressure taps are carefully distributed in a manner to facilitate adequate resolution of the expected pressure gradients. It should be noted that on the upper and lower surfaces, the tap spacing is dense in 5 % of the leading edge region. And the taps at the other locations are distributed uniformly. The diameter of the tap holes is 0.8 mm and the axis of the hole is perpendicular to the airfoil surface. The inner surface of the hole should be smooth, with no burr and chamfer. The distribution of pressure hole is shown is Fig. 5.7.

Fig. 5.5: Final airfoil model.

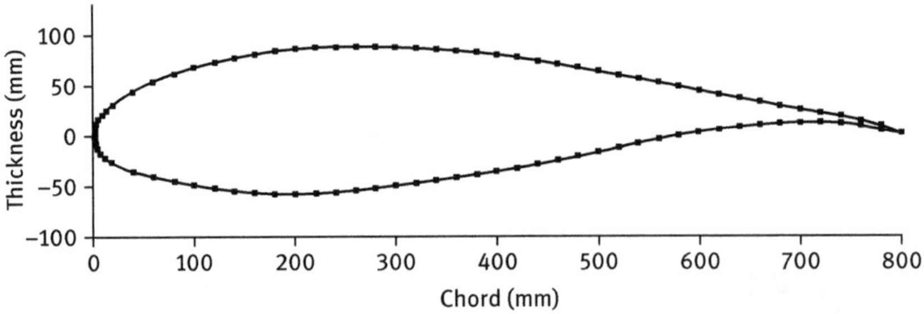

Fig. 5.6: Distribution of pressure taps.

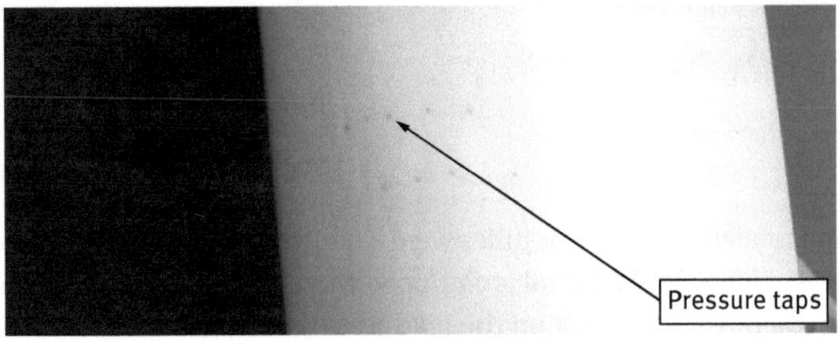

Fig. 5.7: Distribution of pressure taps in the airfoil model.

After drilling the pressure taps, an appropriate rectangular hole was created in the nonmiddle section of the span. Copper wires with diameter 1.2 mm were inserted into the pressure holes from inside to outside. After filling the rectangular space, the section of the wire protruding above the surface was cut back to be flush with the model's surface. The copper wires were led from the inside of the model along a spindle to outside the wind tunnel and then connected to the sensor of a PSI 9816 system by the plastic tubes with external diameter is 2 mm. The tubes were checked for tightness to prevent blocking or leakage of flow.

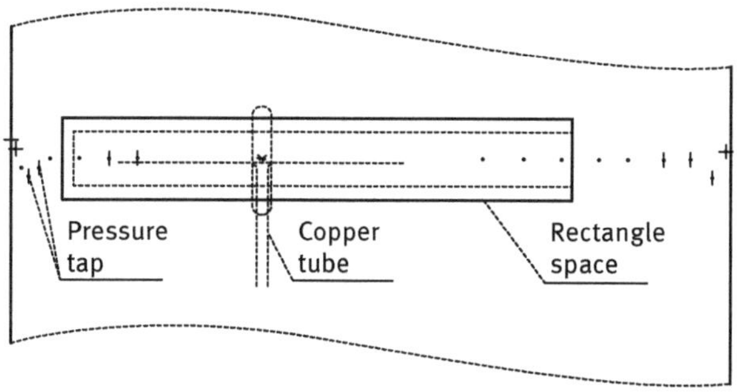

Fig. 5.8: Position of pressure taps on the airfoil section.

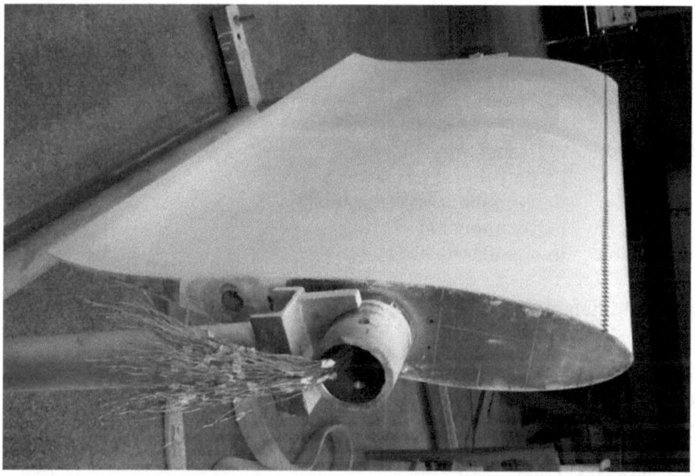

Fig. 5.9: The copper tubes.

5.3 Apparatus, method and data processing of the experiment

5.3.1 Wind tunnel

The experiment was carried out in the two-dimensional (2D) testing section of the large-scale low-velocity wind tunnel named NF-3 at Northwestern Polytechnical University. The NF-3 wind tunnel is comprises anechoic chamber, intake section, static flow section, contraction section, experimental section, the first diffusion section, transition section, the second diffusion section, dynamic section, the third diffusion section, deflector and muffler tower. The whole length of the tunnel is 80 m. The tunnel is made mainly of steel and is equipped with a direct current motor with power of 1120 kW. The layout of the wind tunnel [158] is shown in Fig. 5.10.

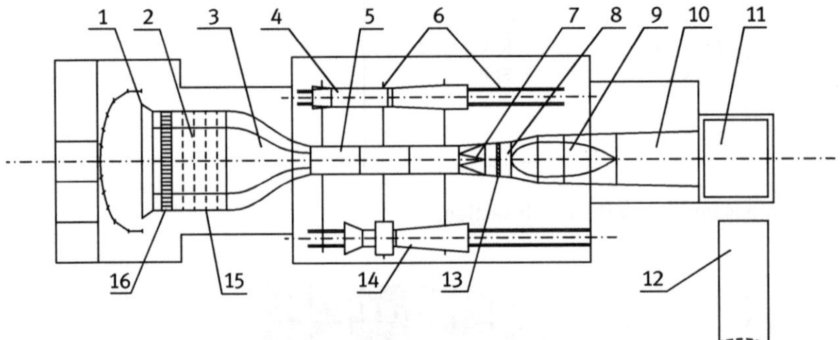

1. Intake section
2. Static flow section
3. Contraction section
4. Two-dimensional section
5. Three-dimensional section
6. Rail
7. Transition section
8. The second diffusion section
9. Dynamic section
10. The third diffusion section
11. Muffler tower
12. Transformer substation
13. Partition curtain
14. Propeller testing section
15. Damping net
16. Cellular devices

Fig. 5.10: The overall layout of wind tunnel NF-3.

The main parameters of the 2D section of the wind tunnel are listed as below.
(1) The size of the 2D section is 1.6 m (height) × 3 m (length) × 8 m (depth).
(2) The maximum wind velocity $V_{\max} = 130$ m/s.
(3) The shrinkage ratio is 20.
(4) The turbulence intensity is less than 4.5 ‰.
(5) The maximum Reynolds number 7×10^6.

5.3.2 Installation of the model

The top and bottom of the model were fixed to the roof and floor of the wind tunnel section to ensure two-dimensional flow over the model. The model was fixed to the rotating disc of the wind tunnel so that the AOA can be accurately controlled. The wake rake is equipped with as many as 187 total-pressure tubes and 9 static-pressure tubes. The wake rake is fixed 1.3 airfoil chords (1.04 m) away from the airfoil trailing edge, with its center approximately at the height of the trailing edge with zero incident angles and behind the center line of the airfoil section. The installation of the airfoil model in the wind tunnel is shown in Fig. 5.11 and 5.12.

5.3.3 Test apparatus

The data obtained from the measurements are inflow velocity, temperature, static pressure, the pressure distribution at the pressure taps on the blade surface and AOA, and the pressure distribution at the wake rake. Inflow velocity is measured with an anemograph, temperature is measured with a thermometer, the air pressure is measured with a barometer and the AOA can be obtained from the scale on the rotating disc. The AOA for the airfoil model is controlled by the control system of the NF-3 wind tunnel. The interface of the AOA control system is shown in Fig. 5.13.

The main components of the AOA measuring system [158], which follows a closed-cycle control strategy, are FANUC motor, servo power, angle transducer, STD/8088 controller and computers. The transmission system for the two-dimensional section is depicted in Fig. 5.14.

Fig. 5.11: The installation of the model in the wind tunnel (free transition).

Fig. 5.12: The installation of the model in the wind tunnel (fixed transition).

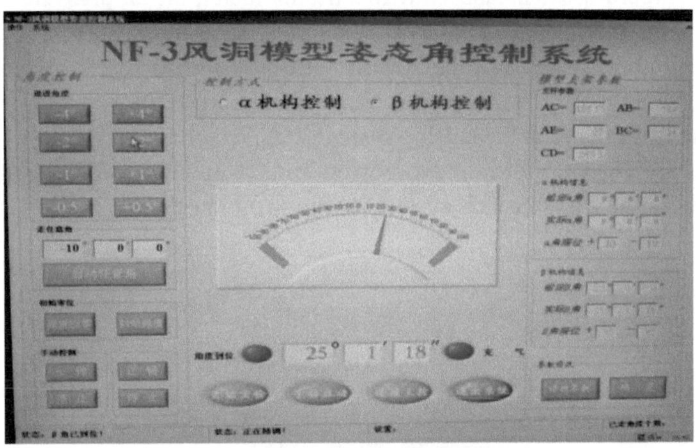

Fig. 5.13: The interface of the AOA control system.

A synchronous sensor is installed on the axis of the rotating disc to ensure that the rotation control error remains within the limit ±30″. Both the top and bottom discs rotate simultaneously. There is an annular sealing tape which can be controlled by a magnetic valve to release air so as to reduce the resistance of rotation when AOA is changing. When the required AOA is reached, the high pressure gas is inflated to prevent air leakage.

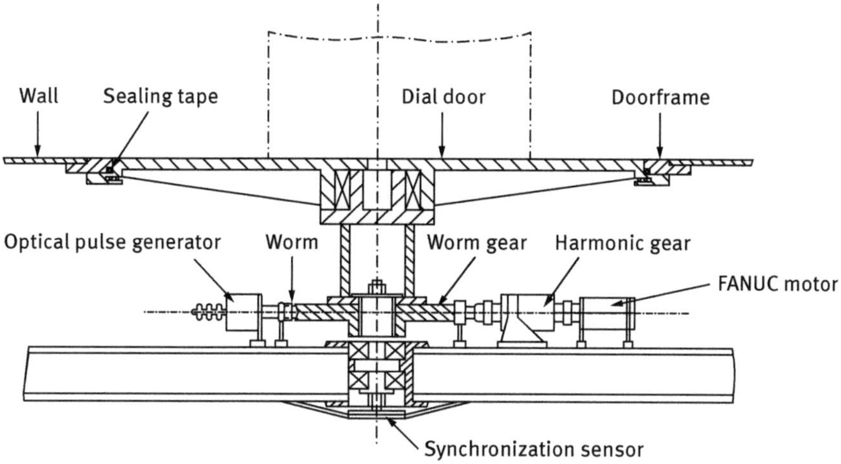

Fig. 5.14: The transmission system for the 2D section.

The wake rake, shown in Fig. 5.15, is equipped with as many as 187 total-pressure tubes and 9 static pressure tubes. The wake rake is fixed 1.3 airfoil chords (1.04 m) away from the airfoil trailing edge, with its center approximately at the height of the trailing edge with zero incident angles and behind the center line of the airfoil section.

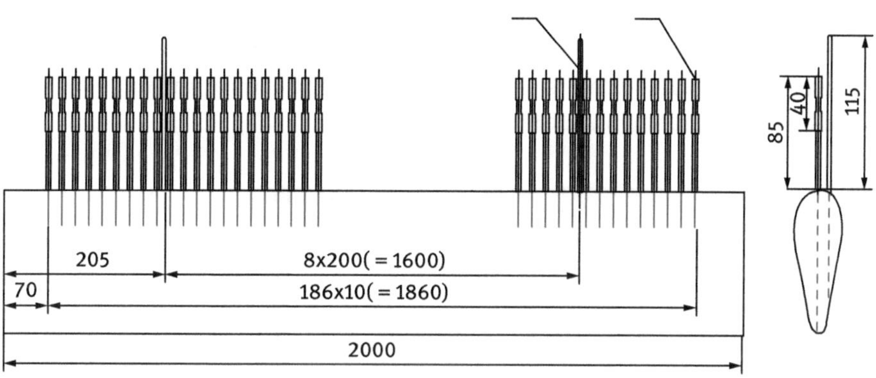

Fig. 5.15: Schematic diagram of the distribution of the wake rake.

The American PSI 9816 pressure data collection system, shown in Fig. 5.16, was used to collect the pressure at the airfoil surface and wake rake. This system, with measurement range of 105 Kpa, is composed of 32 sensor units and each unit is equipped with 16 channels which permit simultaneous measurements for 512 points. The measurements of 94 pressure taps were conducted simultaneously with the collection speed of 100 times/s for every channel and measure accuracy better than 0.05 % FS.

Fig. 5.16: The PSI 9816 pressure data collection system.

The interface of the pressure measuring system is shown in Fig. 5.17. The pressure distribution of the airfoil can be gathered and displayed on a computer. The flow chart of the pressure measuring system is shown in Fig. 5.18. The control and monitoring of airfoil wind tunnel experiments are shown in Fig. 5.19.

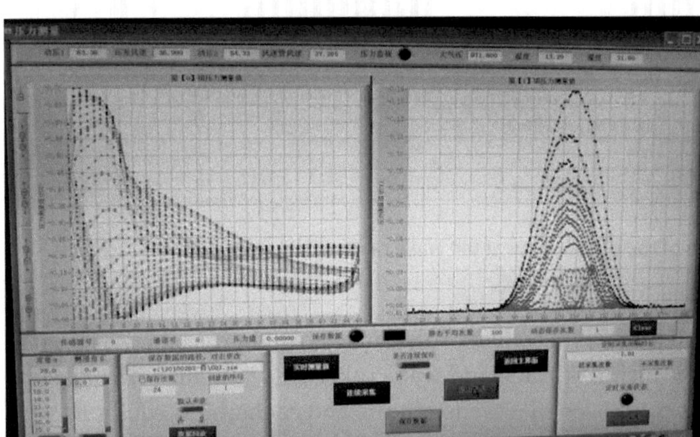

Fig. 5.17: The interface of the pressure measuring system.

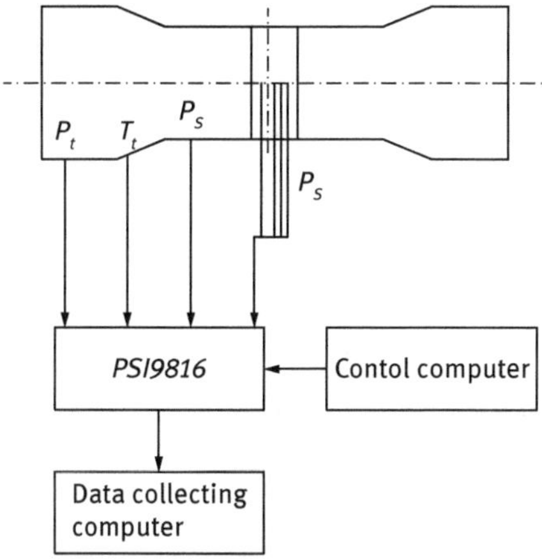

Fig. 5.18: The flow chart of the pressure measuring system.

Fig. 5.19: Controlling and monitoring experiments.

5.3.4 The experiments and data processing

Three Reynolds numbers of 2.0×10^6, 3.0×10^6 and 4.0×10^6 were chosen during the tests. The tests were conducted under free and fixed transition conditions and various Reynolds numbers to study the aerodynamic characteristics of the airfoil. The detailed parameters of the WT180 airfoil experiments are listed in Tab. 5.1.

Tab. 5.1: Parameters of WT180 airfoil experiments.

No.	Transition	Condition	Data	Remarks
1	Free	Changing α, $Re = 2.0 \times 10^6$	Pressure	$T = 12.5$; $V = 37.38$
2	Free	Changing α, $Re = 3.0 \times 10^6$	Pressure	$T = 12.5$; $V = 55.90$
3	Free	Changing α, $Re = 4.0 \times 10^6$	Pressure	$T = 12.5$; $V = 74.83$
4	Fixed	Changing α, $Re = 2.0 \times 10^6$	Pressure	$T = 13.2$; $V = 37.51$
5	Fixed	Changing α, $Re = 3.0 \times 10^6$	Pressure	$T = 13.1$; $V = 56.21$
6	Fixed	Changing α, $Re = 4.0 \times 10^6$	Pressure	$T = 13.6$; $V = 74.89$

Note: T is the temperature inside the wind tunnel (°C), V is the inflow velocity (m/s).

The pressure distribution of an airfoil can be obtained by interpolating the measured data. Then the lift coefficient and drag coefficient can be calculated. The pressure coefficient is defined as

$$C_{\mathrm{p}} = \frac{p - p_\infty}{\frac{1}{2}\rho U_\infty^2}, \tag{5.1}$$

where U_∞ is the inflow velocity, P_∞ is the inflow static pressure, ρ is the air density.

The lift coefficient and drag coefficient after stall can be calculated through the formulas (5.2) and (5.3):

$$C_{\mathrm{L}} = \cos\alpha \int_0^1 (c_{\mathrm{p,l}} - c_{\mathrm{p,u}})\,\mathrm{d}\left(\frac{x}{c}\right) - \sin\alpha \int_{-\frac{t}{2}}^{\frac{t}{2}} (c_{\mathrm{p,f}} - c_{\mathrm{p,r}})\,\mathrm{d}\left(\frac{y}{c}\right), \tag{5.2}$$

$$C_{\mathrm{D}} = \sin\alpha \int_0^1 (c_{\mathrm{p,l}} - c_{\mathrm{p,u}})\,\mathrm{d}\left(\frac{x}{c}\right) + \cos\alpha \int_{-\frac{t}{2}}^{\frac{t}{2}} (c_{\mathrm{p,f}} - c_{\mathrm{p,r}})\,\mathrm{d}\left(\frac{y}{c}\right), \tag{5.3}$$

where c is the chord of the model, t is the maximum thickness, α is the angle of attack, 'u' represents the upper surface, 'l' represents the lower surface, 'f' represents the leading edge, 'r' represents the trailing edge.

Since much error will arise when the drag is calculated with the airfoil surface pressure distribution before stall, the airfoil drag coefficient is usually calculated by measuring the wake pressure distribution measured by the wake rake:

$$C_{\mathrm{D}} = 2\int_W \sqrt{C_{\mathrm{p,o1,i}} + \Delta C_{\mathrm{p,1}}}\left(1 - \sqrt{C_{\mathrm{p,o2,i}} + \Delta C_{\mathrm{p2}}}\right)\mathrm{d}\bar{z}. \tag{5.4}$$

In equation (5.4), the parameters are defined as $C_{\mathrm{p,o2,i}} = (P_{\mathrm{o,i}} - p_\infty)/q_\infty$, $C_{\mathrm{p,o2,i}} = (P_{\mathrm{o,i}} - p_\infty)/q_\infty$, $\Delta C_{\mathrm{p1}} = 1 - \bar{C}_{\mathrm{p,o1}}$, $\Delta C_{\mathrm{p2}} = 1 - \bar{C}_{\mathrm{p,o2}}$, $q_\infty = \mu \cdot (P_1 - P_2)$. $C_{\mathrm{p,o1,i}}$ represents the total pressure coefficient with respect to the average static pressure of wake. $C_{\mathrm{p,o2,i}}$ represents the total pressure coefficient with respect to the static pressure of the reference point (at entrance). $\bar{C}_{\mathrm{p,o1}}$ is the average total pressure coefficient in the wake rake region, $\bar{C}_{\mathrm{p,o2}}$ is the average total pressure coefficient for far down stream.

Tab. 5.2: Accuracy of the experiment and the national standard.

Data source	C_y	C_x	m_z
Experiment	0.0023	0.0004	0.0003
National standard	0.0040	0.0005	0.0012

5.4 Results of the experiments

Three Reynolds numbers, 2.0×10^6, 3.0×10^6 and 4.0×10^6 [163, 164] were chosen for the tests, which were conducted under both free and fixed transition conditions with varying Reynolds numbers.

(1) The range of AOA

Smooth surface: $\alpha = -10 \sim 10°$, $\Delta\alpha = 2°$; $\alpha = 11 \sim 21°$, $\Delta\alpha = 1°$, $\alpha = 23 \sim 25°$, $\Delta\alpha = 2°$.

(2) Reynolds number and wind velocity

Reynolds number: 2.0×10^6, 3.0×10^6 and 4.0×10^6.

Wind velocity: calculated from the local conditions.

(3) Experimental conditions

Fixed transition: Zigzag tape with thickness 0.35 mm (80# emery) is put at 5 % chordwise position on both the upper and lower surfaces.

5.4.1 Free transition conditions

The data from the pressure taps and wake rake were processed and converted to a pressure distribution under each AOA form which the aerodynamic characteristics can be obtained. Due to space limitations, only the main aerodynamic characteristics are shown in the following, such as lift coefficient (C_L), drag coefficient (C_D), moment coefficient (C_M), lift-to-drag ratio (C_L/C_D) and pressure distributions at selected AOAs.

As shown in Fig. 5.20, the lift coefficients of three Reynolds numbers before stall are nearly the same when AOA is smaller than 5°. At regions just before stall, lift coefficients increase with increasing Reynolds number. The maximum lift coefficient are 1.58 ($Re = 2.0 \times 10^6$), 1.69 ($Re = 3.0 \times 10^6$), 1.76 ($Re = 4.0 \times 10^6$). The lift coefficients of the three Reynolds numbers are close to each other after stall starts ($\alpha = 13°$). However, the lift coefficient at Reynolds number of 4.0×10^6 is larger than the other two cases when AOA surpasses 20°. As shown in Fig. 5.21, the drag coefficients of the three Reynolds numbers are almost the same over most of the AOA range. However, the drag coefficient at Reynolds number 4.0×10^6 is obviously larger than the other two cases when AOA is larger than 17°. The minimum drag coefficient decreases with increasing Reynolds number, decreasing from drag coefficient of 0.0064 at 2.0×10^6 to 0.0059 at 4.0×10^6. The conclusions can be drawn that the maximum lift coefficient

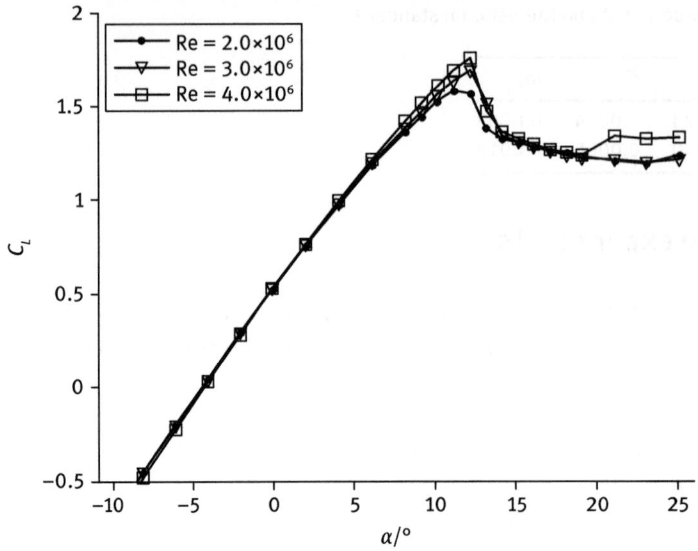

Fig. 5.20: Lift coefficients of airfoil WT180 (free transition).

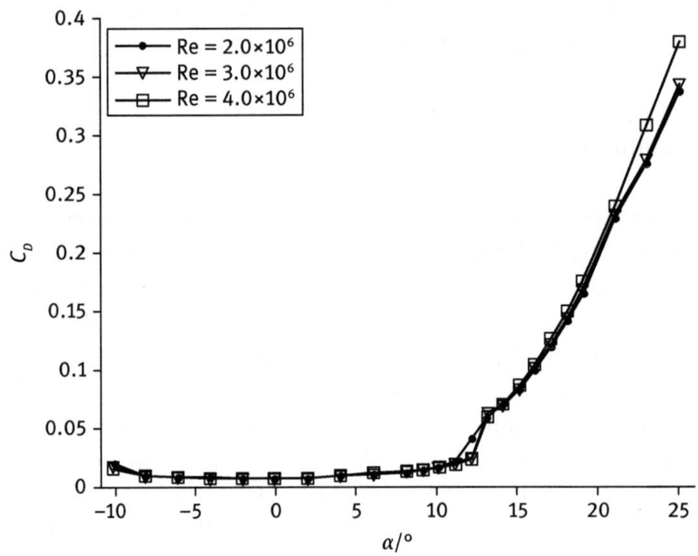

Fig. 5.21: Drag coefficients of airfoil WT180 (free transition).

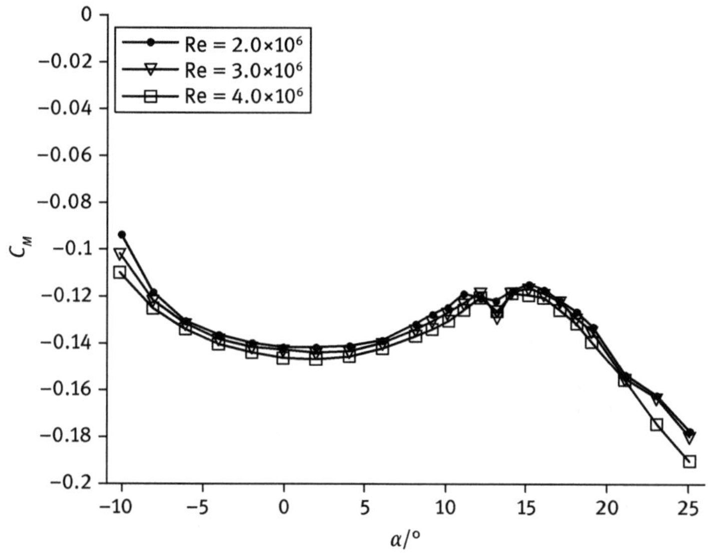

Fig. 5.22: Moment coefficients of airfoil WT180 (free transition).

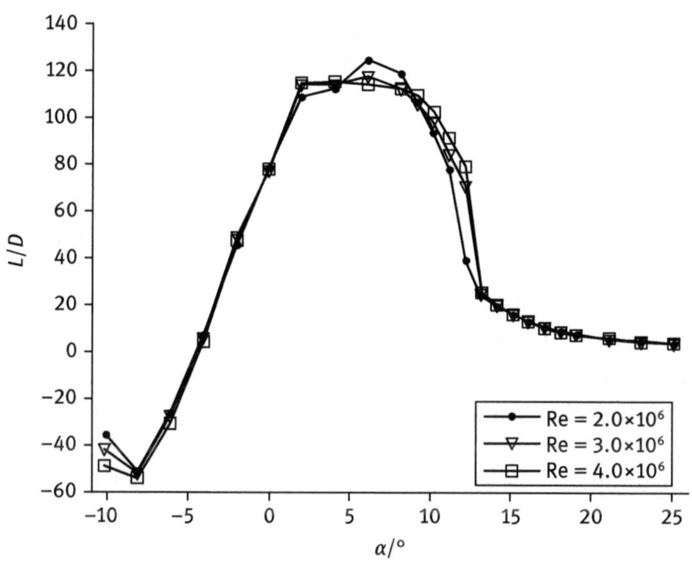

Fig. 5.23: Lift-to-drag ratios of airfoil WT180 (free transition).

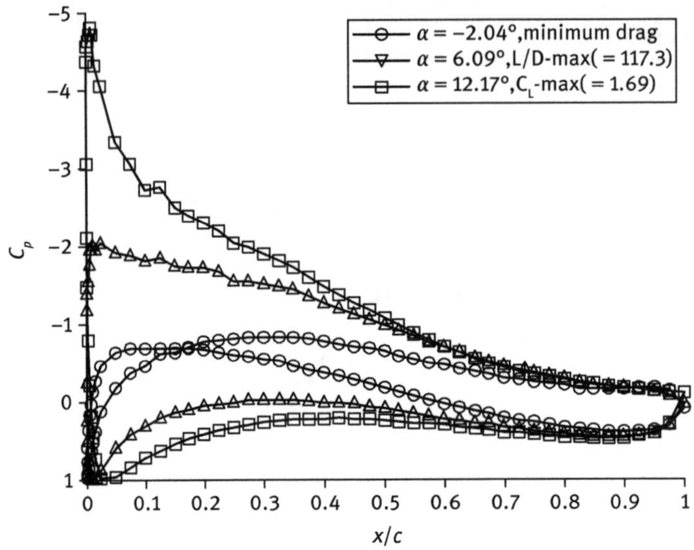

Fig. 5.24: Distribution of pressure coefficients at $Re = 3.0 \times 10^6$ (free transition).

of the WT180 airfoil increases with increasing Reynolds numbers and that the drag coefficient decreases over most of the AOA range at free transition, which corresponds to the basic effects of the Reynolds number.

The moment coefficients of airfoil WT180 under the free transition condition are shown with respect to AOA in Fig. 5.22. The lift-to-drag ratio at three Reynolds numbers (free transitions) are shown in Fig. 5.23. It can be seen that stall starts at AOA of about 12–13° and the maximum lift-to-drag ratios are 124.73 ($Re = 2.0 \times 10^6$), 117.30 ($Re = 3.0 \times 10^6$) and 114.96 ($Re = 4.0 \times 10^6$).

The distribution of pressure coefficients at $Re = 3.0 \times 10^6$ and free transition condition are depicted in Fig. 5.24. The pressure distributions illustrated are obtained at the AOA with the minimum drag ($C_D = 0.0059$), maximum lift-to-drag ratio ($L/D_{max} = 117.3$) and maximum lift coefficient ($C_{L,max} = 1.69$). The maximum lift coefficient is obtained at AOA of 12.17° at which the transition point on the suction surface is located near the leading edge ensuring that the airfoil has low roughness sensitivity.

5.4.2 Fixed transition conditions

To simulate the flow on a rough airfoil that may occur in real wind turbine operating conditions, the fixed transition condition was set utilizing zigzag tape with width 3 mm and granular height 0.35 mm. The zigzag tape is attached at 5 % chordwise position on both the upper and lower surfaces.

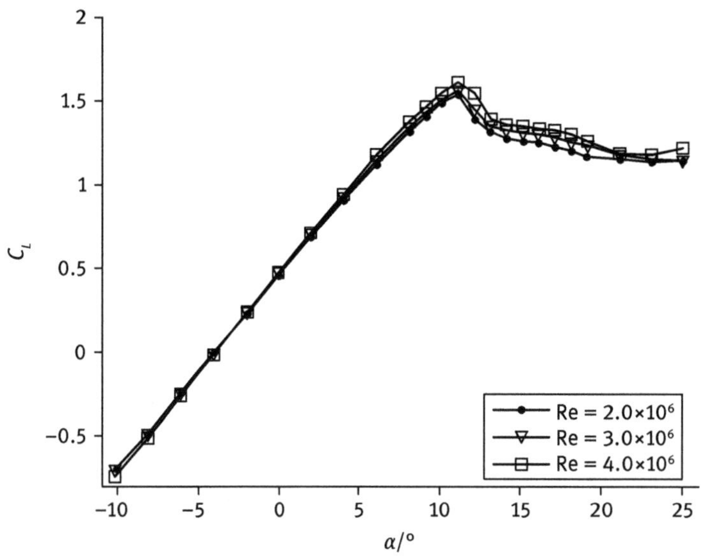

Fig. 5.25: Lift coefficients of airfoil WT180 (fixed transition).

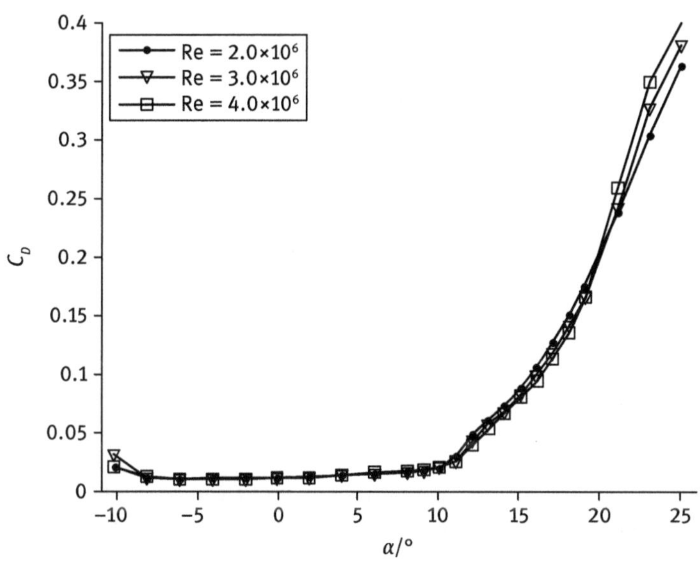

Fig. 5.26: Drag coefficients of airfoil WT180 (fixed transition).

As shown in Fig. 5.25 and 5.26, the maximum lift coefficient increases with increasing Reynolds number from 1.54 ($Re = 2.0 \times 10^6$) to 1.61 ($Re = 4.0 \times 10^6$). The drag coefficient decreases with increasing Reynolds numbers when AOA is less than 20°. Contrarily, when AOA is larger than 20°, the drag coefficients will increase with the Reynolds number.

The moment coefficients of airfoil WT180 under free transition condition are shown with respect to AOA in Fig. 5.27. The variations in lift-to-drag ratio with respect to AOA at three Reynolds numbers (fixed transitions) are shown in Fig. 5.28. It can be seen that stall initiates at AOA of about 12° and the maximum lift-to-drag ratios are 79.50 ($Re = 2.0 \times 10^6$), 83.06 ($Re = 3.0 \times 10^6$), 82.96 ($Re = 4.0 \times 10^6$). The differences in the lift-to-drag ratios between the three Reynolds numbers are small, which implies that the roughness sensitivity of the airfoil is low.

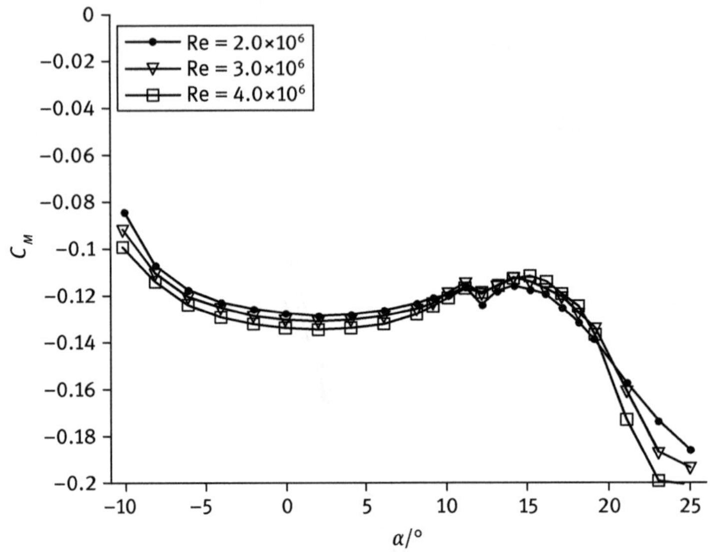

Fig. 5.27: Moment coefficients of airfoil WT180 (fixed transition).

The distribution of pressure coefficients at $Re = 3.0 \times 10^6$ and fixed transition condition are depicted in Fig. 5.29. Compared with putting the zigzag tape on the suction surface of the airfoil, sticking it on the pressure surface has a greater influence on aerodynamic characteristics. The pressure distributions shown are obtained at AOA with the minimum drag ($C_D = 0.0099$), maximum lift-to-drag ratio ($L/D_{max} = 83.06$) and maximum lift coefficient ($C_{L,max} = 1.56$). The maximum lift coefficient is obtained at AOA of 11.15°.

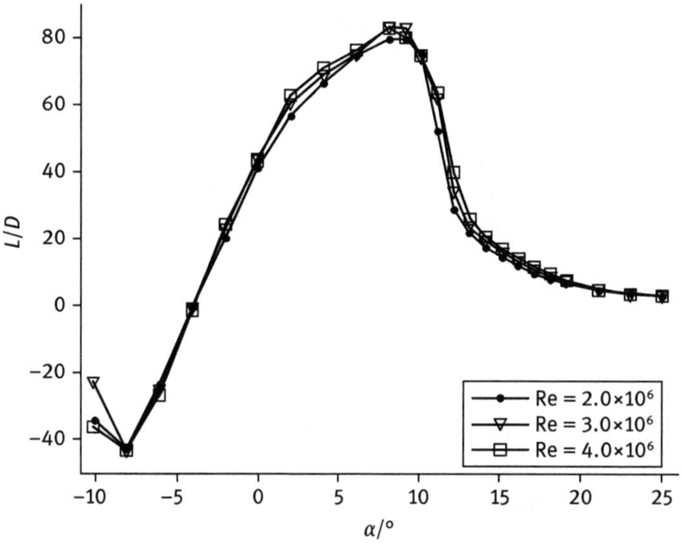

Fig. 5.28: Lift-to-drag ratios of airfoil WT180 (free transition).

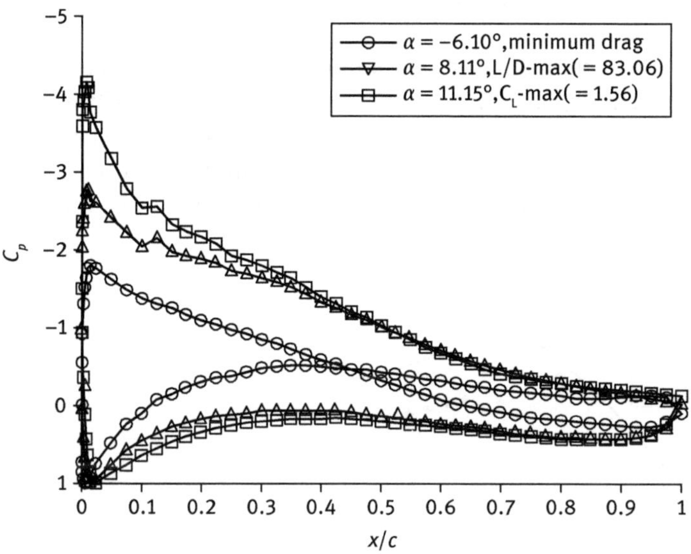

Fig. 5.29: Distribution of pressure coefficients at $Re = 3.0 \times 10^6$ (fixed transition).

5.4.3 Comparison of the results from experiments and RFOIL

The experimental results obtained under Reynolds number of 3.0×10^6 were compared with those from RFOIL calculations and the aerodynamic characteristics of both results were analyzed. The transition point is fixed at 1 % of chord on the suction surface and 5 % of chord on the pressure surface to simulate the fixed transition condition of the experiment. The Reynolds numbers for the RFOIL calculation are the same as those of the experiment, which are $Ma = 55.90/340 = 0.164$ for free transition and $Ma = 56.21/340 = 0.165$ for fixed transition. There is a difference between the design Reynolds numbers and experiment Reynolds numbers, which has little influence on the airfoil's aerodynamic characteristics.

As can be seen from Fig. 5.30 and 5.31, the measured lift and drag coefficients of the WT180 airfoil under smooth condition (free transition) were compared with results from RFOIL. The difference of the maximum C_L between the measured value (1.6902) and the predicted value (1.717) is within 2 % for smooth flow. The shape of C_L versus AOA is linear before reaching the maximum C_L. The minimum C_D was measured as 0.0059, which is in agreement with the predicted value of 0.00592. The values of the measured lift and drag coefficients are different for the calculation in the post-stall region. The reason is that it is hard to accurately predict the aerodynamic performance using the flow solver RFOIL in deep stall.

The deviation between experiment and computation is large near the zone with maximum C_L/C_D as shown in Fig. 5.32. This is because that C_L/C_D are sensitive to the slight change of C_D which is the denominator of C_L/C_D. In the wind tunnel experiment, the height of the wake rake is consistent with that of pressure taps whose arrangement will have an impact on the wake resistance. And the accuracy of the wake rake will influence the test data.

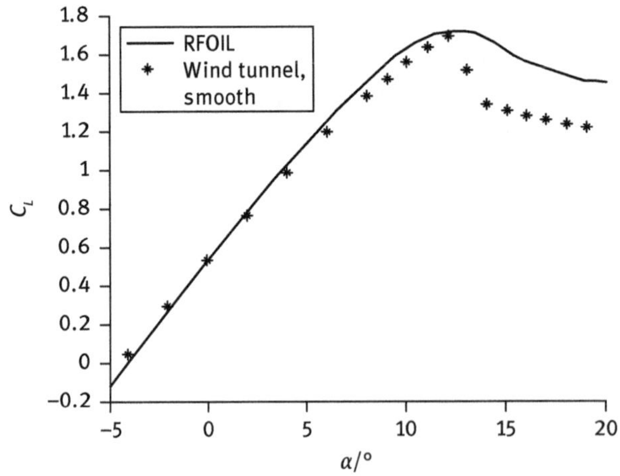

Fig. 5.30: Comparison of lift coefficients.

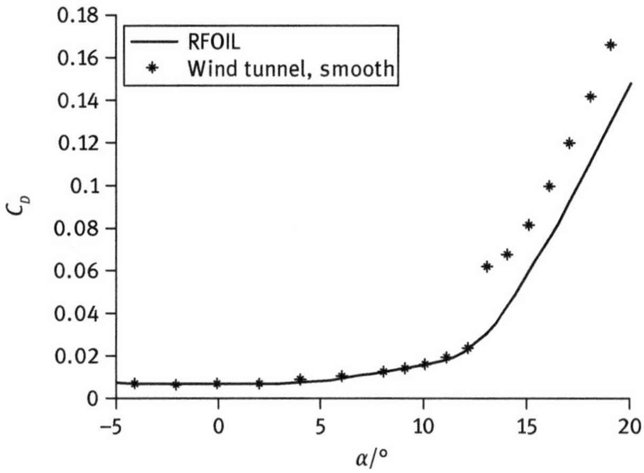

Fig. 5.31: Comparison of drag coefficients.

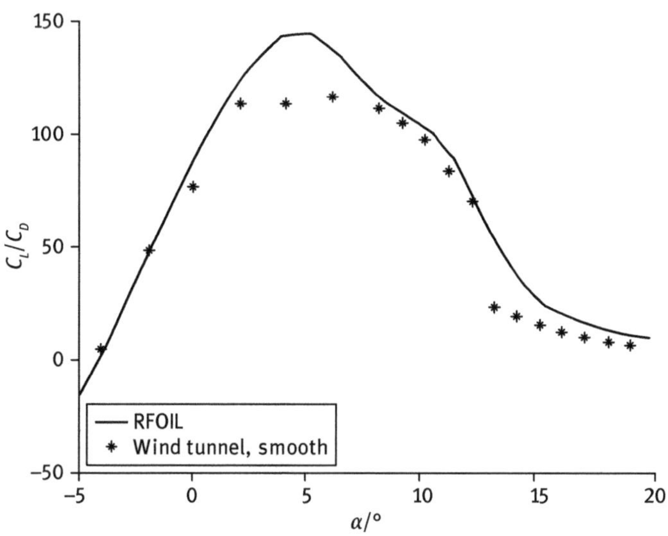

Fig. 5.32: Comparison of lift-to-drag ratios.

The measured and RFOIL predicted pressure distributions at the AOA of the minimum drag (AOA = 2.04°), maximum lift-to-drag ratio (AOA = 6.09°) and maximum lift coefficient (AOA = 12.17°) are shown in Fig. 5.33–5.35. The Reynolds number is $Re = 3.0 \times 10^6$. The measurement and the numerical predictions showed desirable agreement.

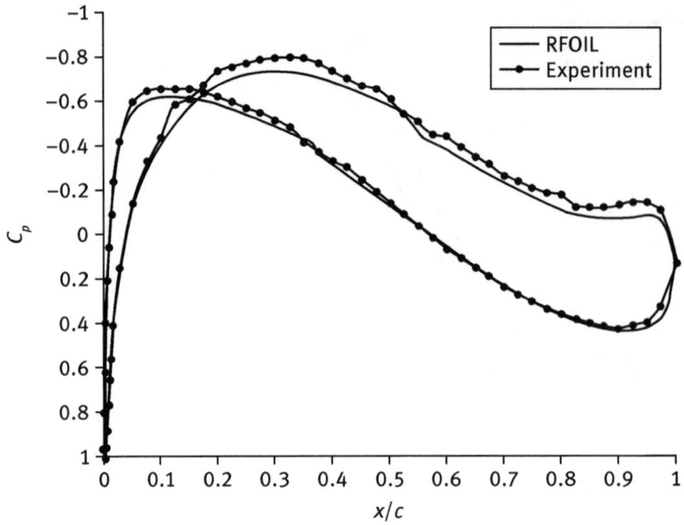

Fig. 5.33: Pressure distribution ($\alpha = 2.04°$).

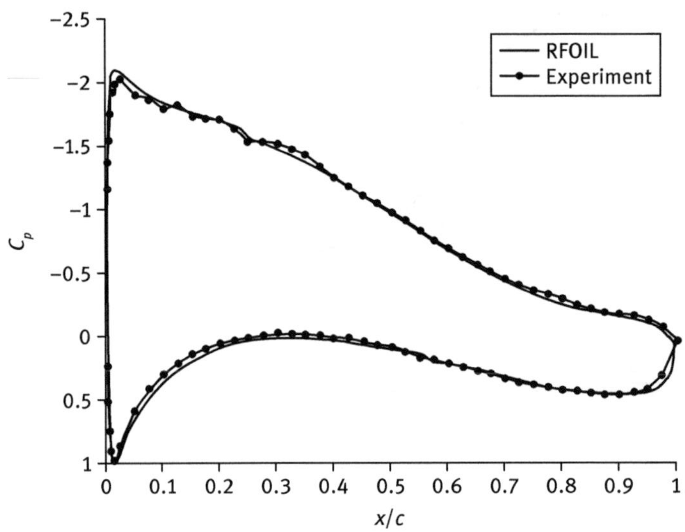

Fig. 5.34: Pressure distribution ($\alpha = 6.09°$).

The measured C_L and C_D of the WT180 airfoil under fixed transition were compared with RFOIL predictions, shown in Fig. 5.36 and 5.37. The difference of the maximum C_L between the measured value (1.56) and the predicted value (1.62) is 0.06. The shape of C_L and C_D versus AOA (before the maximum C_L) match well with the prediction. In RFOIL calculations, transition is initiated on the 1 % chord on the suction surface and

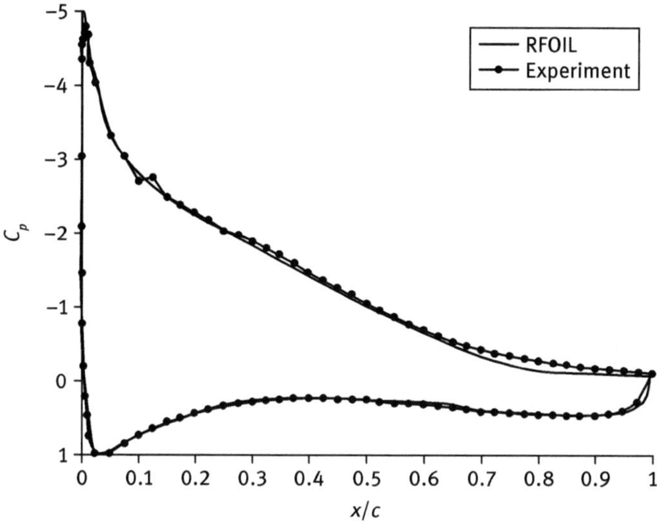

Fig. 5.35: Pressure distribution ($\alpha = 12.17°$).

5 % chord on the pressure surface. However, there must be some differences between the transition computations in RFOIL and the real transition on zigzag tape which leads to the C_D from RFOIL being slightly smaller than that from the experiment. And the slopes of the C_L curve, the maximum C_L and the $C_\mathrm{L}/C_\mathrm{D}$ are slightly larger than those from the experiment.

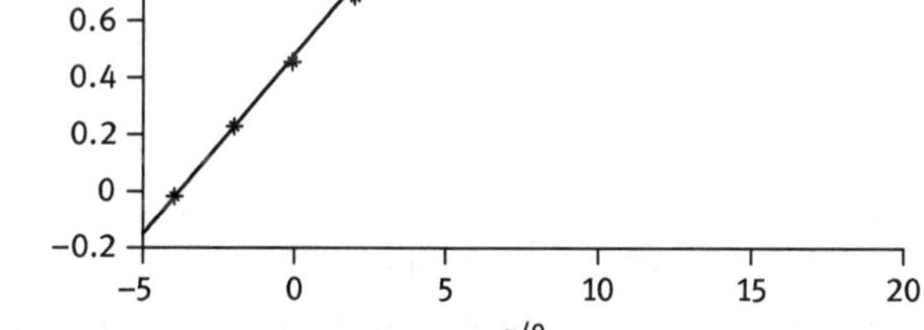

Fig. 5.36: Comparison of lift coefficients.

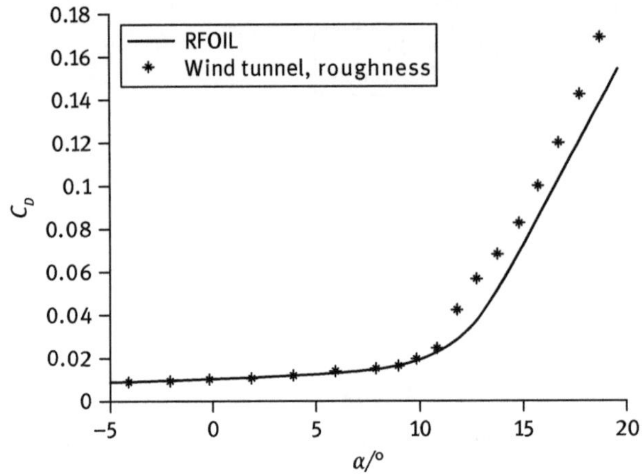

Fig. 5.37: Comparison of drag coefficients.

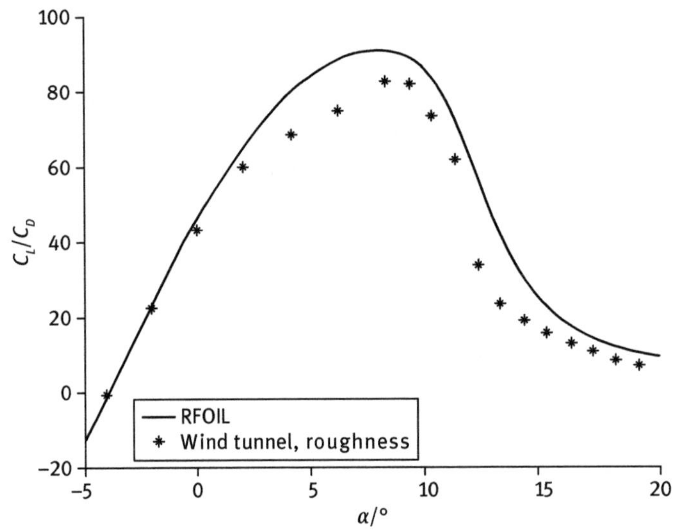

Fig. 5.38: Comparison of lift-to-drag ratios.

The measured and RFOIL predicted pressure distributions of airfoil WT180 ($Re = 3.0 \times 10^6$) at AOA of the minimum drag ($\alpha = -6.10°$), maximum lift-to-drag ratio ($\alpha = 8.11°$) and maximum lift coefficient ($\alpha = 11.15°$) are shown in Fig. 5.39–5.41. The measurement and the numerical prediction showed desirable agreement. This gives more validation to the performance of RFOIL.

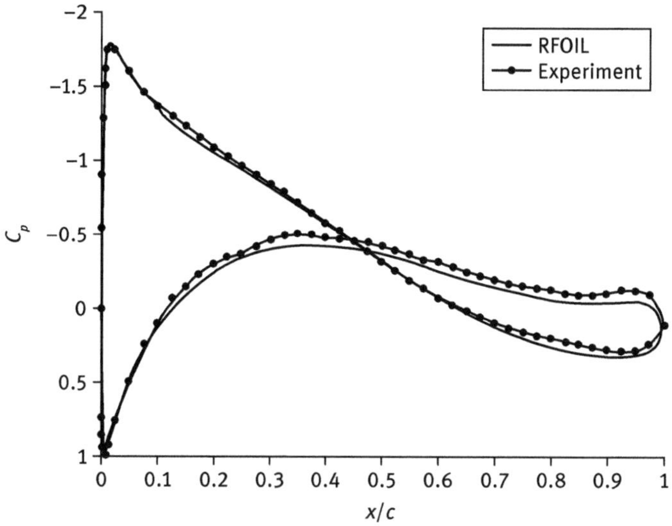

Fig. 5.39: Pressure distribution ($\alpha = -6.10°$).

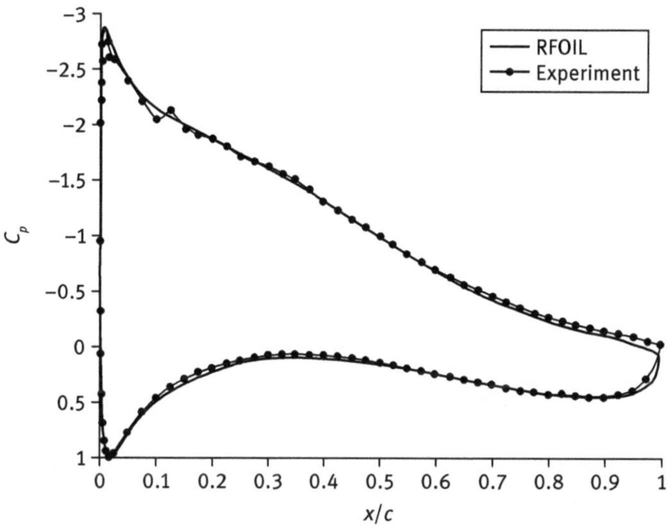

Fig. 5.40: Pressure distribution ($\alpha = 8.11°$).

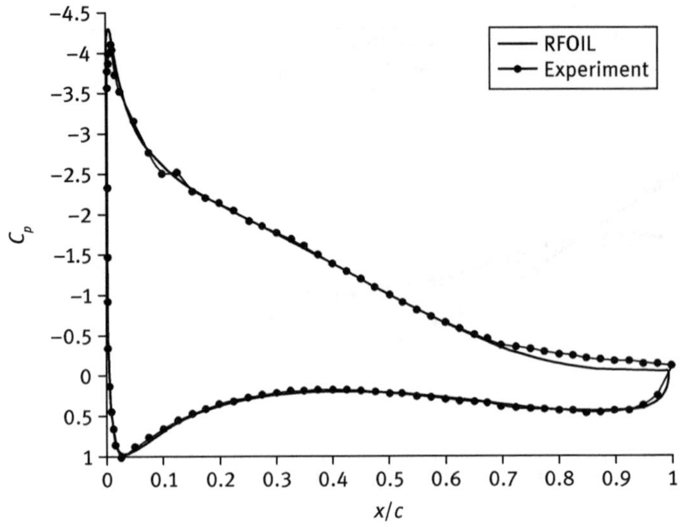

Fig. 5.41: Pressure distribution ($\alpha = 11.15°$).

5.4.4 Comparing different experimental cases

To demonstrate the aerodynamic performance of the designed airfoils, comparisons were made between the new WT180 airfoil and the well-known wind turbine airfoil Risø-A1-18 [106]. Two free transition conditions (nearly the same) were chosen, with WT180 at Reynolds number of 1.6×10^6 and Risø-A1-18 at Reynolds number of $Re = 1.6 \times 10^6$. The comparison of the results are shown in Fig. 5.42–5.44.

The lift coefficients of the WT180 airfoil are larger than those of the Risø-A1-18 airfoil at all AOA as shown in Fig. 5.42. The maximum C_L of the WT180 airfoil is 1.5813 and that of Risø-A1-18 is 1.433. The drag coefficients of WT180 are smaller than those of Risø-A1-18 over the main AOA range ($\alpha < 20°$) as shown in Fig. 5.43. The lift-to-drag ratio of WT180 is larger than that of Risø-A1-18 in the AOA range of $-5° < \alpha < 20°$. In conclusion, the WT180 airfoil shows higher lift coefficient and lift-to-drag ratio than the Risø-A1-18 airfoil for both on-design and off-design operation conditions over the main range of AOA. The detailed comparison of aerodynamic data is listed in Tab. 5.3. In Tab. 5.3, α_1 is the AOA with the maximum lift coefficient, α_2 is the AOA with the minimum drag coefficient, α_3 is the AOA with the maximum lift-to-drag ratio.

Tab. 5.3: Comparison of aerodynamic data.

Airfoil name	$C_{L,max}$	α_1	$C_{D,min}$	α_2	L/D_{max}	α_3
Risø-A1-18	1.433	10.74	0.0095	4.44	106.154	7.69
WT180	1.5813	11.15	0.0064	−2.04	124.73	6.09

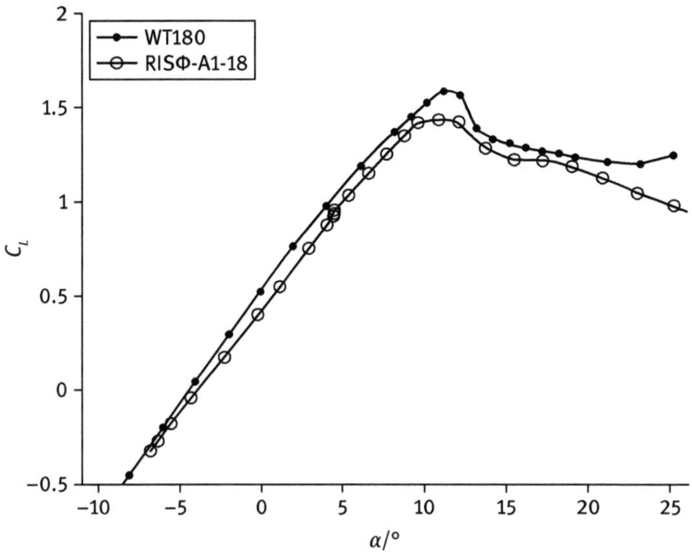

Fig. 5.42: Comparison of lift coefficients.

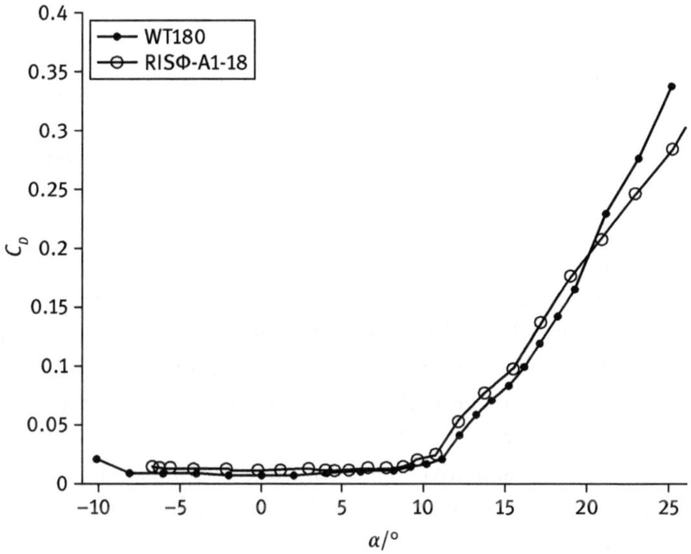

Fig. 5.43: Comparison of drag coefficients.

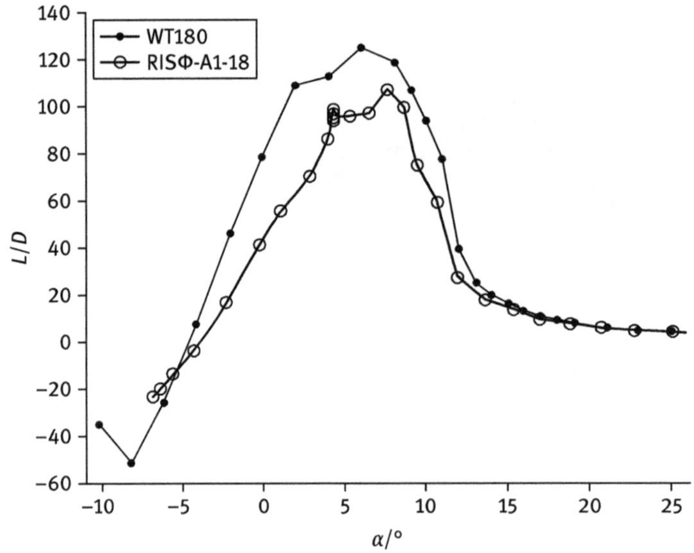

Fig. 5.44: Comparison of lift-to-drag ratios.

5.5 Chapter conclusions

Wind tunnel experiments on the WT180 airfoil were carried out for both smooth and rough surface conditions in the 3.0×1.6 m low-speed wind tunnel, and three Reynolds numbers (2.0×10^6, 3.0×10^6, and 4.0×10^6) were chosen during the tests. Results show that the new airfoil has high lift coefficient and lift-to-drag ratio at both free and fixed transition conditions. It has good characteristics in on-design and off-design operation conditions, which make it suitable for wind turbine applications. A comparison of the RFOIL prediction and experimental results generally showed good agreement, which validates the airfoil boundary layer flow theory used in RFOIL. A comparison was made between the new WT180 airfoil and the well-known wind turbine airfoil Risø-A1-18. Results show that the WT180 airfoil has higher lift coefficient and lift-to-drag ratio than the Risø-A1-18 airfoil. The new airfoil also shows better off-design performance. But the airfoil was found to be a little sensitive to leading edge roughness. Therefore, new airfoils with higher aerodynamic performance and lower roughness sensitivity will be designed in future work.

In conclusion, the novel design theory and methodology for wind turbine airfoils proposed in this book has been verified by the wind tunnel experiments. This approach has shown one way for the development of new wind turbine airfoils and blades.

6 Aerodynamics of wind turbine rotors and tip-loss corrections

6.1 Introduction

Wind turbine blades extract kinetic energy from wind and then transform it into mechanical energy of the rotor. The mechanical energy is then transformed into electrical energy through the transmission system. It can be seen that the aerodynamic research on wind turbines mainly focuses on the rotor.

When the wind is passing through the wind turbine, its velocity cannot change abruptly. Due to energy extraction by the rotor, the wind speed just before it reaches the rotor gradually decreases to a lower level than that of far upstream. The radial velocity before the rotor is not zero because of flow expansion. There will be a suddenly pressure drop across the rotor plane which is explained by basic momentum theory [165–167].

Based on momentum theory and blade element theory, the basic aerodynamic model of a wind turbine rotor is established considering the tip-loss effects. Both theoretical and experimental analyses were conducted to study the aerodynamic characteristics of the wind turbine rotor.

6.2 Aerodynamics of the wind turbine rotor

6.2.1 The momentum theory [92]

The actuator disc (AD) momentum theory, in short, momentum theory for wind turbines, describes the momentum balance on a rotating annular stream tube. The AD model is a simplified model for representing a wind turbine rotor with a circular disc, ignoring the friction. A stream tube with a circular cross section is created that extends from the disc upstream and downstream along the stream lines. A control volume is formed by the cylindrical surface of this stream tube and two cross sections at both ends. Four axial locations are important in the momentum analysis of this stream tube: free stream region or far upstream, just before the rotor, just after the rotor and far wake region or far downstream. The wind flows into the control volume with an axial velocity V_0 and a free stream pressure P_0. Then, it flows across the disc with axial velocity V and pressure drop Δp as shown in Fig. 6.1. Finally, it leaves the control volume with an axial velocity P_1 and a far wake pressure P_0.

The rotor disc is divided into several annuli along the radial direction and it is assumed that every annulus is independent of other annuli. Applying the axial momentum theory for the annulus control volume, the axial force equals the rate of axial

DOI 10.1515/9783110344387-007

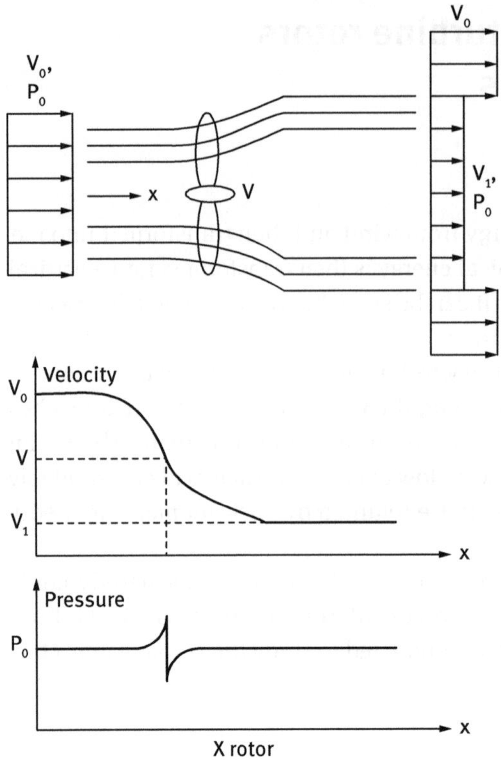

Fig. 6.1: Sketch of flow field.

momentum change of the flow:

$$dT = dm(V_0 - V_1) = \rho V(V_0 - V_1)\,dA, \tag{6.1}$$

where dm is the mass flow rate of the circular control volume, ρ is the density of air, A is the area of the rotor disc.

According to momentum theory, the thrust on the rotor can also be expressed as

$$dT = \Delta p\,dA. \tag{6.2}$$

For incompressible steady flows, Bernoulli's constant can be applied separately to a stream line both before and after the disc. Due to energy extraction by the rotor, the total pressure or Bernoulli constant after the disc remains constant but at a lower level than that of far upstream.

$$p_0 + \tfrac{1}{2}\rho V_0^2 = p + \tfrac{1}{2}\rho V^2, \tag{6.3}$$

$$p - \Delta p + \tfrac{1}{2}\rho V^2 = p_0 + \tfrac{1}{2}\rho V_1^2. \tag{6.4}$$

From equations (6.3) and (6.4), we obtain

$$\Delta p = \tfrac{1}{2}\rho(V_0^2 - V_1^2). \tag{6.5}$$

Combining equations (6.1), (6.2) and (6.5), we obtain that

$$V = \frac{V_0 + V_1}{2}. \tag{6.6}$$

The axial induction factor is defined using the following equations

$$\begin{cases} V = V_0 (1 - a), \\ V_1 = V_0 (1 - 2a). \end{cases} \tag{6.7}$$

Taking equation (6.7) into equation (6.1), the thrust on the circular area with radius increment $\mathrm{d}r$ can be expressed as

$$\mathrm{d}T = 4\pi\rho V_0^2 a(1 - a) r \,\mathrm{d}r. \tag{6.8}$$

When applying the moment of momentum theory on a single AD annulus, the torque is obtained

$$\mathrm{d}M = \mathrm{d}m(\omega r) r = 2\pi\rho V \omega r^3 \,\mathrm{d}r, \tag{6.9}$$

where ω is the tangential induction angular velocity at radius r. Then the tangential induction factor is defined as

$$b = \frac{\omega}{2\Omega}, \tag{6.10}$$

where Ω is the rotational rotor speed.

Combining equations (6.7), (6.9) and (6.10), the torque on a single AD annulus is

$$\mathrm{d}M = 4\pi\rho\Omega V_0 b(1 - a) r^3 \,\mathrm{d}r. \tag{6.11}$$

6.2.2 The blade element theory [92]

Similarly to the assumption above, the blade is also divided into several independent segments, which are called the blade elements. The inflow velocity and force on a blade element are shown in Fig. 6.2.

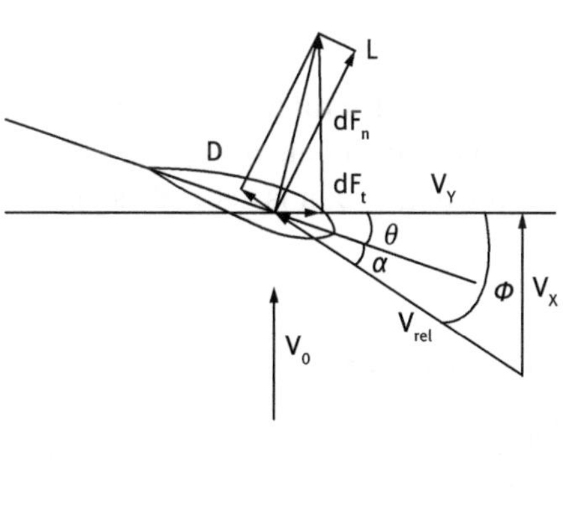

Fig. 6.2: The inflow velocity and force on a blade element.

According to momentum theory and considering the rotation of the wake after the rotor, we obtain

$$\begin{cases} V_x = V_0(1-a), \\ V_y = \Omega r(1+b), \end{cases} \tag{6.12}$$

where V_x, V_y is the component of inflow velocity along x- and y-direction respectively.

The resultant inflow velocity V_{rel} at a blade element can be expressed as

$$V_{\text{rel}} = \sqrt{V_x^2 + V_y^2} = \sqrt{(1-a)^2 V_0^2 + (1+b)^2(\Omega r)^2}. \tag{6.13}$$

The inflow angle and angle of attack at a blade element are shown below:

$$\varphi = \arctan \frac{(1-a)V_0}{(1+b)\Omega r}, \tag{6.14}$$

$$\alpha = \varphi - \theta, \tag{6.15}$$

where θ is the twist angle of the blade element.

The aerodynamic force on the blade element with the relative inflow velocity V_{rel} can be expressed as

$$\begin{cases} \mathrm{d}F_{\text{n}} = \dfrac{1}{2}\rho c V_{\text{rel}}^2 C_{\text{n}}\,\mathrm{d}r, \\ \mathrm{d}F_{\text{t}} = \dfrac{1}{2}\rho c V_{\text{rel}}^2 C_{\text{t}}\,\mathrm{d}r, \end{cases} \tag{6.16}$$

where c is the chord; C_{n}, C_{t} are the normal force coefficient and tangential force coefficient, respectively.

With blade element theory applied on a blade segment, the thrust and torque are derived:

$$\mathrm{d}T = \tfrac{1}{2}B\rho c V_{\text{rel}}^2 C_{\text{n}}\,\mathrm{d}r, \tag{6.17}$$

$$\mathrm{d}M = \tfrac{1}{2}B\rho c V_{\text{rel}}^2 C_{\text{t}} r\,\mathrm{d}r, \tag{6.18}$$

where B is the number of blades.

6.2.3 The blade element momentum theory [92]

The blade element momentum (BEM) theory, consisting of actuator disc momentum theory and blade element theory, is widely used in aerodynamic design and power prediction for wind turbine rotors. Firstly, the solidity of the rotor is defined as follows

$$\sigma = \frac{Bc}{2\pi r}. \tag{6.19}$$

Combining equations (6.8) and (6.17) above, it is derived that

$$\frac{a}{1-a} = \frac{\sigma C_{\text{n}}}{4\sin^2\varphi}. \tag{6.20}$$

Similarly, combining equations (6.11) and (6.18) above, it is derived that

$$\frac{b}{1+b} = \frac{\sigma C_\mathrm{t}}{4\sin\varphi\cos\varphi}. \tag{6.21}$$

The simplest way to get the solution of the BEM equations (6.14), (6.10) and (6.21) is to use the fixed point iterative method. To start the iterative method, initial values of the unknown parameters are guessed. After the initial solution is given, iterations are processed until convergence is achieved under a given precision.

6.3 The tip-loss correction model

There must be some differences between the physics of an actuator disc with an infinite number of blades and an actual wind turbine with a finite number of blades, so Prandtl [168] introduced the concept of tip loss:

$$F = \frac{2}{\pi}\arccos\left[\exp\left(-\frac{B}{2}\times\frac{R-r}{r\sin\varphi}\right)\right]. \tag{6.22}$$

6.3.1 The tip-loss correction model of Glauert [169]

In the tip-loss correction model of Glauert, the axial and tangential induction factors are corrected as follows:

$$a = \frac{1}{4F\sin^2\varphi/(\sigma C_\mathrm{n})+1}, \tag{6.23}$$

$$b = \frac{1}{4F\sin\varphi\cos\varphi/(\sigma C_\mathrm{t})-1}. \tag{6.24}$$

6.3.2 The tip-loss correction model of Wilson and Lissaman [170]

In the tip-loss correction model of Wilson and Lissaman, the tangential induction factor is the same as that of the Glauert tip-loss correction model. For the axial induction factor, however, the mass flux is corrected:

$$\frac{(1-aF)aF}{(1-a)^2} = \frac{\sigma C_\mathrm{l}\cos\varphi}{4\sin^2\varphi}, \tag{6.25}$$

$$\frac{bF}{1+b} = \frac{\sigma C_\mathrm{l}}{4\cos\varphi}. \tag{6.26}$$

6.3.3 The tip-loss correction model of De Vries [171]

De Vries thought that the induced velocity and the relative velocity at the blade element in Wilson and Lissaman's model do not satisfy the orthogonality [172]. So the axial induction factor is then modified as

$$\frac{(1-aF)aF}{(1-a)^2} = \frac{\sigma C_l \cos\varphi}{4\sin^2\varphi}, \tag{6.27}$$

$$\frac{bF(1-aF)}{(1+b)(1-a)} = \frac{\sigma C_l}{4\cos\varphi}. \tag{6.28}$$

6.3.4 The tip-loss correction model of Shen [172]

Based on the above mentioned models, Shen made a modification to the induction factors through introducing corrections to the normal and tangential force coefficients. This correction factor [172] is

$$F_1 = \frac{2}{\pi}\cos^{-1}\left[\exp\left(-g\frac{B(R-r)}{2r\sin\varphi}\right)\right], \tag{6.29}$$

where R is the radius of the rotor, $g = \exp[-0.125(B\lambda - 21)] + 0.1$ is a coefficient determined from experimental data [172], $\lambda = \Omega R/V_0$ is the tip speed ratio. Then, the normal and tangential force coefficients after correction are

$$\begin{cases} C_{\mathrm{n,correction}} = F_1 C_{\mathrm{n}}, \\ C_{\mathrm{t,correction}} = F_1 C_{\mathrm{t}}. \end{cases} \tag{6.30}$$

The other definitions are the same as De Vries's correction model.

6.4 The BEM model with Shen's tip-loss correction [172]

When applying the momentum theory on a single AD annulus and considering the tip-loss correction, thrust and torque on a rotor annulus are

$$\mathrm{d}T = 4\pi\rho V_0^2 aF(1-aF)r\,\mathrm{d}r, \tag{6.31}$$

$$\mathrm{d}M = 4\pi\rho\Omega V_0 bF(1-aF)r^3\,\mathrm{d}r. \tag{6.32}$$

With the blade element theory and Shen's tip-loss correction, the thrust and torque on a blade segment can also be expressed as

$$\mathrm{d}T = \frac{1}{2}B\rho c V_{\mathrm{rel}}^2 F_1 C_{\mathrm{n}}\,\mathrm{d}r, \tag{6.33}$$

$$\mathrm{d}M = \frac{1}{2}B\rho c V_{\mathrm{rel}}^2 F_1 C_{\mathrm{t}} r\,\mathrm{d}r. \tag{6.34}$$

Combining the equations (6.29)–(6.32), the axial and tangential induction factor can be obtained as

$$a = \frac{2 + Y_1 - \sqrt{4Y_1(1-F) + Y_1^2}}{2(1 + FY_1)}, \tag{6.35}$$

$$b = \frac{1}{(1 - aF)Y_2/(1-a) - 1}, \tag{6.36}$$

where $Y_1 = 4F\sin^2\varphi/(\sigma C_\mathrm{n} F_1)$ and $Y_2 = 4F\sin\varphi\cos\varphi/(\sigma C_\mathrm{t} F_1)$.

When axial induction factor a is larger than a critical value of a_c, the wind turbine enters into a turbulent wake state and the axial momentum theory equation (6.31) is invalid. The empirical correction relationship between the axial induction factor and thrust should be utilized instead:

$$\mathrm{d}T = 4\pi\rho V_0^2[a_\mathrm{c}^2F^2 + (1 - 2a_\mathrm{c}F)aF]r\,\mathrm{d}r, \tag{6.37}$$

and the corrected axial and tangential induction factor is

$$a = \frac{2 + (1 - 2a_\mathrm{c}F)Y_1 - \sqrt{(1 - 2a_\mathrm{c}F)^2Y_1^2 + 4Y_1(1 - 2a_\mathrm{c}F + a_\mathrm{c}^2F)}}{2}, \tag{6.38}$$

$$b = \frac{1}{(1 - aF)Y_2/(1-a) - 1}, \tag{6.39}$$

where $a_\mathrm{c} = \frac{1}{3}$.

The BEM equations (6.14), (6.38) and (6.39) can be solved using the fixed point iterative method. To start the iterative method, initial values of the unknown induction factors a and b are guessed. Normally, they are set to be 0. Then the inflow angle can be calculated:

$$\varphi = \arctan\frac{(1-a)V_0}{(1+b)\Omega r}. \tag{6.40}$$

Then the angle of attack can be obtained:

$$\alpha = \varphi - \theta. \tag{6.41}$$

The lift and drag coefficients can be obtained through looking at aerodynamics tables of airfoils. So the normal and tangential force coefficients of a blade element are

$$\begin{cases} C_\mathrm{n} = C_\mathrm{l}\cos\varphi + C_\mathrm{d}\sin\varphi, \\ C_\mathrm{t} = C_\mathrm{l}\sin\varphi - C_\mathrm{d}\cos\varphi. \end{cases} \tag{6.42}$$

Taking the newly computed C_n, C_t and φ into equations (6.35) and (6.35) again, the newest value of induction factors a and b are obtained. Iterations are processed until convergence is achieved under a given precision. Here the iterations errors limit for induction factors a and b are set to be 0.001.

6.5 Experimental validation

The aerodynamic model is validated against the wind tunnel measurements obtained from the MEXICO (Model Experiment in Controlled Conditions) Project [173–177]. In this European Commission sponsored project, a three-bladed wind turbine rotor with a diameter of 4.5 m has been tested in the Large Scale Low Speed Facility (LLF) of the German-Dutch Wind Tunnel Organization (DNW) with a test section of $9.5 \times 9.5\,\mathrm{m}^2$.

The designed tip speed of the rotor was set to 100 m/s, corresponding to an angular speed of 424.5 rpm. The rotor was designed for an optimal tip speed ratio of 6.7, which is obtained at a tunnel wind speed of 15 m/s. The blades consist of three airfoils: DU 91-W2–250 (from 20 to 45 % span), Risø-A2-21 (from 55 to 65 % span) and NACA 64418 (from 70 to 100 % span). The experimental pressure data points are obtained at 25 %, 35 %, 60 %, 82 % and 92 % blade span. The parameters of blade shape are listed in Tab. 6.1.

Tab. 6.1: Shape parameters of MEXICO blades.

Rotor radius (m)	Chord (m)	Twist (°)	Relative thickness (%)
0.225	0.100	0	25
0.45	0.240	16.4	25
0.675	0.207	12.1	25
0.9	0.178	8.3	25
1.125	0.150	7.0	23
1.35	0.142	4.8	21
1.575	0.125	3.8	19
1.8	0.116	2.6	18
2.205	0.088	0.5	18
2.25	0.084	0	18

For the sake of comparing with the experimental data, the lift of the airfoil is corrected for 3D effects. When the blade is rotating, there will be radial flow on the suction side of blade. And due to the Coriolis force, there will be a chordwise pressure gradient which leads to decreasing thickness of the boundary layer, the postponement of separating position and stall. The 3D corrections model of Snel was utilized here:

$$(C_l)_{3D} = (C_l)_{2D} + 3(c/r)^2[(C_l)_p - (C_l)_{2D}], \tag{6.43}$$

where $(C_l)_p = 2\pi(\alpha - \alpha_0)$, α_0 is the angle of attack when lift is zero, c is the chord, r is the spanwise position.

Utilizing the aerodynamic model of Section 6.4 with the pitch angle of blade at 0.7° and the tower height at 6.5 m, the aerodynamic performance of the MEXICO rotor under an inflow velocity of 15 m/s is shown in Fig. 6.3–6.6.

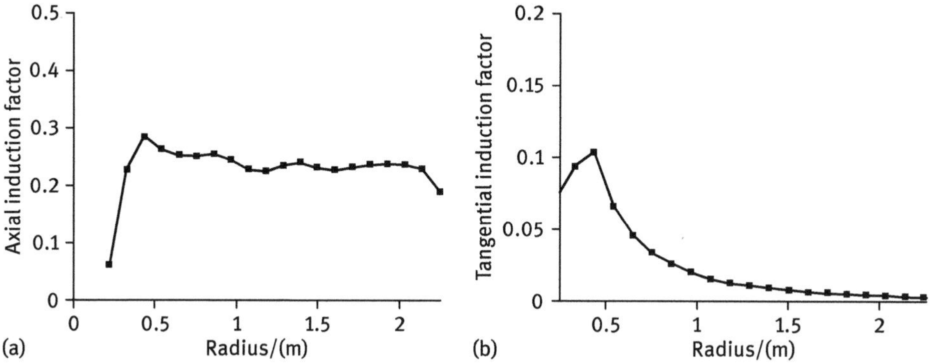

Fig. 6.3: The induction factors of the rotor.

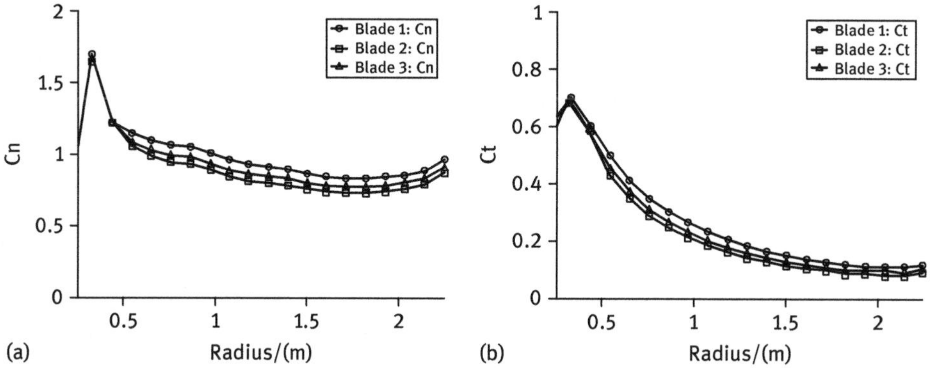

Fig. 6.4: The normal and tangential force coefficients.

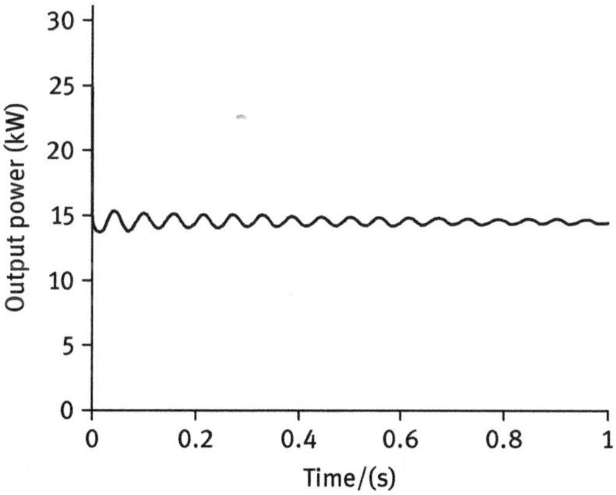

Fig. 6.5: Rated power of the rotor.

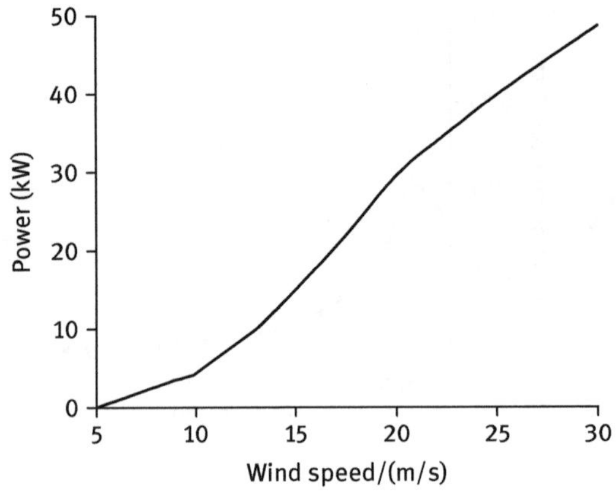

Fig. 6.6: Power characteristic of the rotor.

From the pressure distributions on the five cross sections (at 25 %, 35 %, 60 %, 82 % and 95 % span), the normal force coefficient C'_n and the tangential force coefficient C'_t related to the airfoil chord can be calculated. Then the axial force coefficient C_n and the tangential force coefficient C_t are computed from the equations below:

$$\begin{cases} C_n = C'_n \cos\theta' - C'_t \sin\theta', \\ C_t = C'_n \sin\theta' + C'_t \cos\theta', \end{cases} \tag{6.44}$$

where the angle $\theta' = \theta + \theta_{pitch}$ is the sum of pitch and local twist angles. As the normal force coefficient C_n and tangential force coefficient C_t have already been obtained, the normal force and tangential force are easily obtained through equations (6.31) and (6.32).

The axial and the tangential forces are plotted for the flow past the MEXICO rotor at a wind speed of 10 m/s and pitch angles of –2.3°. A comparison of the results from computations with Shen and Glauert tip-loss correction models are shown in Fig. 6.7 (a). The results show that Shen's tip-loss correction model predicts the force (near the blade tip) better than the traditional tip-loss correction model of Glauert. Similar to the situation above, the comparison at a wind speed of 10 m/s and pitch angles of 0.7° is shown in Fig. 6.7 (b). The comparison at a wind speed of 15 m/s (tip speed ratio of 6.7) and pitch angles of –2.3° and 0.7° are shown in Fig. 6.8 (a) and (b) respectively. Results show that Shen's tip-loss correction model agrees somewhat better with the experimental data than the traditional tip-loss correction model of Glauert. Comparisons at a wind speed of 24 m/s (tip speed ratio of 4.2) and pitch angles of –2.3° and 0.7° are shown in Fig. 6.9 (a) and (b) respectively. Although the difference between computed and experimental data is larger, Shen's tip-loss correction model performs better than the correction model of Glauert.

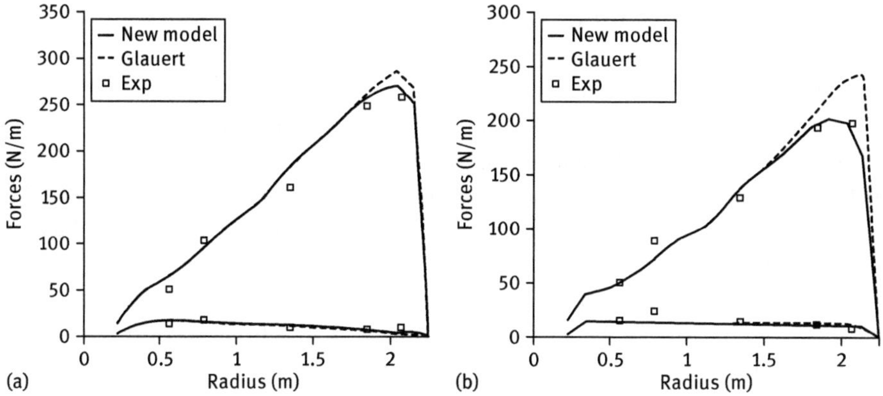

Fig. 6.7: Normal and tangential force of the MEXICO rotor at wind speed of 10 m/s.

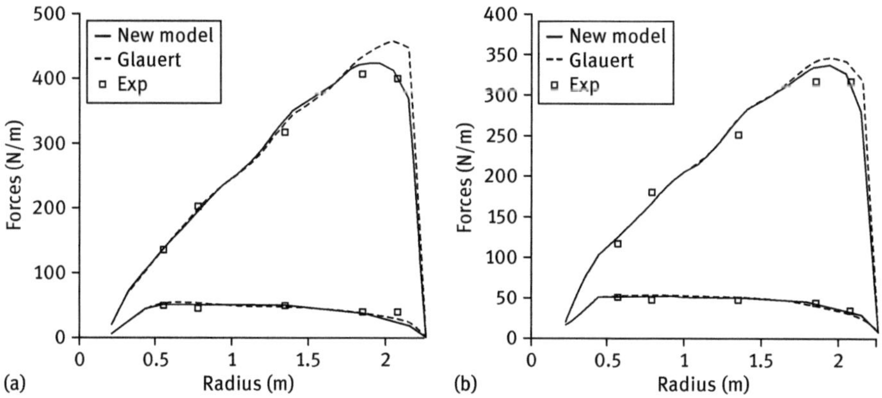

Fig. 6.8: Normal and tangential force of the MEXICO rotor at wind speed of 15 m/s.

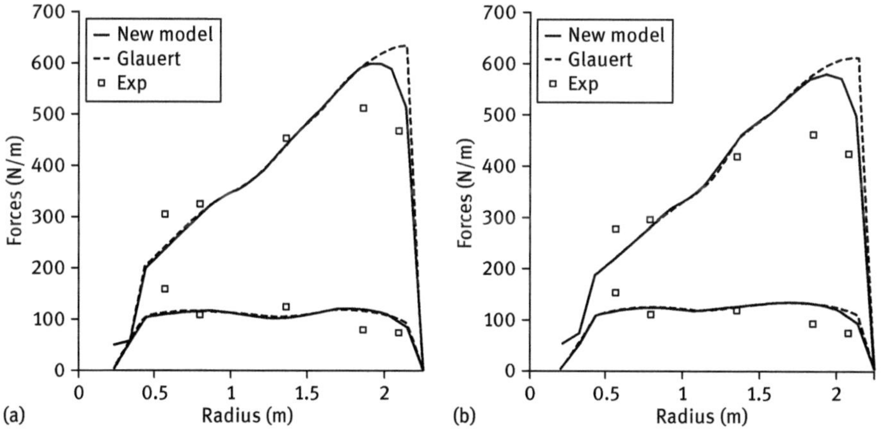

Fig. 6.9: Normal and tangential force of the MEXICO rotor at wind speed of 24 m/s.

6.6 Chapter conclusions

The one-dimensional blade element momentum theory of wind turbines was described. Brief comparisons and descriptions of tip-loss correction models by Glauert, Wilson and Lissaman, De Vries, and Shen were made. The traditional BEM theory and tip-loss correction of Shen were used to calculate the aerodynamic performance of the MEXICO rotor. The axial induction factor, tangential induction factor, normal force coefficient, tangential force coefficient and power output were obtained. The comparison of BEM with Shen's tip-loss correction and BEM with Glauert's tip-loss correction were made under some design cases in the MEXICO experiment. Shen's tip-loss correction model show higher ability simulating the blade force, the advantage is more obvious especially near the blade tip. The material in this chapter also provides a foundation for the shape optimization and aeroelastic analysis of wind turbine blades.

7 Integrated representations for wind turbine blade shapes

7.1 Introduction

Traditionally, airfoils were expressed by coordinates and blades were expressed by many airfoil sections discretely distributed along the blade span with the corresponding chord and twist. Optimizations for the blade were made on the basis of these discrete parameters. Considering the requirements of structure strength, stiffness and stability of wind turbine blades as well as the manufacturing process and costs, the shape of the optimized wind turbine blade still needs to be adjusted properly, such as the linearization treatment of chord distribution and twist distribution along the blade span [12, 34, 148, 178]. Normally, the wind turbine airfoil design, design of blade spanwise distribution, and the internal structural design of the blades are completed one by one, which is called serial design, without considering the coupling relationships between these steps.

Based on functional analysis methods, a versatile integrated representation (shape function) for 3D blade surface has been established by analyzing the distribution characteristics of chord and twist along the blade span and the 3D characterizations of various kinds of blades. The geometry of any kind of blade can be expressed through this shape function, such as the distribution of airfoil, the distribution of chord and twist of blade. The integrated representation has provided the foundations for parametric finite element modeling and the parallel design of wind turbine blades.

7.2 The integrated representations of 3D blade surface

Based on the integration and optimization of 2D wind turbine airfoils, versatile expressions for the curved shape of 3D blades were proposed. For a curved face, the points $P(x, y, z)$ on the surface should satisfy the function of a surface. In other words, the surface can be represented by the coordinates (x, y, z) [179]:

$$F(x, y, z) = 0. \tag{7.1}$$

Due to the imperfection of traditional expressions of airfoils, the versatile integrated expressions for the shapes of airfoils and blades cannot be fulfilled. Based on the integrated expressions of airfoils, which was inspired by the Joukowsky transformation and series theory, the integrated representations for 3D profiles of wind turbine blades were studied.

DOI 10.1515/9783110344387-008

7.2.1 Integrated expressions for 3D flat blades

The integrated expression $\rho(\theta) = C_0 + C_1\theta + C_2\theta^2 + C_3\theta^3 + \cdots + C_k\theta^k + \cdots$ (shape function) of airfoils has been described in equation (3.33). Then the coordinates of a 2D airfoil in the Cartesian coordinate system were obtained after conformal transformation:

$$\begin{cases} x = a\left(\rho + \dfrac{1}{\rho}\right)\cos(\theta), \\ y = a\left(\rho - \dfrac{1}{\rho}\right)\sin(\theta). \end{cases}$$

In addition to representing the curved surface by a three-variable equation $F(x, y, z) = 0$, a parametric equation system can also be used with two variables, s and t [179]:

$$\begin{cases} x = x(s, t), \\ y = y(s, t), \quad s \in I_1,\ t \in I_2. \\ z = z(s, t), \end{cases} \tag{7.2}$$

The parametric equation system was used to fulfill the integrated representations of wind turbine blades. In the blade coordinate system (as shown in Fig. 7.1), the relative spanwise location is defined with u ($u = r/R$). For a blade section with spanwise location of u_0, the coordinates of the airfoil section are

$$\begin{cases} x = a\left(\rho(\theta) + \dfrac{1}{\rho(\theta)}\right)\cos(\theta), \\ y = a\left(\rho(\theta) - \dfrac{1}{\rho(\theta)}\right)\sin(\theta), \\ z = u_0 R. \end{cases}$$

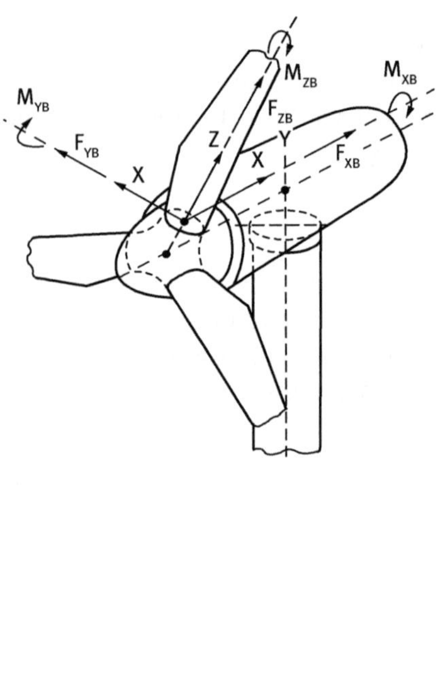

Fig. 7.1: Schematic drawing of blade coordinates.

If the blade section is extended spanwise (stretched along the z-axis), a flat blade can be formed with the same chord at all sections and without twist:

$$\begin{cases} x = a\left(\rho(\theta) + \dfrac{1}{\rho(\theta)}\right)\cos(\theta), \\ y = a\left(\rho(\theta) - \dfrac{1}{\rho(\theta)}\right)\sin(\theta), \\ z = u \cdot R. \end{cases} \tag{7.3}$$

Selecting the same airfoil ($\rho = \rho_0$) along the blade span and setting the radius of the blade as 10 m, the length of chord as $c_0 = 0.15R = 1.5$ m and the blade span as $u \in [0, 1]$, the profile of blade is obtained as shown in Fig. 7.2.

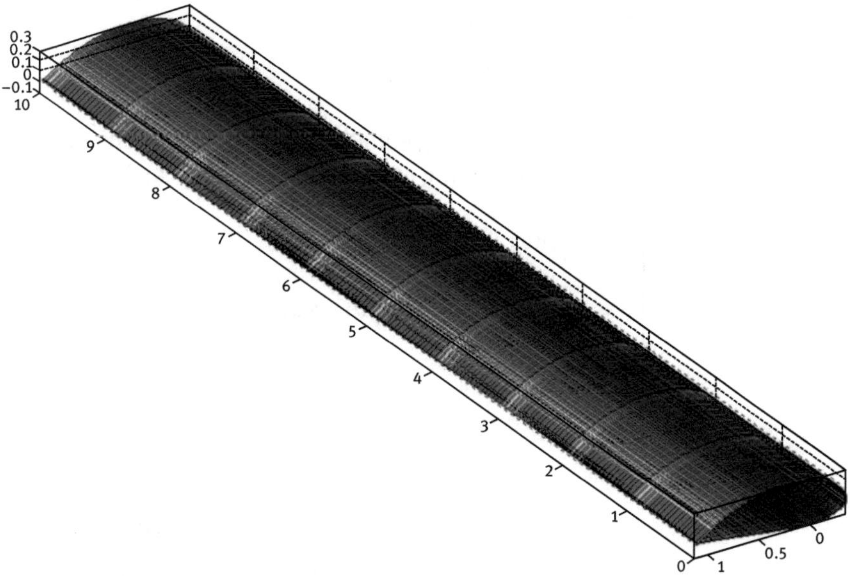

Fig. 7.2: 3D flat blade represented by shape function.

7.2.2 Integrated expressions on 3D blade with chord variation

Fig. 7.3 shows the variation of chord with respect to local tip speed ratio under ideal conditions. It can be seen from the figure that the chord is gradually decreasing from blade root to blade tip. Considering the manufacturing factors, many blades were designed with linear spanwise distribution of chord, such as the blade of the ART-600 kW wind turbine from NREL and the Tjæreborg 2 MW wind turbine. And some blades were designed with nonlinear distribution of chord to enhance efficiency, like blade of the MEXICO 25kw rotor and the E101 wind turbine from ENERCO. Based on the research

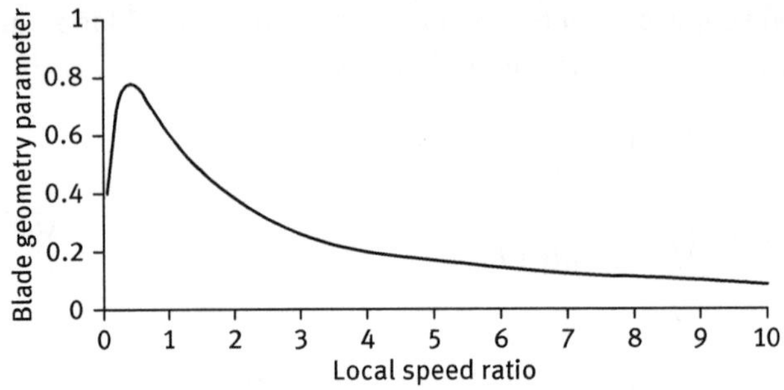

Fig. 7.3: Variation of chord with respect to local tip speed ratio under ideal conditions.

on chord distributions of various wind turbine blades and series theory, an easy and general chord distribution function is proposed:

$$c(u) = \sum_{0}^{n} a_i(1-u)^i, \tag{7.4}$$

where n is the order of the polynomial. Taking equation (7.4) into equations (7.3), the integrated expressions of a blade with varying chord along the span can be derived:

$$\begin{cases} x = a\left(\rho(\theta) + \dfrac{1}{\rho(\theta)}\right)\cos(\theta)\cdot c(u), \\ y = a\left(\rho(\theta) - \dfrac{1}{\rho(\theta)}\right)\sin(\theta)\cdot c(u), \\ z = u\cdot R. \end{cases} \tag{7.5}$$

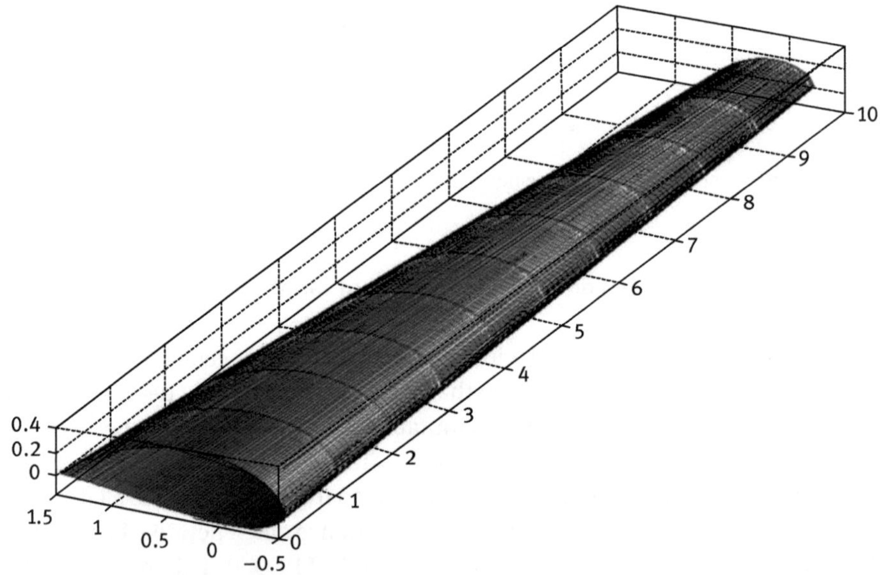

Fig. 7.4: 3D blade with varying chord represented by shape function.

Selecting the same airfoil ($\rho = \rho_0$) along the blade span and setting the radius of blade as 10 m, the variation of chord as $c = c(u)$ and the blade span as $u \in [0, 1]$, the profile of blade is obtained as shown in Fig. 7.4.

7.2.3 Integrated expressions on 3D blade with chord and twist variations

Fig. 7.5 shows the variation of inflow angle with respect to local tip speed ratio under ideal conditions. It can be seen that the inflow angle is large at the blade root which may cause stall. If the design angle of attack at each airfoil section is known, the twist of the blade can be easily obtained through $\beta = \varphi - \alpha$.

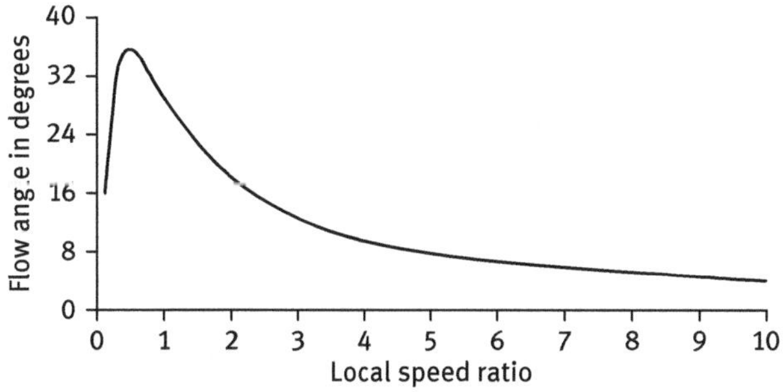

Fig. 7.5: Variation of inflow angle with respect to local tip speed ratio under ideal conditions.

Similar to the variation of chord, the twist also decreases gradually from blade root to blade tip. An easy and general twist distribution function is proposed:

$$\beta(u) = \sum_{i=0}^{n} b_i (1 - u)^i, \tag{7.6}$$

where n is the order of the polynomial. The blade is twisted about the axis passing the aerodynamic center of each blade section (the aerodynamic center locates at 1/4 chord position). Taking equation (7.6) into equations (7.5), the integrated expressions of a blade with varying chord and twist along the span can be obtained

$$\begin{cases} x = \left\{ \left[a\left(\rho(\theta) + \dfrac{1}{\rho(\theta)}\right)\cos\theta - X_M \right] \cos\beta(u) \right. \\ \qquad \left. + a\left(\rho(\theta) - \dfrac{1}{\rho(\theta)}\right)\sin\theta \sin\beta(u) \right\} c(u), \\ y = \left\{ \left[-a\left(\rho(\theta) + \dfrac{1}{\rho(\theta)}\right)\cos\theta - X_M \right] \sin\beta(u) \right. \\ \qquad \left. + a\left(\rho(\theta) - \dfrac{1}{\rho(\theta)}\right)\sin\theta \cos\beta(u) \right\} c(u), \\ z = u \cdot R, \end{cases} \tag{7.7}$$

where $\rho(\theta)$ is the shape function of the airfoil; X_M is the chordwise position of the aerodynamic center, normally adopted as $X_M = 0.25$; $c(u)$ is the chord distribution function; $\beta(u)$ is the twist distribution function.

The equation system (7.7) is the integrated expression of blade shape and is called the shape function of blades. The geometry (distribution of chord, twist and airfoils) of an arbitrary blade can be represented by this shape function. Fig. 7.6 shows a typical blade with varying chord and twist represented by shape function.

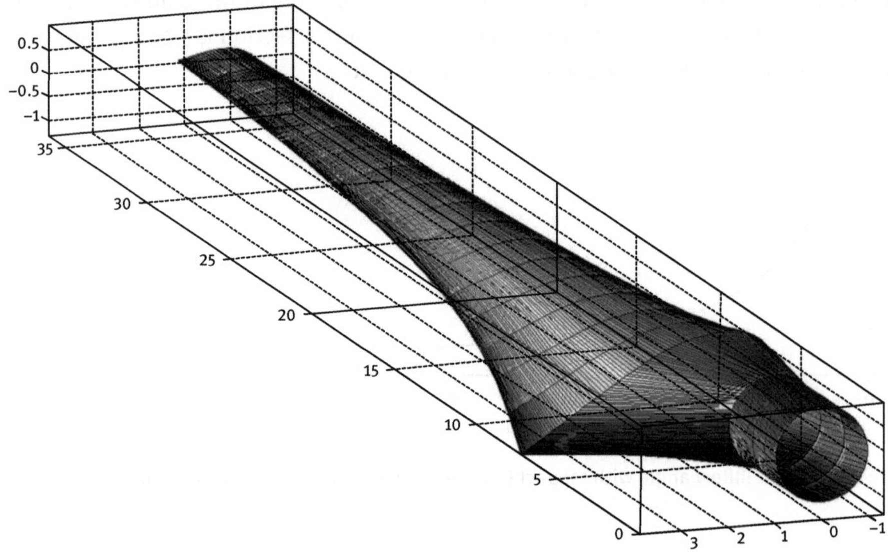

Fig. 7.6: 3D blade with varying chord and twist represented by shape function.

7.3 Integrated representations for blades of an ART-2B rotor

ART-2B [180, 181] (Advanced Research Turbines) from NREL was chosen as the object in this integrated expression research. ART series rotors are research rotors which are designed on the foundation of the Westinghouse WWG-0600 rotor. The ART-2B rotor was basically designed for 3D aerodynamics performance tests of the full-sized rotor.

The rated power of the ART-2B rotor is 667kw. Three airfoils from the NREL S airfoil series are used on the blade: s818, s825 and s826. The relative thickness of s818 (distributed at 15–25 % of blade span) is 24 %. The relative thickness of s825 (distributed at 45–75 % of blade span) is 17 %. And the relative thickness of s826 (distributed at 80–100 % of blade span) is 14 %. Fig. 7.7 shows the shapes of the three airfoils with x/c and y/c representing the relative position with respect to chord.

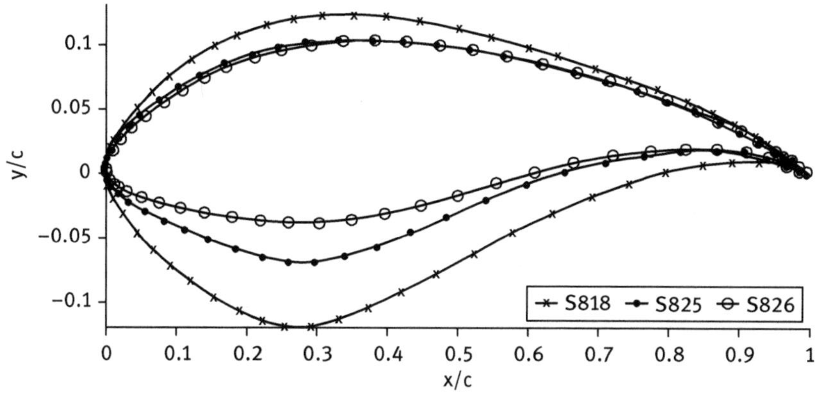

Fig. 7.7: Shapes of airfoils used on ART-2B blades.

The integrated expression for one airfoil used on the ART-2B blade can be achieved by using the 11th order polynomial, and the coefficients of the shape function are shown in Tab. 7.1.

Tab. 7.1: Coefficients of the shape function.

	Coefficients					
	C_0	C_1	C_2	C_3	C_4	C_5
ρ_1	1.004	−8.718e−2	1.647	−3.412	3.243	−1.536
ρ_2	1.001	6.707e−2	6.475e−1	−1.057	4.277e−1	3.744e−1
ρ_3	0.9994	1.434e−1	1.877e−1	−1.31e−1	−4.333e−1	7.755e−1
	C_6	C_7	C_8	C_9	C_{10}	C_{11}
ρ_1	3.017e−1	2.893e−2	−2.689e−2	5.525e−3	−5.12e−4	1.851e−5
ρ_2	−4.83e−1	2.294e−4	−5.865e−2	8.521e−3	−6.631e−4	2.15e−5
ρ_3	−5.62e−1	2.227e−1	−5.195e−2	7.127e−3	−5.326e−4	1.674e−5

Linear distribution of chord was applied in the ART-2B blade and two variables can be used to fit the chord distribution. Combined with the least squares method, the twist distribution was obtained using the 4th order polynomial. The integrated representation result of the ART-2B blade is shown in Fig. 7.8. The standard deviation between the integrated representation and the original blade shape is 0.0331. The integrated equations of blade chord and twist are expressed as follows:

$$\begin{aligned} c &= 2.004(1-u) + 0.543, \\ \beta &= 41.5931(1-u)^4 - 24.773(1-u)^3 + 7.3645(1-u)^2 \\ &\quad + 0.7698(1-u) - 0.4621. \end{aligned} \tag{7.8}$$

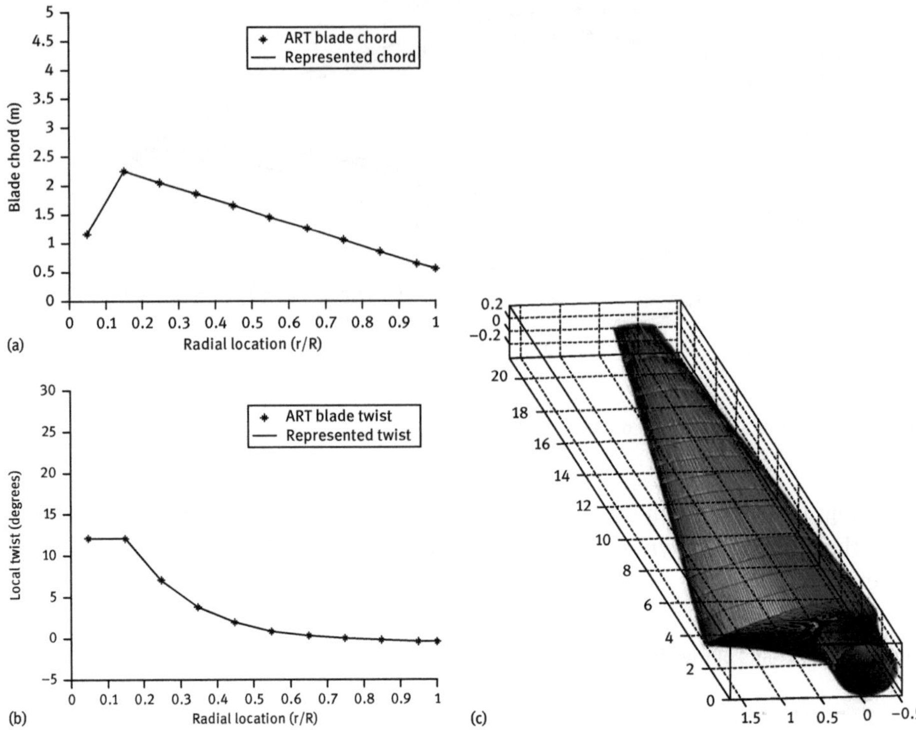

Fig. 7.8: ART-2B rotor represented by shape function.

After further derivation, the uniform shape functions of the blade were obtained.

$$\begin{cases} x = \left\{ \left[a\left(\rho_i + \frac{1}{\rho_i}\right)\cos\theta - X_M \right] \cos\beta(u) + a\left(\rho_i - \frac{1}{\rho_i}\right)\sin\theta\sin\beta(u) \right\} c(u), \\ y = \left\{ \left[-a\left(\rho_i + \frac{1}{\rho_i}\right)\cos\theta - X_M \right] \sin\beta(u) + a\left(\rho_i - \frac{1}{\rho_i}\right)\sin\theta\cos\beta(u) \right\} c(u), \\ z = u \cdot R, \end{cases}$$

where

$$\rho_i = \begin{cases} \rho_1 & \mu \in [0.15, 0.35], \\ \rho_2 & \mu \in [0.45, 0.75], \\ \rho_3 & \mu \in [0.8, 1]. \end{cases}$$

In order to compare the aerodynamic performance between the new represented blade and the original blade, the traditional BEM (Blade Element Momentum) theory and Shen tip-loss correction model were utilized. According to [172], the tip-loss correction model of Shen (Technical University of Denmark) fits the experimental results better than the Glauert tip-loss correction model. The power characteristics of the original blades and the new representation under variable speed control and fixed speed control

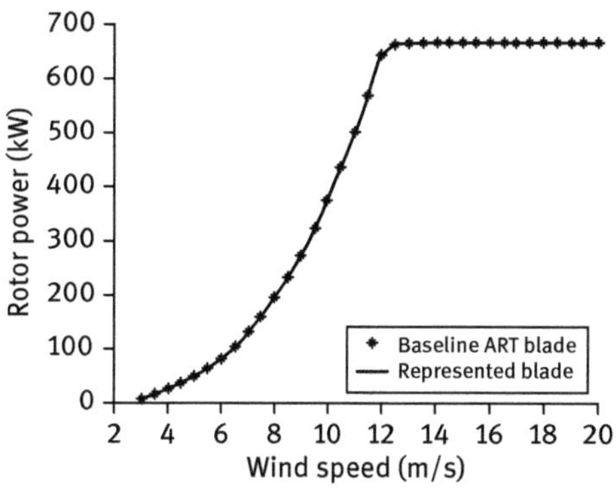

(a) Variable speed control

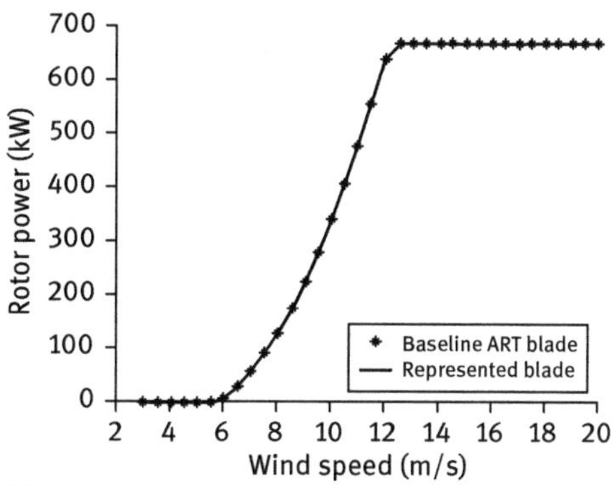

(b) Fixed speed control

Fig. 7.9: Power performance of the two blades.

are shown in Fig. 7.9. The maximum error of power is 0.5 % smaller (0.35 % for variable speed control and 0.44 % for fixed speed control), which shows that the shape function can represent the blade shape perfectly.

7.4 Chapter conclusions

Based on the integrated expression for airfoils, a versatile integrated representation for a 3D blade was established by analyzing the spanwise distribution characteristics of chord and twist and the 3D characterizations of various kinds of blades. This method can represent the geometry of any kind of blade, such as the distribution of airfoil, the distribution of chord and twist. The integrated representation has been applied on the ART-2B blade and the results show the high ability of this method. The integrated method in this chapter provides the foundation for the blade optimization conducted in the rest of the book and provides references for the representation of other sophisticated 3D surfaces.

8 Shape optimization of wind turbine blades

8.1 Introduction

With increasing rotor diameter of a wind turbine, it becomes more important to accurately predict the energy output and aerodynamic load characteristics of the rotor. The design of a wind turbine is a comprehensive engineering project which includes aerodynamic design, structural design and control system design, etc. The design tasks are independent but correlated with each other. Aerodynamic design focuses on the geometric shape of the wind turbine blade and the spanwise distribution of chord, twist angle and relative thickness to ensure a high power coefficient of the rotor [92]. When working on the aerodynamic design of a wind turbine, the relevant technical parameters, such as blade number, rotor diameter, rated wind speed and tip speed ratio, etc. should be set.

Blade element momentum theory is frequently used in the design of the aerodynamic shape of the blades. Appropriate adjustment to the shape of the wind turbine blades is needed after optimization, considering structural strength, rigidity and stability requirements, manufacturing process and cost. The tip-loss correction model of Shen was adopted when calculating the normal force and tangential force in the optimization design of the wind turbine. The new calculation model of Shen for axial and tangential induction factors of a wind turbine is briefly introduced. Firstly, the key parameters that have a great influence on the performance of the wind turbine are illustrated, like tip speed ratio, rotor diameter, Reynolds number, airfoil aerodynamic characteristics and blade root load. In addition, the relationship between the cost of the wind turbine and the energy output is discussed and a mathematical model for the cost of energy is established. An optimization model with the target of decreasing the cost of energy and the variables of chord, twist, and relative thickness of blade is proposed. Within the whole span of the blade, the newly-designed wind turbine airfoil group was adopted. The reduction in blade mass can lead to an increase in fatigue life and a decrease in the cost of wind turbine blades. However, wind turbine blades are made of composite materials so that their density changes along with the sophisticated curve surface. So it is relatively hard to establish the mass calculation mathematical model using traditional composite laminate theory. Therefore, when other conditions are consistent (thickness of blade composite shell and density, etc.), transferring mass calculation to area calculation of a curved surface is proposed. In order to minimize the mass, the area of the curved surface should be minimized. For the design of blades for a pitch-regulated wind turbine, the newly-designed CQU-A airfoil series was adopted. In addition, the multiobjective optimization model with targets for maximum power coefficient and minimal blade area is proposed. Taking chord and twist as the design parameters and fully considering the constraints of blade root load, a 2 MW wind turbine blade was designed and optimized with the use of improved

DOI 10.1515/9783110344387-009

particle swarm algorithm. Comparative analysis between the optimization result and the initial blades as well as the EU Tjæreborg experimental blades was conducted in order to verify the feasibility of this optimization design.

8.2 Influences of key parameters on the performance of rotors

Based on the aerodynamic model for wind turbine rotors in Chapter 6, the thrust and torque on blades can be calculated so that power and annual energy production (AEP) under different tip speed ratios can be obtained. Variable-speed constant-frequency generators are adopted in most pitch-regulated multi-megawatt level (MW-level) wind turbines. In order to study the influence of key parameters such as Reynolds number, lift-to-drag ratio, tip speed ratio and rotor radius on power characteristics of pitch-regulated wind turbines, two cases will be discussed. The first case is wind turbine blades with three kinds of rotor radius and the same airfoil group. The second is wind turbine blades with the same rotor radius and different airfoil groups.

8.2.1 Three rotors with different power

Rotor diameter directly determines the power of a wind turbine. The longer the blades are, the larger the power is and the higher the AEP will be. Wind turbine blades are gradually becoming longer and the diameter of MW-level wind turbine rotors has already reached as long as 130 m. Additionally, Reynolds number on the blades will change along with the size of the wind turbine. The largest Reynolds number of a traditional 5 MW wind turbine blade may reach above 8.0×10^6. Fig. 8.1 and 8.2 respectively show how the Reynolds numbers of traditional 2 MW and 5 MW wind turbine blades change with the rotor speed along the blade span. With increasing rotor speed, the Reynolds number will increase. Additionally, the maximum Reynolds numbers under different rotor speeds usually appear at 70 % spanwise location on blades. With traditional 2 MW wind turbine blades, the rotor speed generally changes from 20 rpm to 25 rpm and the maximum Reynolds number varies from 3.0×10^6 and 6.0×10^6. The maximum Reynolds number of traditional 5 MW wind turbine blades generally changes from 7.0×10^6 to 1.0×10^6 when the rotor speed changes from 10 rpm to 15 rpm.

The change of Reynolds number will directly influence the aerodynamic performance of wind turbine airfoil sections. The variations in maximum lift-to-drag ratio of the newly-designed wind turbine CQU-A airfoil series with respect to Reynolds numbers are shown in Fig. 8.3. This airfoils group contains seven airfoils with relative thickness from 15 % to 40 %. The maximum lift-to-drag ratios of four relative thinner airfoils, which have a greater influence on the aerodynamic performance of blades, were compared. From Fig. 8.3 it can be seen that the maximum lift-to-drag ratio of these

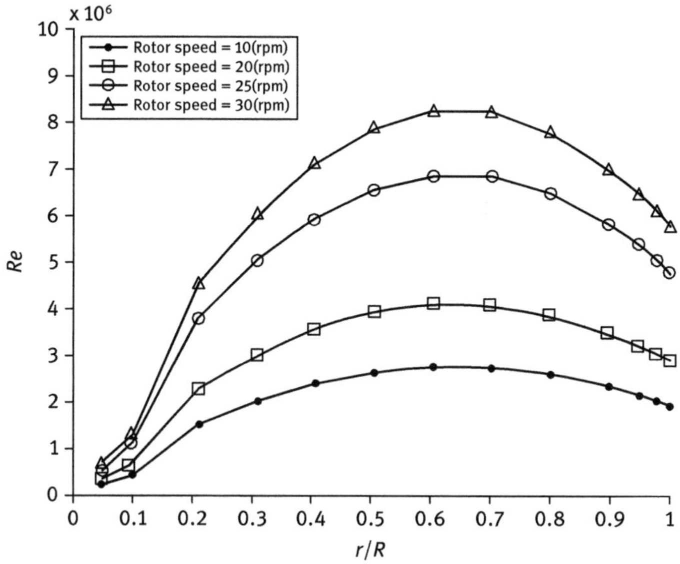

Fig. 8.1: Variation of Reynolds number on blades of a 2 MW wind turbine.

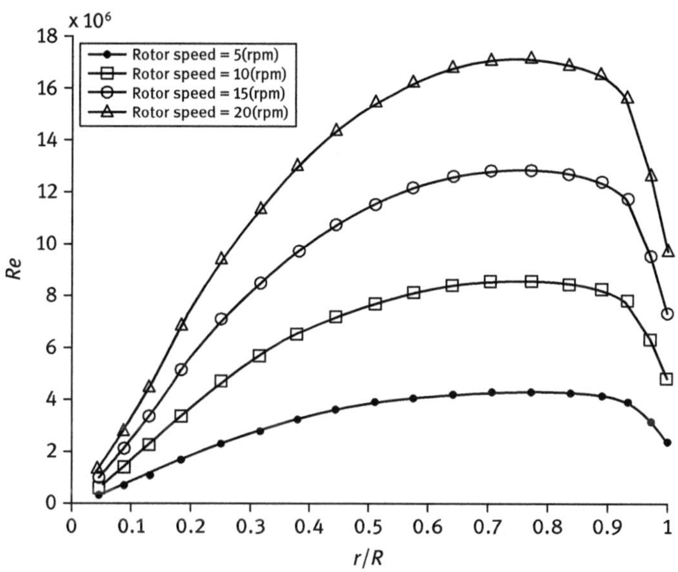

Fig. 8.2: Variation of Reynolds number on blades of a 5 MW wind turbine.

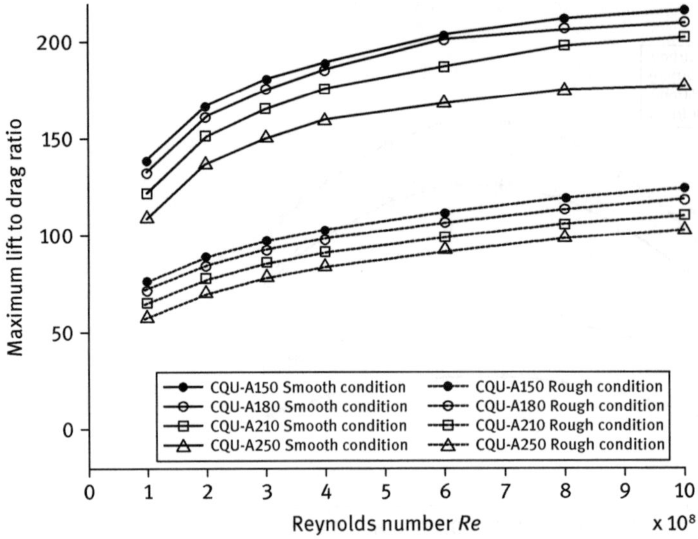

Fig. 8.3: Variation of maximum lift-to-drag ratio of rough and smooth airfoils.

airfoils increases with the Reynolds number under smooth and rough conditions. The change in the maximum lift-to-drag ratio is much larger with Reynolds numbers in the range of 1.0×10^6 to 3.0×10^6, which indicates that the maximum lift-to-drag ratio is more sensitive to Reynolds number within this scope.

In conclusion, the length of the blade has a great influence on the Reynolds number and the latter will directly influence the aerodynamic performance of the airfoil. In order to understand more about the relationships among these parameters, three wind turbines (660 kW, 2 MW and 5 MW) are selected in this section. In order to get rid of the influence from using different airfoils, the same strategy for airfoil distribution is adopted. NACA 644 airfoil series are distributed at spanwise location between 85 % and 100 %; DU airfoil series are distributed at spanwise location between 15 % and 75 %. In total, seven standard airfoils with different thickness were adopted and the transitional airfoils were obtained through interpolation. The relevant aerodynamic characteristics of the above three airfoils were treated as follows. Firstly, the selections of Reynolds numbers are shown. For blades of the 660 kW (small and medium sized) wind turbine, the Reynolds number does not change greatly and is set at 2.0×10^6 of the whole blade; for blades of the 2 MW wind turbine, the rotor speed of 20 rpm was selected. The Reynolds number is set to be 2.0×10^6, 3.0×10^6, 3.5×10^6 and 2.5×10^6 at four spanwise regions (from 15 % to 25 %, from 25 % to 45 %, from 45 % to 85 % and from 85 % to 100 %). For blades of the 5 MW wind turbine, the rotor speed of 12 rpm was selected. The Reynolds number is set to be 2.0×10^6, 5.0×10^6, 7.0×10^6, 8.5×10^6 and 6.5×10^6 at four spanwise regions (from 15 % to 25 %, from 25 % to 45 %, from 45 % to 60 %, from 60 % to 90 % and from 90 % to 100 %). Secondly, the selections of Mach number are shown. As Mach number has little influence on the aerodynamic performance of blades, we select the Mach number of 0.15. Thirdly, all the aerodynamic characteristics of airfoils are calculated by RFOIL software and then they are extended to angles of attack within ±180° through extrapolation. The aerodynamic characteristics of the airfoil after extrapolation are shown in Fig. 8.4.

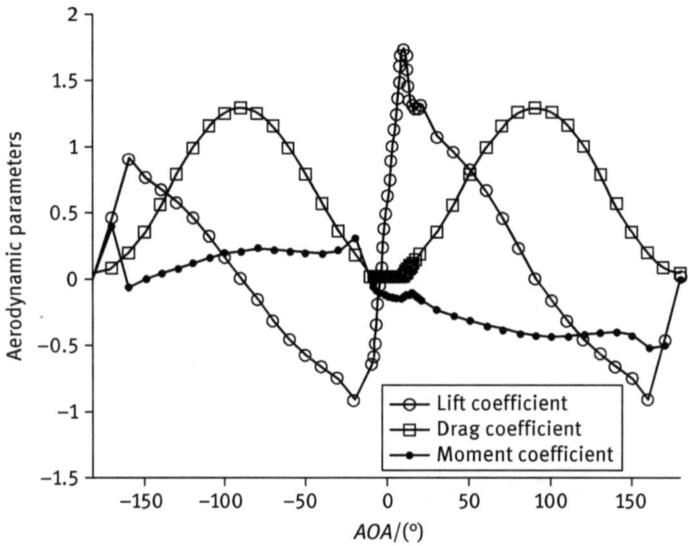

Fig. 8.4: Aerodynamic coefficients of CQU-A210 airfoil.

The variations in rotor power coefficients for the three wind turbines with respect to tip speed ratio are shown in Fig. 8.5 (under smooth conditions) and Fig. 8.6 (under rough conditions), respectively. The maximum power coefficients of the three rotors and the corresponding optimal tip speed ratios are listed in Tab. 8.1. It can found that the maximum power coefficient and the optimal tip speed ratio will increase with increasing rotor diameter both under smooth or rough conditions. The phenomenon mainly rests on the increase of the rotor diameter which leads to an increase in the Reynolds number (i.e., Fig. 8.1 and 8.2) and the improvement in aerodynamic performance of the airfoil (the lift-to-drag ratio of the airfoil directly correlates with the power coefficient).

8.2.2 Two rotors with the same power and different airfoil series

Two blades from different airfoil series, but with the same chord, twist and relative thickness distribution, were compared to study the influence of using different airfoil series on the aerodynamics characteristics of rotors. The original airfoils on the Tjæreborg 2 MW rotor were replaced with newly-designed airfoils (the maximum relative

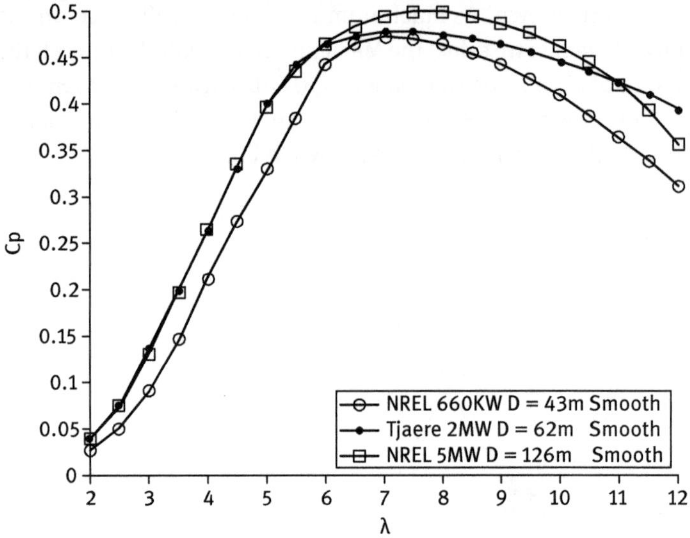

Fig. 8.5: Power coefficient characteristics of three wind turbines (smooth condition).

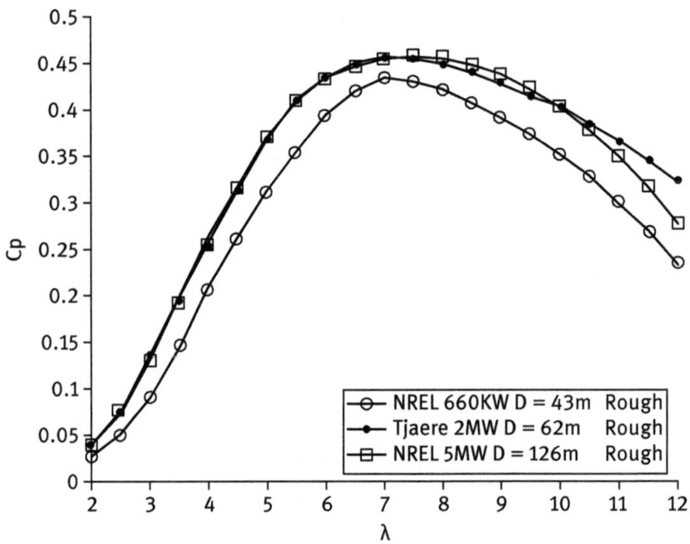

Fig. 8.6: Power coefficients of three wind turbines (rough condition).

Tab. 8.1: Performance of three wind turbines.

		NREL 660 kW	Tjæreborg 2 MW	NREL 5 MW
Rotor diameter		43.0 m	62.0 m	126.0 m
Smooth condition	Maximum power coefficient	0.47152	0.47658	0.4931
	Optimal tip speed ratio	7.0	7.5	8.0
Rough condition	Maximum power coefficient	0.43302	0.45385	0.45545
	Optimal tip speed ratio	7.0	7.5	8.0

thickness varies from 15 % to 40 %) which are shown in Fig. 8.7 and 8.8. The new blade is named CQU-A. A comparison of the aerodynamic characteristics of the new and original rotors was made. The variations in power coefficients with respect to tip speed ratio and the distribution of annual energy production with respect to wind speed are shown in Fig. 8.9 and 8.10 respectively. The detailed performance of the two rotors is listed in Tab. 8.2. According to the table, the maximum power coefficient of the rotor, distribution of annual energy production and AEP have been improved both under smooth and rough conditions. However, the bending moment at that root is larger than that of the original rotor. The increase in the bending moment at the root means higher requirements for strength and fatigue life of the blades. It is necessary to optimize the chord and twist distribution of the CQU-A blade to increase the power coefficient with the reduction of the root loads.

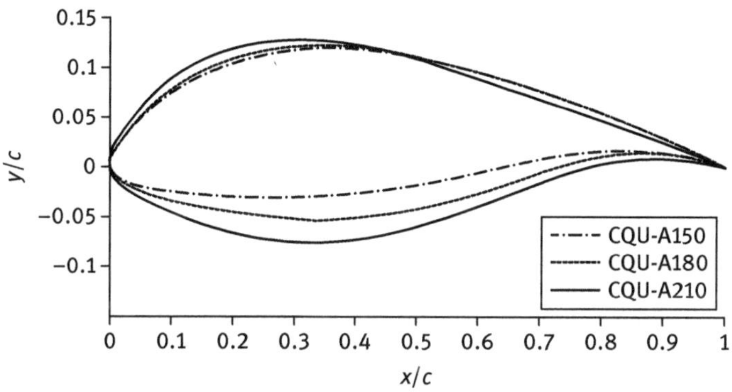

Fig. 8.7: Thinner airfoils of CQU-A series.

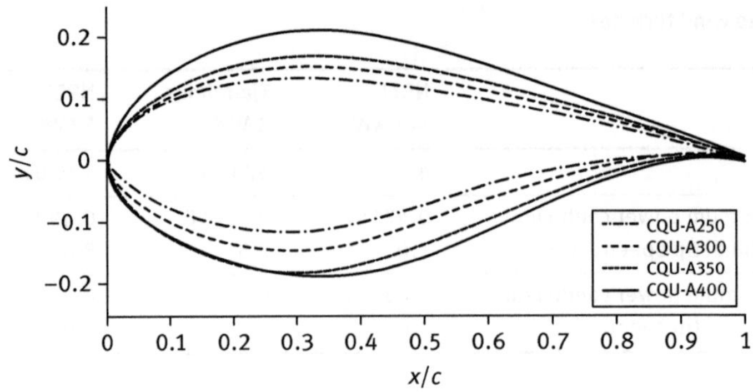

Fig. 8.8: Thicker airfoils of CQU-A series.

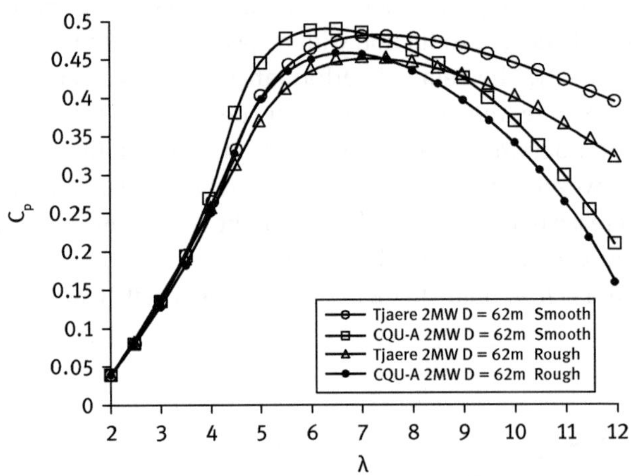

Fig. 8.9: Variation of the power coefficient.

Tab. 8.2: Detailed performance of the two rotors.

	Rotor type	Optimal tip speed ratio	Maximum power coefficient	AEP (kWh)	Bending moment at root (kN m)
Smooth condition	Tjæreborg 2 MW	7.5	0.47658	4.224×10^6	4755
	CQU-A 2 MW	7.0	0.48866	4.321×10^6	5443
Rough condition	Tjæreborg 2 MW	7.5	0.45385	4.071×10^6	4541
	CQU-A 2 MW	7.0	0.45758	4.116×10^6	5178

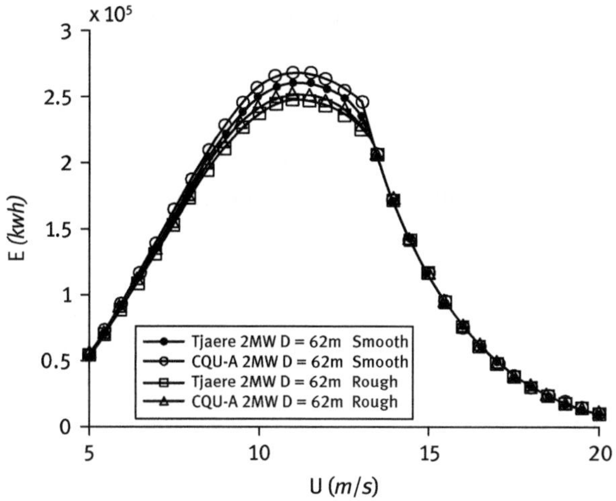

Fig. 8.10: Distribution of annual energy production.

8.3 Optimization model of wind turbine blades based on COE

Wind turbine blades will deform under inertial force, wind loads and gravities which directly affect their aerodynamic performance. Meanwhile, changing the aerodynamic properties of the blades can cause variations in the output power of a wind turbine. So the process optimization of blades should consider these effects that actually happen on real blades. The optimization model of wind turbine blades was established coupling the dynamics model and the wind turbine blade aerodynamics model.

8.3.1 Optimization objective function

A cost model for wind turbine optimization normally includes the capital costs of foundation, tower, rotor blades, gearbox and generator, plus the costs of operation and maintenance. Estimating the cost of a wind turbine is an important and difficult task, but also crucial for the success of an optimization. The total cost of a wind turbine can be expressed as

$$C = \sum_{i=1}^{N} C_i = \sum_{i=1}^{N} R_i \left(b_i + (1 - b_i)w_i\right), \tag{8.1}$$

where C_i is the cost of the i-th component of the wind turbine and N is the number of main components, R_i is the initial cost of the i-th component determined from a reference rotor, b_i is the fixed part of the i-th component that accounts for manufacturing and transport, $(1 - b_i)$ is the variable part of the i-th component, and w_i is the weight parameter of the i-th component.

As the costs for operation and maintenance often can be considered a small percentage of the capital cost, reducing the capital cost becomes an essential task for designing wind turbines. Moreover, a well-designed wind turbine with a low energy cost always has an aerodynamically efficient rotor. Therefore, the rotor design plays an important role for the whole design procedure of a wind turbine. Here, we restrict our objective to the cost of the rotor. Thus, the objective function is defined as

$$f(x) = \mathrm{COE} = \frac{C_{\mathrm{rotor}}}{\mathrm{AEP}}, \tag{8.2}$$

where COE is the cost of energy produced by a wind turbine rotor and C_{rotor} is the total cost for producing, transporting and erecting a wind turbine rotor. The fixed part of the cost of a wind turbine rotor b_{rotor} is chosen to be 0.1. Therefore, the total cost of a rotor C_{rotor} is a relative value defined as

$$C_{\mathrm{rotor}} = b_{\mathrm{rotor}} + (1 - b_{\mathrm{rotor}})w_{\mathrm{rotor}}, \tag{8.3}$$

where w_{rotor} is the weight parameter of the rotor. The weight parameter is calculated from the chord and mass distributions of the blades. Supposing that a blade can be divided into n cross sections, w_{rotor} is estimated as

$$w_{\mathrm{rotor}} = \sum_{i=1}^{n} \frac{m_i \times c_{i,\mathrm{opt}}}{M_{\mathrm{tot}} \times c_{i,\mathrm{or}}}, \tag{8.4}$$

where m_i is the mass of the i-th cross section of the blade; $c_{i,\mathrm{opt}}$ is the averaged chord of the i-th cross section of the optimized blade; $c_{i,\mathrm{or}}$ is the averaged chord of the i-th cross section of the original blade; M_{tot} is the total mass of the blade.

The power curve is determined with the BEM method. In order to compute the AEP, it is necessary to combine the power curve with the probability density of wind (i.e., the Weibull distribution). The function defining the probability density can be written in the following form:

$$f(V_i < V < V_{i+1}) = \exp\left(-\left(\frac{V_i}{A}\right)^k\right) - \exp\left(-\left(\frac{V_{i+1}}{A}\right)^k\right), \tag{8.5}$$

where A is the scale parameter, k is the shape factor and V is the wind speed. The shape factor is chosen to be $k = 2$, corresponding to the Rayleigh distribution.

The output power of a wind turbine has a relationship to wind speed, as shown in Fig. 8.11. If a wind turbine operates for about 8700 hours per year, its AEP can be evaluated as

$$\mathrm{AEP} = \sum_{i=1}^{N-1} \frac{1}{2}(P(V_{i+1}) + P(V_i)) \times f(V_i < V < V_{i+1}) \times 8700. \tag{8.6}$$

The $P(V_i)$ is the power at wind speed V_i which can be calculated through

$$P(V_i) = C_{\mathrm{P}}(\lambda, \beta)P_{V_i} = \frac{1}{2}\rho V^3 A_{\mathrm{S}} C_{P_i}(\lambda, \beta) \tag{8.7}$$

where ρ is the air density, A_{S} is the area of rotor plane, C_{P} is the power coefficient and β is the pitch angle.

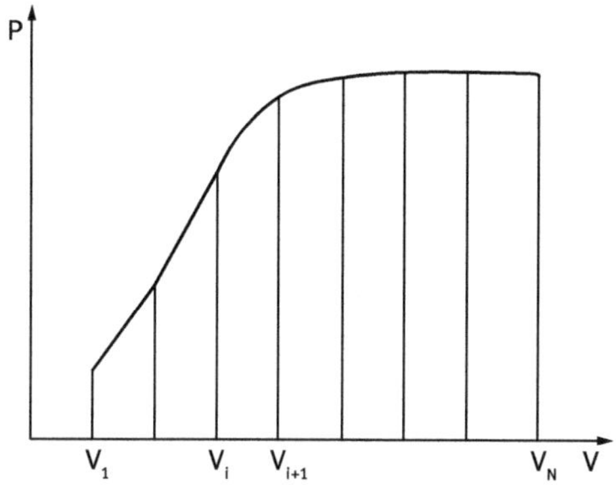

Fig. 8.11: The power curve with respect to inflow speed.

8.3.2 Design variables and constraints

To obtain a reliable optimization of a wind turbine blade, the geometry of the blade needs to be represented as far as possible. This requires a great number of design variables. On the other hand, the selection of more design variables in the optimization procedure requires more calculation time. The design variables are often chosen to be the parameters controlling rotor shape and airfoil characteristics, as well as regulating rotor speed and pitch angle. The rotor shape is controlled by the rotor diameter, chord, twist, relative thickness and shell thickness.

To illustrate our optimization model, we focus on two different rotors: one 2 MW rotor and one 5 MW virtual rotor. The optimization is based on the original rotors. The rotor diameter and the rotor speed are chosen to be the same. As a consequence of the choice, chord, twist angle, relative thickness and tip pitch angle are chosen as design variables. The distributions of chord, twist angle and relative thickness are very important for the aerodynamic performance of a rotor. Based on a general chord distribution, a cubic polynomial is used to control the chord distribution. Because of the multiple distribution characteristics, a spline function is used to control the distributions of twist angle and relative thickness. The constraints of the design variables are

$$X_{i,\min} \le X_i \le X_{i,\max}, \quad i = 1, 2, 3. \tag{8.8}$$

Since the cost of a rotor depends on the lifetime of the blades, which is influenced by the maximum load and the fatigue load; the maximum output power, the maximum shaft torque and the maximum thrust need to be restricted. These three variables are also constrained in the optimization process. The axial and tangential loads on the blades are calculated from the lift and drag forces as

$$\begin{cases} F_\mathrm{n} = F_\mathrm{l} \cos\varphi + F_\mathrm{d} \sin\varphi, \\ F_\mathrm{t} = F_\mathrm{l} \sin\varphi - F_\mathrm{d} \cos\varphi, \end{cases} \tag{8.9}$$

where F_l is the lift force, F_d is the drag force and φ is the flow angle. The total thrust of the rotor is obtained from the axial force and is constrained as

$$T \leq T_{max}, \tag{8.10}$$

where T_{max} is the maximum thrust that is taken to be the maximum thrust of the original rotor. A similar constraint for the shaft torque is also imposed. The tangential force F_t contributes mostly to the shaft torque and the output power. A bigger shaft torque will increase the load of the transmission system and reduce the lifetime of the gearbox. Therefore, the shaft torque distribution on the blades is constrained as

$$M \leq M_{max}, \tag{8.11}$$

where M_{max} is the maximum shaft torque that is taken to be the maximum thrust of the original rotor.

8.3.3 Optimization program and method

The aerodynamic model discussed in Section 6.4 was utilized in this optimization. As is a common procedure for optimization problems, we have one objective function and multiple constraints. To achieve the optimization, the *fmincon* function in MATLAB is used. The optimization process is shown in Fig. 8.12.

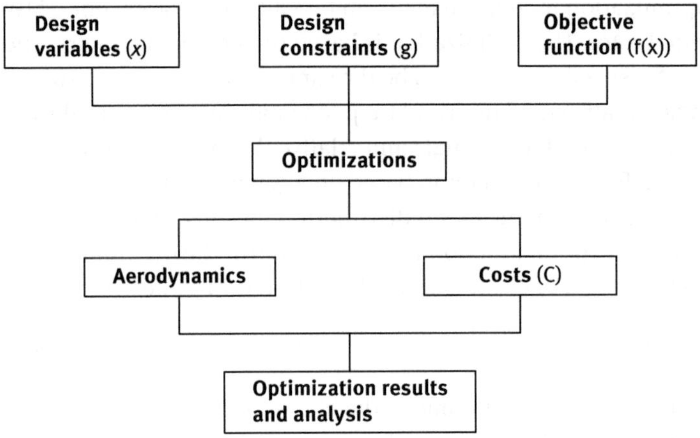

Fig. 8.12: Optimization process.

Here, only the cost of the rotor is taken into account in relation to the annual energy production. To demonstrate the capabilities and the reliability of the optimization method, optimizations for two wind turbine rotors with different sizes will be carried out consecutively in the following, and the performance of the optimized rotors will be compared in detail with the original rotors for the same design load cases.

8.3.4 Optimization results

8.3.4.1 Optimization results for the 2 MW rotor

One 2 MW rotor is chosen as the first test case of the optimization model. The turbine is a three-bladed rotor of radius 30.56 m. The shape parameters and structural parameters of the blades are listed in Tab. 8.3 and 8.4 respectively. In the BEM computations, 20 uniformly distributed blade elements are used. The optimization design is performed from a radial position at a radius of 6.46 m to the tip of the blade. In the optimization process, the lower limits for chord, twist angle and relative thickness are 0 m, 0° and 12.2 %, respectively, and the upper limits are 3.3 m, 8° and 100 %, respectively. As before, the maximum values of the design variables for the optimized rotor are chosen to be the same as the original rotor. To reduce computational time, four points along the blade are used to control the shape of the blade. The rotor diameter, the rotor speed

Tab. 8.3: The shape parameters of the blades.

Rotor radius (m)	Chord (m)	Twist (°)	Relative thickness (%)
1.46	2.00	0	100
2.96	2.00	0	100
6.46	3.30	8	30.58
9.46	3.00	7	24.10
12.46	2.70	6	21.13
15.46	2.40	5	18.70
18.46	2.10	4	16.81
21.46	1.80	3	15.46
24.46	1.50	2	14.38
27.46	1.20	1	13.30
30.56	0.02	0	12.20

Tab. 8.4: The structural parameters of the blades.

r (m)	EI_1 (MN m^2)	EI_2 (MN m^2)	m (kg/m)
1.46	14480.00	14480.00	3460.00
2.96	1901.60	1963.60	442.64
6.46	566.57	1400.30	337.19
9.46	235.68	851.64	275.64
12.46	119.16	525.11	229.02
15.46	59.83	329.37	191.40
18.46	29.76	207.20	166.92
21.46	13.80	120.99	159.47
24.46	5.39	59.94	84.52
27.46	1.51	24.54	47.88
30.56	0.16	7.22	19.99

and the airfoil shapes remain unchanged. Note from our optimizations that the final solution is more or less independent of the choice of spline function.

In the optimization, chord, twist angle and relative thickness are forced to decrease from the starting point to the blade tip. With these constraints, the optimization converges after about 30 iterations. In order to save optimization time, four control points were chosen for interpolating for the chord, twist and relative thickness distributions.

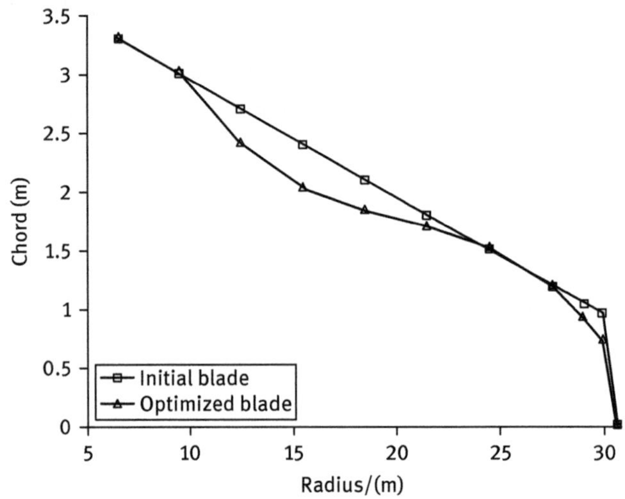

Fig. 8.13: Comparison of chord distribution.

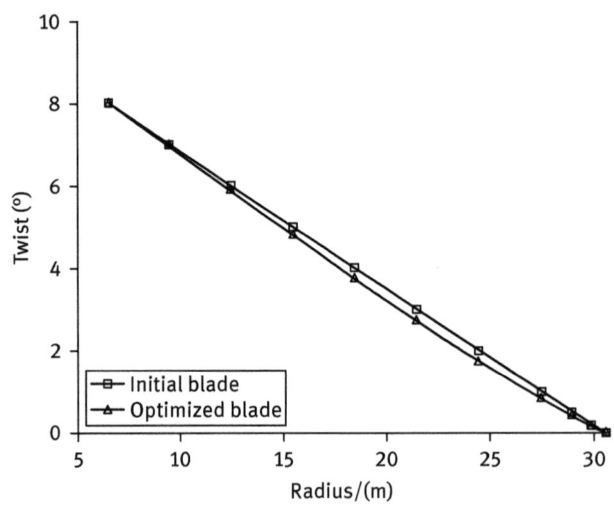

Fig. 8.14: Comparison of twist distribution.

The chord and twist distributions of the original and the optimized rotor are shown in Fig. 8.13 and 8.14 respectively. From Fig. 8.13 it can be seen that the optimized blade attains a remarkable reduction in chord length in the region between 10 and 23 m, as compared to the original rotor. The optimized blade still has a linear chord distribution. At a radius of 15 m, the chord reduction reaches a maximum value of about 16 %. From a position at a radius of 23 m to a position at a radius of 28 m, the optimized chord has almost the same value as the original distribution. This is because the original chord was already almost optimal in this region. The twist angle (Fig. 8.14) is a little bit smaller than the original distribution.

8.3.4.2 Optimization results for the 5 MW rotor

One 5 MW rotor is chosen as the first test case of the optimization model. The turbine is a three-bladed rotor of radius 30.56 m. The shape parameters and structural parameters of the blades are listed in Tab. 8.5 and 8.6 respectively. The rotor is equipped with three blades of 63 m. In the BEM computations, 40 uniformly distributed blade elements are used. The optimization design is performed from a position at a radius of 15 m to the tip of the blade. In the optimization process, lower limits for chord, twist angle and relative thickness are 0 m, 0° and 18 %, respectively, and the upper limits are 5 m, 15° and 100 %, respectively. For the same reason as for the previous rotor, the maximum values of the design variables for the optimized rotor are chosen to be the same as the original rotor. The rotor diameter, the rotor speed and the airfoil shapes remain unchanged in the optimization.

To reduce computational time, four points along the blade are used to control the shape of the blade. Using the same ideas as for the 2 MW rotor, the optimization converges after about 36 iterations. The chord and twist angle distributions of the original and the optimized rotor are shown in Fig. 8.15 and 8.16 respectively. From Fig. 8.15 it can be seen that the original rotor attains an almost linear chord distribution, whereas the optimized blade has a sharp decrease after 30 m. In the region between 30 and 48 m, the optimized rotor has a smaller chord than the original rotor. The chord reduction reaches a maximum value of about 8.2 % at a radius of 40 m. Between 48 and 60 m on the blade, the optimized rotor has a slightly smaller chord than the original rotor. From Fig. 8.16 it can be seen that the twist angle is reduced significantly in the region between 25 and 35 m, whereas it is almost the same on the rest of the blade.

Tab. 8.5: The shape parameters of a 5 MW blade.

Rotor radius (m)	Chord (m)	Twist angle (°)	Relative thickness (%)
2.87	3.542	13.308	40
5.60	3.854	13.308	40
8.33	4.167	13.308	40
11.75	4.557	13.308	40
15.85	4.652	11.48	35
19.95	4.458	10.162	35
24.05	4.249	9.011	30
28.15	4.007	7.795	25
32.25	3.748	6.544	25
36.35	3.502	5.361	21
40.45	3.256	4.188	21
44.55	3.010	3.125	18
48.65	2.764	2.319	18
52.75	2.510	1.526	18
56.17	2.313	0.863	18
58.90	2.086	0.37	18
63	0.5	0	18

Tab. 8.6: The structural parameters of the blades.

r (m)	EI_1 (MN m^2)	EI_2 (MN m^2)	m (kg/m)
2.87	1.90967e10	1.95485e10	767.89
5.60	1.12329e10	1.53515e10	607.25
8.33	5.81482e9	8.45976e9	409.23
11.75	4.65455e9	7.17288e9	425.85
15.85	2.54191e9	5.03389e9	352.32
19.95	2.02233e9	4.46923e9	338.17
24.05	1.54902e9	3.95249e9	320.56
28.15	1.0514e9	3.37782e9	293.02
32.25	6.40991e8	2.68491e9	260.56
36.35	3.78233e8	2.16981e9	234.83
40.45	2.15099e8	1.48632e9	192.46
44.55	1.18041e8	1.11418e9	160.55
48.65	8.39593e7	7.55915e8	134.48
52.75	5.49758e7	4.84947e8	102.81
56.17	3.71712e7	3.75799e8	86.87
58.90	2.5448e7	2.73536e8	67.77
63.00	170000	5.01e6	10.32

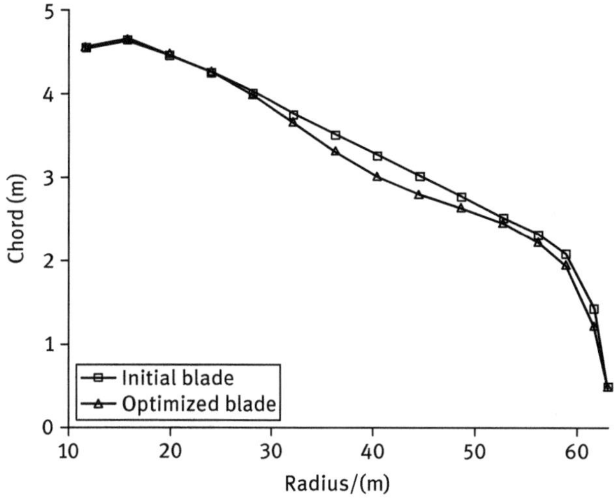

Fig. 8.15: Comparison of chord distributions.

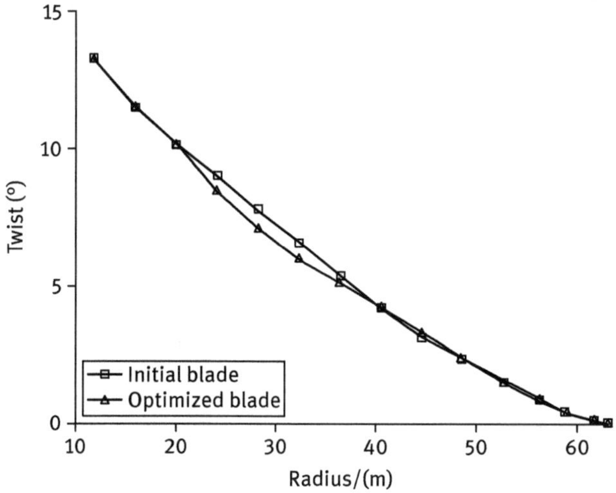

Fig. 8.16: Comparison of twist distributions.

8.3.5 Comparison of rotor performance

8.3.5.1 Performance of original 2 MW rotor and optimized rotor

The power output of the original 2 MW and the optimized rotor are listed in Tab. 8.7. It can be seen that the optimized power output is slightly smaller than that of the original rotor while the cost of the optimized rotor is reduced. At a rated wind speed of 14 m/s, the output power of the optimized rotor is reduced by about 2.6 % compared with the original rotor. The AEP of the optimized rotor is reduced by about 4 %, whereas the cost of the optimized rotor has been reduced by about 7.1 %. Thus, the cost of energy produced by the rotor is reduced by about 3.4 %.

Tab. 8.7: Power output of original and optimized rotor.

Inflow velocity (m/s)	Power of original rotor (MW)	Power of optimized rotor (MW)
5	0.057	0.048
7	0.27	0.26
9	0.61	0.60
10	0.82	0.83
11	1.07	1.06
12	1.34	1.32
13	1.62	1.63
14	1.91	1.86
15	2	2

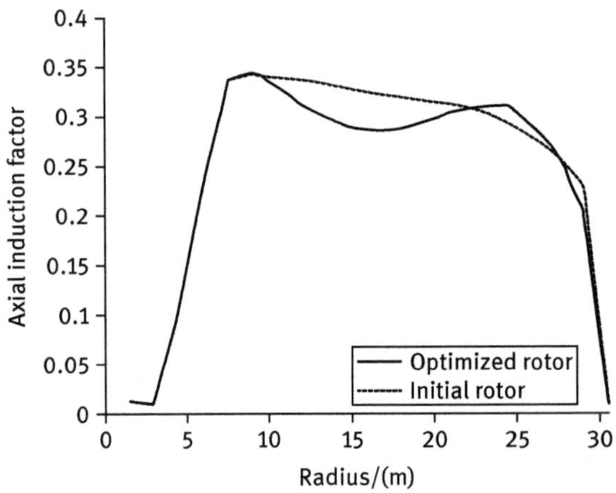

Fig. 8.17: Axial induction factors.

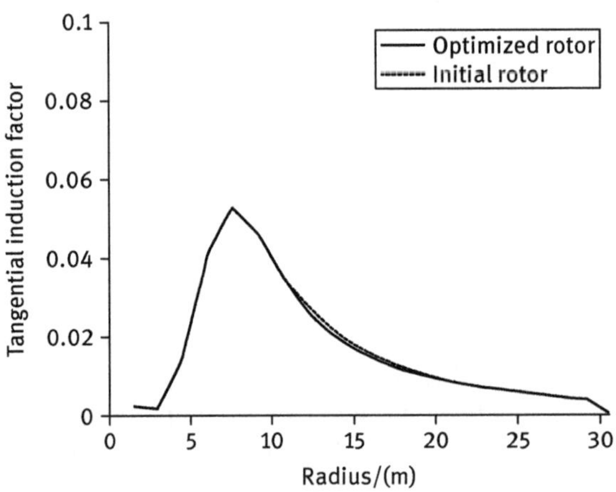

Fig. 8.18: Tangential induction factors.

In order to investigate why the optimized rotor with a smaller chord in the middle part of the blade produces almost the same power as the original rotor, the axial and tangential induction factors of the two rotors at wind speed of 10 m/s are shown in Fig. 8.17 and 8.18 respectively. In Fig. 8.17, the axial induction factor of the optimized rotor is about 0.3 in the region from 10 to 22 m, whereas the original rotor has a value around $\frac{1}{3}$. As is known, the maximum power coefficient (Betz limit) occurs at an axial induction factor of $\frac{1}{3}$. Hence, the original rotor has almost reached the maximum power coefficient. However, since the optimized rotor has an important chord reduction in this region and a slightly smaller power coefficient (a slightly smaller tangential force), the optimized rotor is economically more efficient. In the area close to 25 m, the axial induction factor reaches almost $\frac{1}{3}$, and therefore, the optimized rotor has locally the highest power coefficient. The tangential induction factor is almost the same for both rotors which is shown in Fig. 8.18. The angle of attack is plotted in Fig. 8.19. It can be seen from Fig. 8.19 that the angle of attack on most of the blade, from a radial position of 10 m to the blade tip, is between 5° and 10° for both rotors. Moreover, the angle of attack of the optimized rotor is higher than that of the original rotor, especially in the region with a smaller chord.

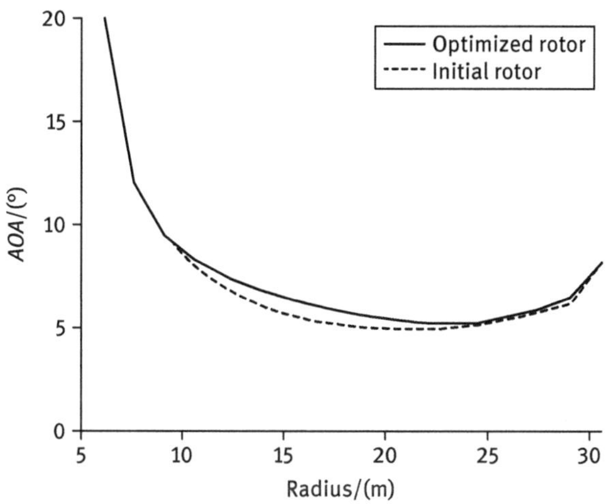

Fig. 8.19: AOA at inflow speed of 10 m/s.

The structural performance of the original and the optimized rotor is investigated at a wind speed of 14 m/s. The flapwise and edgewise deflections of the blade on the two rotors are plotted in Fig. 8.20 and 8.21 respectively. From the figure, we can see that the deflections of the optimized rotor are slightly smaller along the whole blade in both the flapwise and the edgewise directions than those of the original rotor. The comparison of axial force and tangential force between the original and the optimized rotor is shown in Fig. 8.22 and 8.23. The axial forces and the tangential forces of the optimized rotor are reduced significantly in a region from 10 to 23 m radius. The total thrusts of the two rotors are listed in Tab. 8.8. From the table, it can be seen that the thrust has been reduced for all wind speeds below rated power, and that the maximum reduction in thrust (about 5 %) is reached at a wind speed of 14 m/s. Small deflections and reduced forces increase the lifetime of the rotor and thus further reduce the cost of the rotor.

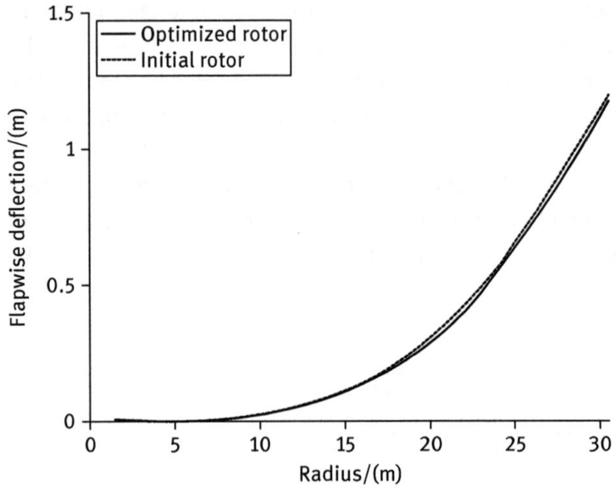

Fig. 8.20: Flapwise deflections.

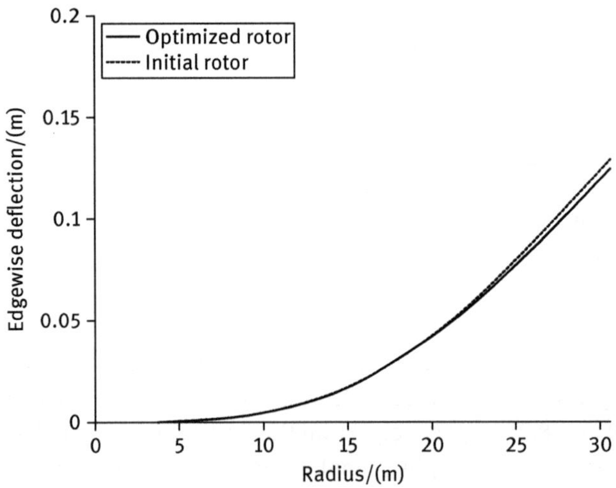

Fig. 8.21: Edgewise deflections.

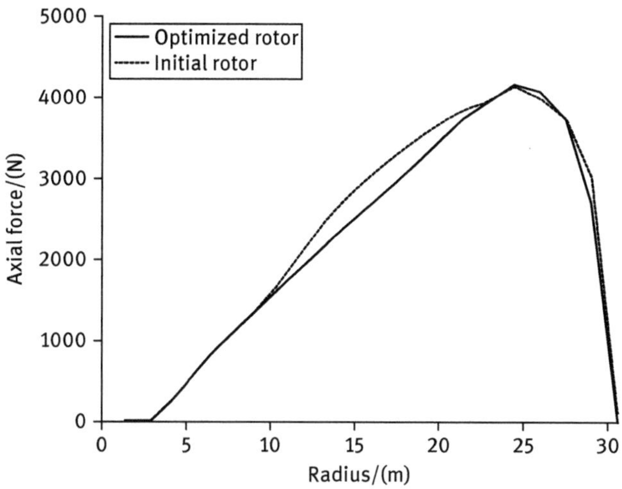

Fig. 8.22: Axial force.

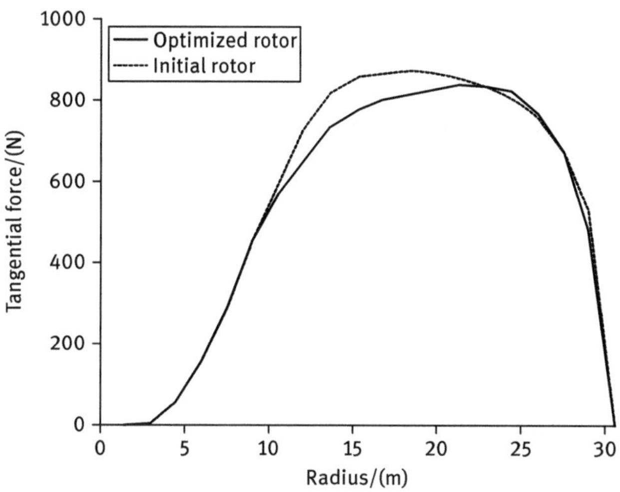

Fig. 8.23: Tangential force.

8.3.5.2 Performance of original 5 MW rotor and optimized rotor

The power output of the original 5 MW and the optimized rotor are listed in Tab. 8.9. From the table, it can be seen that the power of the optimized rotor is almost the same as the original rotor, while the cost of the optimized rotor is reduced. The AEP of the optimized rotor is only reduced by about 0.1 %, whereas the cost of the optimized rotor has been reduced by about 2.7 %. Thus, the cost of energy for the original rotor is reduced by about 2.6 %.

Tab. 8.8: Total thrusts of the two rotors.

Inflow velocity (m/s)	Thrust of original rotor (kN)	Thrust of optimized rotor (kN)
5	54.54	54.25
7	84.91	83.75
9	123.62	120.89
10	143.24	139.66
11	162.14	157.11
12	179.33	173.22
13	195.62	187.78
14	209.99	200.45
15	200.27	205.15
16	184.00	179.82
17	166.66	164.38
18	158.76	157.09
19	149.45	149.37
20	141.57	141.78

Tab. 8.9: Power output of original and optimized rotor.

Inflow velocity (m/s)	Power of original rotor (MW)	Power of optimized rotor (MW)
5	0.23	0.22
7	0.87	0.91
9	2.44	2.45
10	3.43	3.42
11	4.56	4.54
11.5	5	5
12	5	5
13	5	5
15	5	5

In order to investigate why the optimized rotor with a smaller chord produces almost the same power as the original rotor, the axial and tangential induction factors of the two rotors at wind speed of 9 m/s are shown in Fig. 8.24 and 8.25 respectively. From Fig. 8.24, the axial-induced velocity interference factor of the new rotor is found to be about $\frac{1}{3}$ in the region from 30 to 50 m, whereas the original rotor has a linearly changing value from 0.3 to 0.4. As it is known that the maximum power coefficient (Betz limit) occurs at an axial induction factor of $\frac{1}{3}$, the optimized rotor has almost reached the maximum power coefficient and is expected to be more optimal. On the other hand, since the optimized rotor has an important chord reduction in this region, the optimized rotor is economically more efficient. The almost same tangential force for the optimized rotor is achieved with a smaller twist angle at a radial position between

20 and 40 m and with a smaller axial induction factor (a bigger angle of attack) at a radial position between 40 and 60 m. The tangential induction factor (Fig. 8.25) is almost the same for both rotors. The angle of attack is plotted in Fig. 8.26. From the figure, the angle of attack on most of the blade from 20 to 58 m is below 5° for both rotors. However, the angle of attack of the optimized rotor is higher than that of the original rotor. Thus, the optimized rotor produced a higher lift coefficient and it needs only a smaller chord to produce the same tangential force.

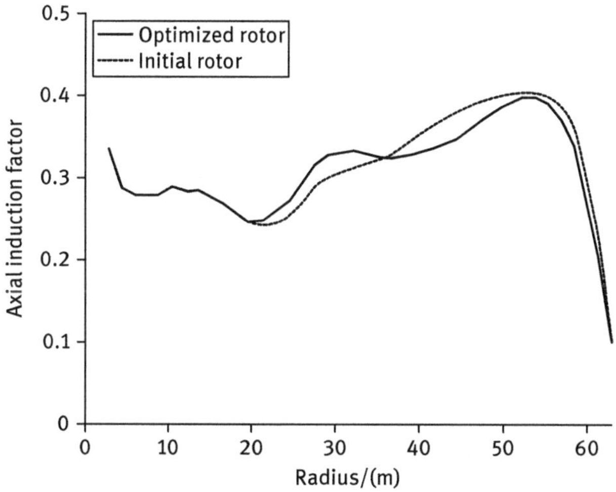

Fig. 8.24: Axial induction factors.

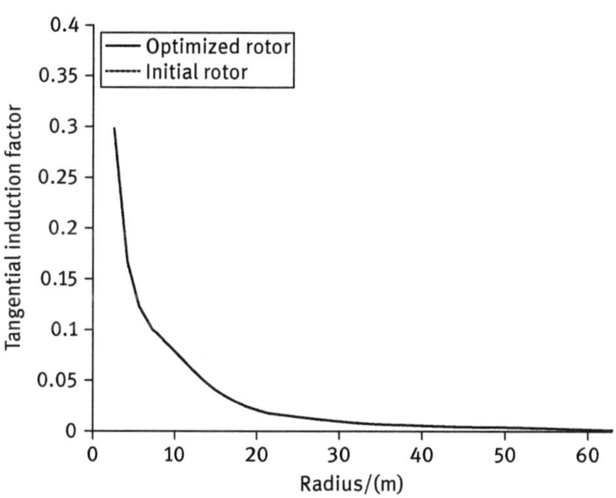

Fig. 8.25: Tangential induction factors.

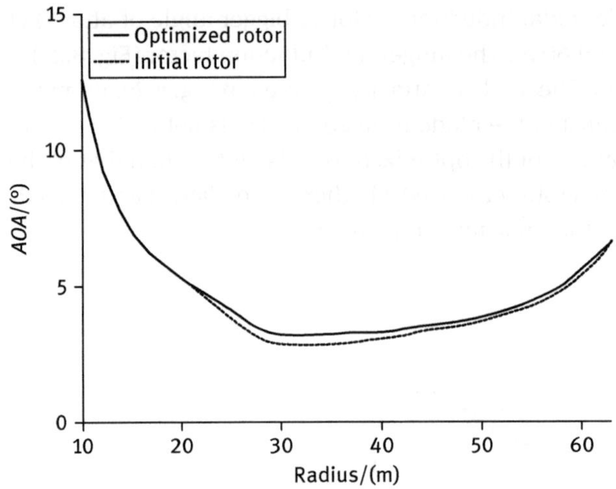

Fig. 8.26: AOA at inflow speed of 10 m/s.

The comparison of axial force and tangential force at a wind speed of 11.5 m/s between the original and the optimized rotor is shown in Fig. 8.27 and 8.28. The axial forces and the tangential forces of the optimized rotor are reduced especially in a region from 35 to 55 m radius. The flapwise and edgewise deflections of the blade on the two rotors (at a wind speed of 11.5 m/s) are plotted in Fig. 8.29 and 8.30 respectively. It can be see that the deflections in both directions of the optimized rotor are smaller on the outer part of the blade than those of the original rotor. The flapwise deflection at tip decreased from 3.6 m to 3.3 m with a reduction of 8.3 %. Similarly, the edgewise deflection at tip decreased from 0.22 m to 0.2 m with a reduction of 10 %.

Tab. 8.10: Total thrusts of the two rotors.

Inflow velocity (m/s)	**Thrust of original rotor (kN)**	**Thrust of optimized rotor (kN)**
5	214.669	213.315
7	355.819	350.076
9	505.221	496.675
10	593.156	582.410
11	680.006	666.377
11.5	713.128	697.407
12	596.238	584.729
13	517.300	509.187
15	451.105	446.771

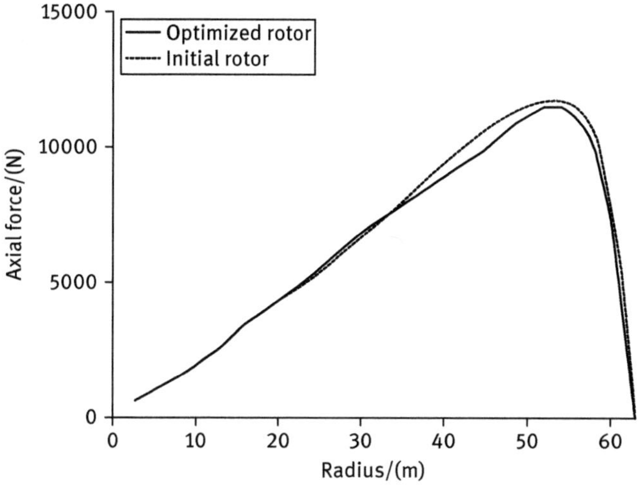

Fig. 8.27: Axial force.

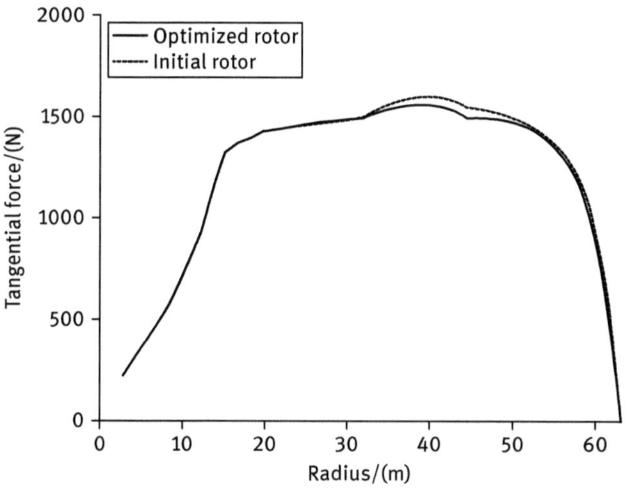

Fig. 8.28: Tangential force.

The total thrusts of the two rotors are listed in Tab. 8.10. From the table, it can be seen that the thrust has been reduced for the optimized rotor with the same pitch setting. At the rated wind speed of 11.5 m/s, the maximum total thrust of the rotor has been reduced by 2.3 %. Smaller deflections and thrust forces can increase the lifetime of the rotor and at the same time reduce the cost of the rotor.

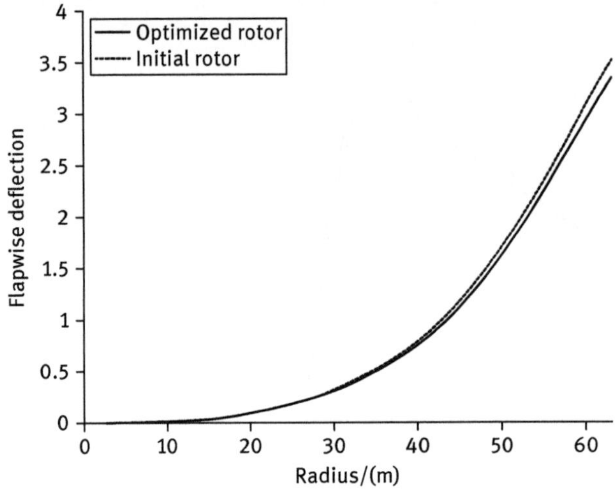

Fig. 8.29: Flapwise deflections.

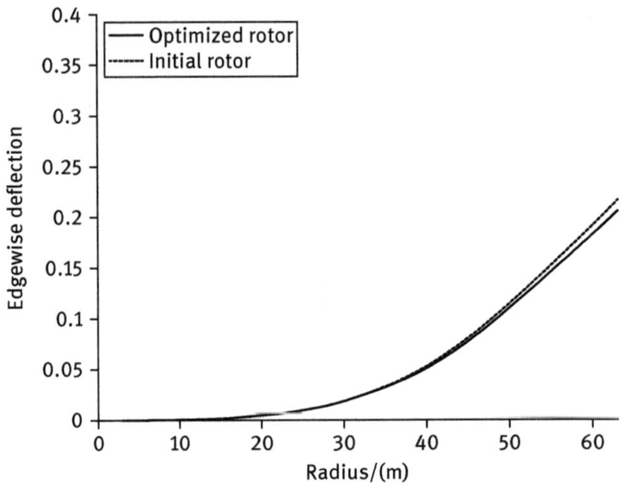

Fig. 8.30: Edgewise deflections.

Tab. 8.11: Basic parameters in blade design.

Rated power	Rated wind speed	Number of blades	Rotor rotation direction	Rotor radius	Designed tip speed ratio
2 MW	12.5 m/s	3	clockwise	31 m	6–8

8.4 Optimization of blades for 2 MW wind turbines

8.4.1 Design of new wind turbine blades

The airfoil series CQU-A, shown in Fig. 8.7 and 8.8, has been used for optimization. The CQU-A series, with varying maximum relative thickness between 15 % and 40 %, has high lift coefficient and lift-to-drag ratio. The basic parameters of the designed initial wind turbine blades are shown in Tab. 8.11. The spanwise distribution of airfoils refers to a 2 MW wind turbine. Reynolds number and Mach number are determined according to the position of the airfoils under rated operating conditions. The distribution of airfoil and boundary layer condition along the span are shown in Tab. 8.12. The transitional airfoil between two neighboring standard airfoils can be obtained through interpolation. As the rotor speed has great influence on Reynolds number, as has been shown in Fig. 8.1, Tab. 8.12 demonstrates several Reynolds numbers at some key blade sections when the rotor speed reaches 20 rpm. The change in Reynolds number will directly influence the change in lift-to-drag ratio of the airfoil. Fig. 8.3 shows that the Reynolds number has the greatest influence on the maximum lift-to-drag ratio when it is between 1.0×10^6 and 3.0×10^6 under the smooth condition, which is basically consistent with the variation in maximum lift-to-drag ratio with Reynolds number in the literature [6]. Since Mach number has little influence on the aerodynamic characteristics, we take the Mach number of the whole blade as 0.15.

Tab. 8.12: Distribution of airfoil and boundary layer condition along blade span.

$\mu = r/R$	Relative thickness	Airfoil name	Boundary layer conditions
0.04–0.08	100 %	Circle	$Re = 1.0 \times 10^6$, $Ma = 0.15$
0.10–0.20	100–40 %	Transitional airfoil 1	$Re = 1.0 \times 10^6$, $Ma = 0.15$
0.20–0.25	40 %	CQU-A400	$Re = 1.5 \times 10^6$, $Ma = 0.15$
0.25–0.30	40–35 %	Transitional airfoil 2	$Re = 1.5 \times 10^6$, $Ma = 0.15$
0.30–0.35	35 %	CQU-A350	$Re = 2.5 \times 10^6$, $Ma = 0.15$
0.35–0.40	35–30 %	Transitional airfoil 3	$Re = 2.5 \times 10^6$, $Ma = 0.15$
0.40–0.45	30 %	CQU-A300	$Re = 2.5 \times 10^6$, $Ma = 0.15$
0.45–0.50	30–25 %	Transitional airfoil 4	$Re = 2.5 \times 10^6$, $Ma = 0.15$
0.50–0.60	25 %	CQU-A250	$Re = 3.0 \times 10^6$, $Ma = 0.15$
0.60–0.65	25–21 %	Transitional airfoil 5	$Re = 3.0 \times 10^6$, $Ma = 0.15$
0.65–0.70	21 %	CQU-A210	$Re = 3.0 \times 10^6$, $Ma = 0.15$
0.70–0.75	21–18 %	Transitional airfoil 6	$Re = 3.0 \times 10^6$, $Ma = 0.15$
0.75–0.80	18 %	CQU-A180	$Re = 2.5 \times 10^6$, $Ma = 0.15$
0.80–0.85	18–15 %	Transitional airfoil 7	$Re = 2.5 \times 10^6$, $Ma = 0.15$
0.85–1.00	15 %	CQU-A150	$Re = 2.5 \times 10^6$, $Ma = 0.15$

8.4.2 Establishing the multiple-objects optimization model

8.4.2.1 Objective function

As for pitch-regulated wind turbines, the rotor speed can be continuously adjusted so as to make the wind turbine run at the optimization point $C_{p,max}$. Therefore, in this section, the maximum power coefficient of the wind turbine is chosen as the objective of the optimization:

$$F_1(X) = \max(C_P). \tag{8.12}$$

The definition of the power coefficient of the wind turbine can be written as

$$C_p = \frac{P}{\rho V^3 A/2}. \tag{8.13}$$

Another equation for the power coefficient is obtained through a series of transformations based on BEM theory

$$C_p = \frac{8\lambda^2}{R^4} \int_0^R a'(1-a)r^3 \, dr. \tag{8.14}$$

Wind turbine blades are made of composite materials whose blade density changes along with the sophisticated curve surface. So it is hard to establish the mass calculation mathematical model with the use of traditional composite laminate theory. Therefore, when other conditions are consistent (blade layer thickness and density, etc.), the strategy of transferring the mass calculation to area calculation of curved surface was adopted. Fig. 8.31 shows the schematic diagram for the blade area calculation model. The minimization of blade curve area is an optimization objective:

$$F_2(X) = \min(A_{\text{blade}}), \tag{8.15}$$

$$A_{\text{blade}} = B \int_0^R f(x,y) \, dr, \tag{8.16}$$

where $f(x, y)$ is the circumference of the airfoil section located at r spanwise location.

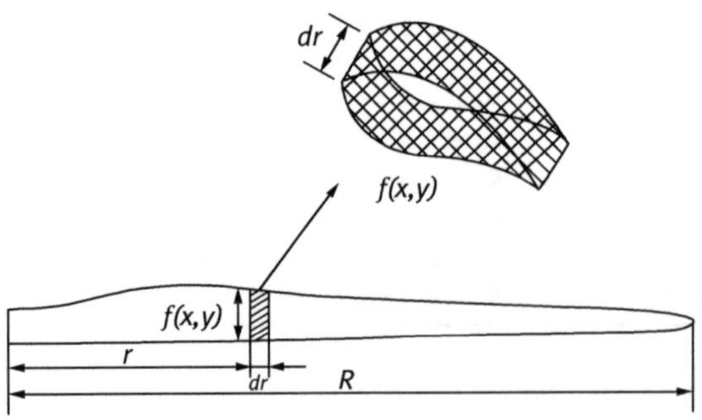

Fig. 8.31: Schematic diagram of blade area calculation model.

This function is expressed as follows

$$f(x, y) = \sum_{i=1}^{m} \sqrt{(x_{i+1} - x_i)^2 + (y_{i+1} - y_i)^2}, \tag{8.17}$$

where x_i, y_i are the coordinates of the airfoil section on the twisted blade. If there were enough points, $f(x, y)$ would be close to the curve length. The coordinate relationship between the airfoil section on the twisted blade and the original airfoil is

$$\begin{cases} x_i = c(x_{0,i} \cos\beta + y_{0,i} \sin\beta), \\ y_i = c(x_{0,i} \sin\beta + y_{0,i} \cos\beta), \end{cases} \tag{8.18}$$

where $x_{0,i}$ and $y_{0,i}$ are coordinates of the airfoil with unit chord length before putting a twist on the blade; c is the blade's chord at spanwise location r; and β is the twist at spanwise location r. Combining equations (8.16)–(8.18), the area of the whole blade can be calculated.

From the two target functions above, this optimization process is substantially a nonlinear bi-object optimization mathematical model. For convenience of solution, the multiple-objects optimization is converted into a single-object optimization function by setting weight coefficient

$$F(X) = \max\{\mu_1 F_1(X) + (1 - \mu_1)[-F_2(X)/\text{ratio}]\}, \tag{8.19}$$

where μ_1 is the weight coefficient ($\mu_1 \in [0, 1]$). In order to set the two target functions in the same order of magnitude, we give a scale factor "ratio". The maximum power coefficient is generally close to 0.5, so $F_2(X)/\text{ratio}$ should also be in the same order of magnitude.

8.4.2.2 Design variables and constraints

The chord, twist and relative thickness directly determine the shape of the blades. The radius of a rotor is set to be 31 m. The thickness distribution of the blade can be obtained through interpolation from Tab. 8.12. Therefore, the chord and twist of the blade were chosen as optimization variables. The distributions of chord and twist of the EU 2 MW Tjæreborg blade were chosen to design the initial blade for this optimization

In order to ensure good aerodynamic characteristics and continual surface smoothness of the blade, eight control points were chosen at the exact spanwise locations of standard airfoils to control the chord distribution. The same strategy is used for twist distribution. The constraint boundaries of the design variables are shown in Tab. 8.13.

$$X_{i,\min} \le X_i \le X_{i\,\max}, \quad i = 1, 2. \tag{8.20}$$

Furthermore, the loads on the blade, which can obtained through BEM theory, should also be constrained in optimization design. Based on the aerodynamics of the wind

Tab. 8.13: Constraint boundaries of the design variables.

	Chord *c* (m)	Twist *θ* (°)	Relative thickness *δ* (%)
Maximum	3.2	12.0	40
Minimum	0.1	−0.1	15

turbine, the normal force coefficient C_n and the tangential force coefficient C_t of a blade section can be expressed as follows:

$$\begin{cases} C_n = L\cos\varphi + D\sin\varphi, \\ C_t = L\sin\varphi - D\cos\varphi, \end{cases} \tag{8.21}$$

where L is the lift coefficient of the airfoil, D is the drag coefficient of the airfoil.

When the rotor is working, the bending moment at the blade root is huge, which will directly influence strength and the fatigue life of the blade. And the bending moments at the blade root are mainly flapwise bending moments and edgewise bending moments. The former are caused by thrust and the latter are caused by the in-plane tangential force distribution. The flapwise bending moment is mainly influenced by aerodynamic loads [183, 184], which should be controlled below the limit value

$$M_{\text{flap}} = \frac{1}{2}\rho B \int_0^R \frac{V_0^2(1-a)^2}{\sin^2\varphi} c C_n r \, \mathrm{d}r \le M_{\text{flap,max}}. \tag{8.22}$$

Furthermore, the torque of the rotor is formed due to the tangential force on the blade. However, over-sized torque is harmful to the mechanical transmission system. Therefore, the torque of a wind turbine should be controlled and restricted:

$$M_{\text{T}} = \frac{1}{2}\rho B \int_0^R \frac{V_0(1-a)wr(1+a')}{\sin\varphi\cos\varphi} c C_t r \, \mathrm{d}r \le M_{\text{T,max}}. \tag{8.23}$$

8.4.3 Optimization result

Based on the aerodynamic model for wind turbine rotors in Chapter 6, the thrust and torque on blades and the power coefficient of the rotor can be calculated. Improved multiple-objects particle swarm algorithm was used in the optimization. For some basic parameters in the particle swarm algorithm, we take 0.9 as the inertia weight w, 0.5 and 0.5 as the learning factors C1 and C2, 16 as the variable dimensions, 50 as the size of population and 200 as the maximum number of iteration, 0.25 as the weight coefficient μ_1 and 500 as the scale factor “ratio”. The flow chart of optimization is shown in Fig. 8.32. Fitness value is calculated based on BEM theory and surface area. The particle swarm parameters are updated according to fitness value. The optimized value can be

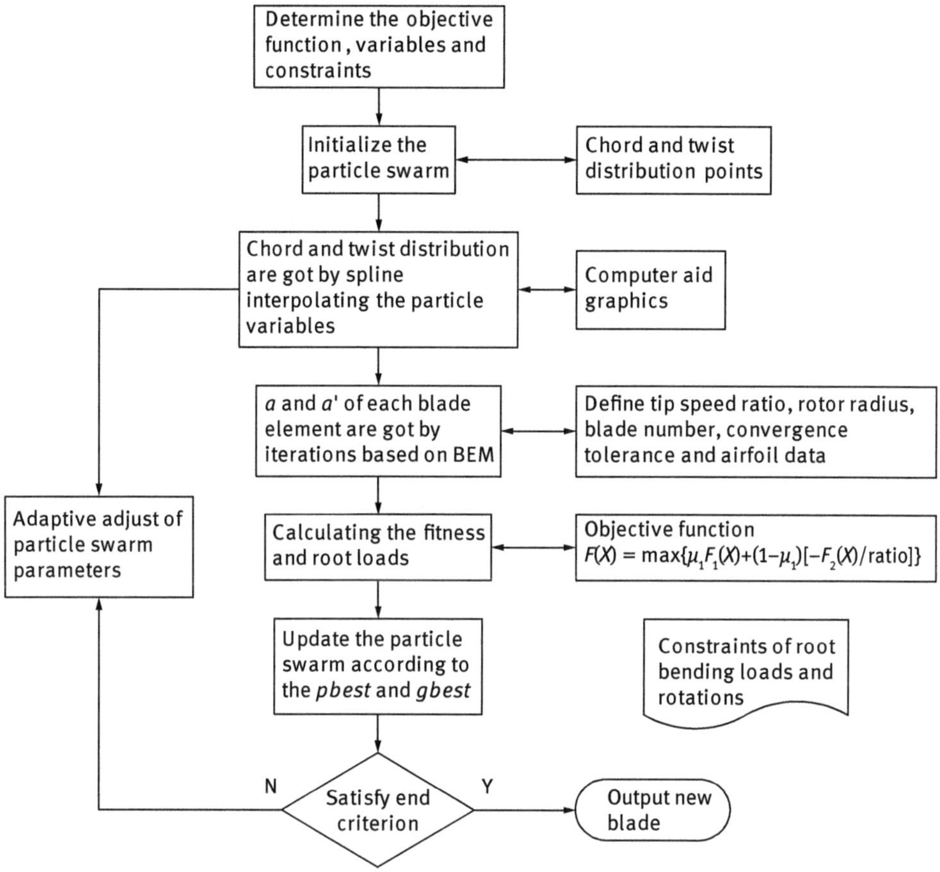

Fig. 8.32: Flow chart of blade optimization.

output if the fitness value meets the iteration termination condition. Otherwise, the iteration will continue until the criterion is satisfied. The aerodynamic characteristics of airfoils are calculated with RFOIL software and then they are extended to angles of attack within ±180° through extrapolation.

Fig. 8.33 and 8.34 show the comparison of the distributions of chord and twist of the blade before and after optimization. From Fig. 8.33 it can be seen that the spanwise distribution of chord of the newly-designed blade is nonlinear. The chord of the optimized blade is smaller than that of the initial blade, which is especially evident in the middle part of the blade. This is because the minimization of the blade area is taken as the target function. Therefore the algorithm will tend to reduce the area and mass of the blade. From Fig. 8.34 it can be seen that the twist of the newly-designed blade is also distributed in a nonlinear way. The twist in the blade root is evidently larger than that of the initial blade while the difference at the blade tip is not as evident as that at the root. Moreover, increasing twist of the blade will decrease the load at the

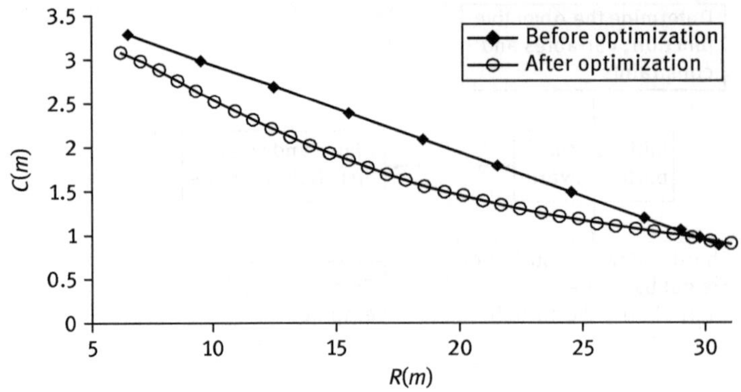

Fig. 8.33: Chord distribution.

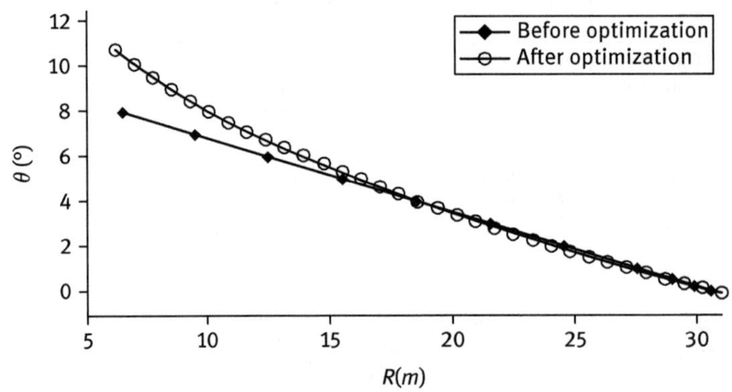

Fig. 8.34: Twist distribution.

blade root, which is beneficial to blade life and the cost of blade will be reduced. The newly-designed blade after optimization is shown in Fig. 8.35. It can be seen that the blade chord is distributed in a smooth and continuous way.

The variation in the power coefficients with respect to tip speed ratio for the Tjæreborg blade, original blade and optimized blade are shown in Fig. 8.36. The power coefficient of the initial blade under design tip speed ratio (2–7.5) is evidently larger than that of the EU Tjæreborg experimental blade, which is caused by the different aerodynamic performance of different airfoil series. The power coefficient of the newly-designed blade under the design tip speed ratio (6–12) is evidently larger than that of the initial blade, which is mainly caused by different shape parameters of the blade. The variations in the distribution of annual energy production and power coefficients of the three rotors with respect to wind speed are shown in Fig. 8.37 and 8.38. The improved distribution of annual energy production and power of the newly-designed rotor is obvious at the rated wind speed ($V \leq 12.5$ m/s). When the wind speed arrives

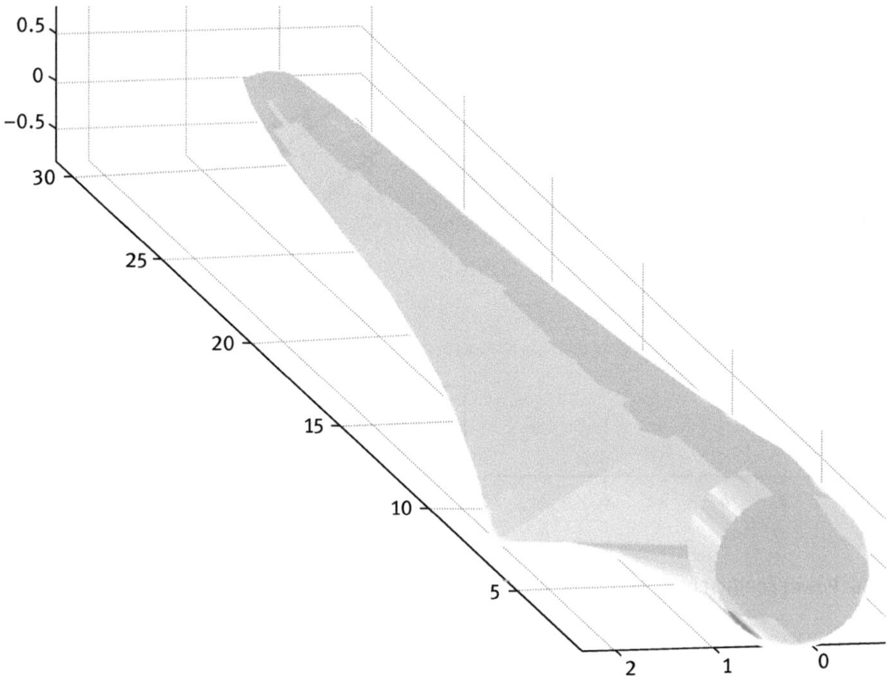

Fig. 8.35: The shape of the optimized blade.

at the rated value, the distribution of annual energy production and power will remain unchanged by keeping power at 2 MW by using the pitch-regulated system.

The comparison of the performance of the Tjæreborg rotor, the initial rotor and the optimized rotor is shown in Tab. 8.14. In terms of maximum power coefficient and AEP, the initial blade is better than that of the Tjæreborg rotor. However, in terms of the root load and rotor torque, the initial blade is worse than that of the Tjæreborg rotor. The maximum power coefficient of the optimized rotor is 0.50830 (increased by 4.019 % compared with the initial rotor), and AEP is 4.406×10^6 kWh (increased by 1.967 % compared with the initial rotor), and the newly-designed rotor area (from $0.2R$ to $1.0R$) is 278.906 m^2 (decreased by 19.005 % compared with the initial rotor). The mass of the rotor is greatly decreased, blade fatigue life span is increased and the cost is accordingly decreased. The new high-performance wind turbine airfoils adopted in the initial blade lead to an increase in blade root loads. After optimization, both the bending moment and torque at the blade root decrease. The reduction in the bending moment is more obvious, which indicates that this design can control the blade root load effectively.

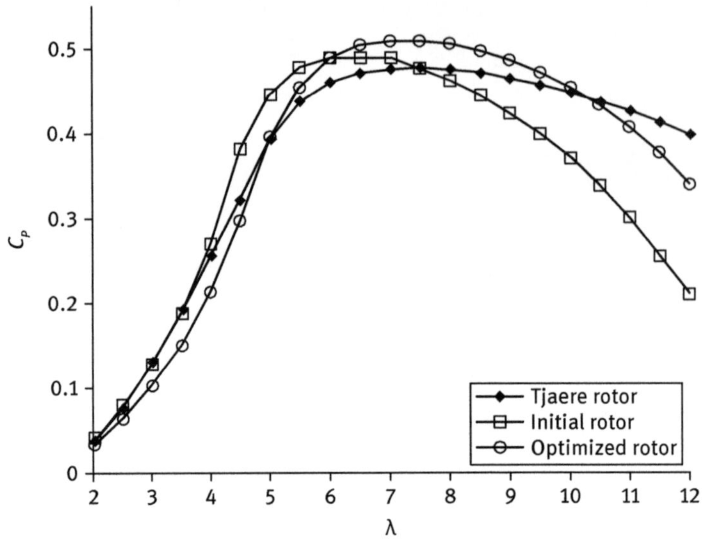

Fig. 8.36: Power coefficient.

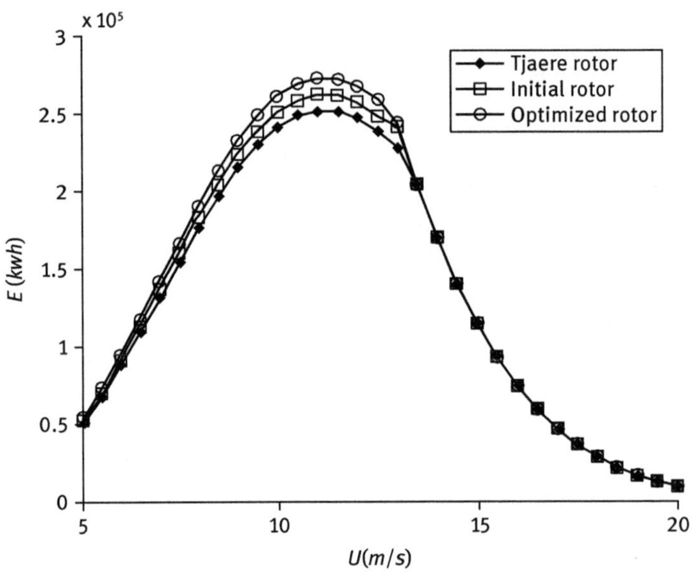

Fig. 8.37: Distribution of annual energy production.

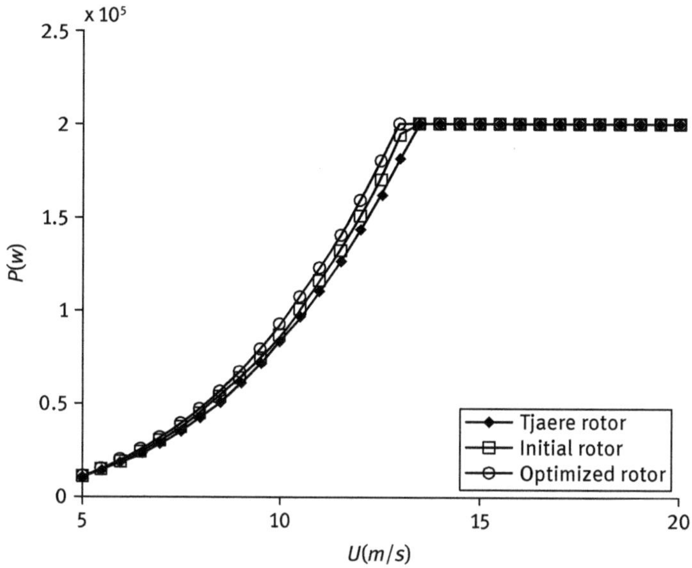

Fig. 8.38: Power characteristics.

Tab. 8.14: Comparison of the performance of the three rotors.

Rotor name	λ	$C_{p,max}$	AEP (kWh)	Sweep area of rotor (m^2)	Root bending moment (kNm)	Rotor torque (kNm)
Tjæreborg	7.0	0.47658	4.224×10^6	357.847	4755	560
Initial	7.0	0.48866	4.321×10^6	344.348	5311	584
Optimized	7.0	0.50830	4.406×10^6	278.906	4542	580

Note: $C_{p,max}$ is the maximum power coefficient, λ is the tip speed ratio.

8.5 Chapter conclusions

In the design of blades, key parameters like tip speed ratio, rotor diameter, Reynolds number, aerodynamic performance of airfoil and blade root load, etc. have a great influence on the performance of a wind turbine. The mutual influences between Reynolds number and airfoil aerodynamic performance were thoroughly analyzed for pitch-regulated wind turbines based on wind turbine aerodynamics. The influence of the key parameters (like rotor diameter, Reynolds number, lift-to-drag ratio and tip speed ratio) on wind turbine performance has been studied for three rotors with the same airfoil group and with different diameters. For blades of two rotors with the same diameter, the influence of different airfoils on rotor performance has also been studied. The results indicates that airfoils with higher performance will definitely raise the power coefficient and AEP of the rotor, as well as the root load.

The optimization mathematical model, taking chord, twist, and relative thickness as design variables and COE as the objectives, was established. Optimizations of blades for 2 MW and 5 MW rotors have been carried out and the comparative analyses of the shape parameters, structural characteristics and aerodynamic characteristics have been made for initial and optimized rotors. The COE of 2 MW and 5 MW rotors have been successfully reduced by 3.45 % and 2.6 % respectively. Meanwhile, the blade load and vibration deformation has been decreased to extend the service life of the blades.

A multiple-objects optimization model was proposed to solve the problem that the loads at the root will increase with the aerodynamic force. Based on modified wind turbine aerodynamics theory, a new type of wind turbine blade design and optimization model was established. The maximization of the power coefficient and the minimization of the blade area were taken as the objectives; chord and twist were taken as design variables. The airfoil series CQU-A has been adapted to the blade during optimization. The improved multiple-objects particle swarm algorithm was utilized. A comparative analysis has been made between optimized blade, original blade, and Tjæreborg blade. The optimized result shows that the newly-designed 2 MW blade has been improved, compared with original blade and the Tjæreborg blade, in terms of maximum power coefficient and AEP. The service life of the blade has been extended and the mass and cost of the blade have been decreased. Meanwhile, the root load is reduced. This research provides references for designing blades of wind turbines with high performance and low cost.

9 Structural optimization of composite wind turbine blades

9.1 Introduction

The blade is one of the most important components of a wind turbine. As rotor size has increased, blades are now increasingly made of composite materials. Composite materials have properties such as lower weight and higher stiffness, which can help the blade to endure larger loads. The structural characteristics of composite blades are the focus of many scholars worldwide [41–46, 49, 185–188].

Firstly, the basics of the mechanics of composite materials was introduced in this chapter. The geometric modeling and initial layup have been designed for the newly-designed 2 MW blade introduced in Chapter 8. A faithful parametric finite element model of composite wind turbine blades has been established. Based on the modified Blade Element Momentum theory in Chapter 6, a new one-way fluid-structure interaction method is introduced. Then the pressure coefficients for each airfoil can be acquired by RFOIL. Then, the pressure distribution for each blade section can be expressed by a high order polynomial which can be applied to the finite element model. Finally, a procedure combining finite element analysis and particle swarm algorithm to optimize composite structures of the wind turbine blade is developed. The thicknesses of unidirectional laminate, biaxial laminate, triaxial laminate, etc. are taken as design variables. Two optimization schemes are proposed and their optimization results analyzed.

9.2 Basics of the mechanics of composite materials

Composite materials refer to multidirectional materials (macroscale) which are composed of two or more kinds of materials with different properties [41–46, 49, 185–188]. Furthermore, composite materials are mainly composed of two parts: one part is the reinforcement (fiber, particle or sheet, etc.), which bear the various loads of the structure; the other part is the matrix (organic polymer, metal, porcelain and carbon, etc.) which can bond the reinforcements, pass along the stress and reinforce the toughness of the composite material. Compared with metal materials and the products made from metal materials, composite materials and the products made from composite materials can greatly reduce the material consumption and assembly workload. They can also greatly simplify the production cycle, enhance the performance of products and extend their service life, etc. Due to the advantages listed above, composite materials have already been widely utilized in the structural design of wind turbine blades.

DOI 10.1515/9783110344387-010

9.2.1 Classification of fiber reinforcement composite materials

Composite blades are normally made of glass fiber reinforced epoxy resin or carbon fiber reinforced epoxy resin which belong to the fiber reinforced composite materials. The classification and characteristics of fiber reinforced composite materials will be introduced in this section.

Fiber materials can mainly be classified as glass fiber, boron fiber, carbon fiber, aramid fiber, etc. Their fundamental characteristics are shown in Tab. 9.1.

Tab. 9.1: Fundamental characteristics of various fiber materials.

Materials	Diameter (μm)	Relative density	Tensile strength (MPa)	Elastic modulus (GPa)	Specific strength (MPa)	Specific modulus (GPa)
E glass fiber	10	2.55	3500	74	1370	29
Boron fiber	10	2.65	3500	410	1320	155
Carbon fiber	100	1.75	3500–7000	350–580	2000–4000	129–130
Aramid fiber	6	1.47	2830	134	1930	91

Matrix materials are mainly classified into thermosetting resin and thermoplastic resin. The common thermosetting resins are epoxy resin, phenolic resin and polyester resin. The main characteristic of epoxy resin is the strong adhesiveness, good wettability into fiber reinforcement, small shrinkage rate of cured resin and higher heat resistance. The commonly used thermoplastic resins are polyethylene resin, polystyrene resin, polypropylene resin, etc. The characteristics of common matrix resins are shown in Tab. 9.2.

Tab. 9.2: Characteristics of common matrix resins.

Name	Relative density	Tensile strength	Modulus (GPa)	Tensile strength (MPa)	Bending strength (MPa)
Epoxy	1.1–1.3	60–95	3–4	90–110	100
Phenolic	1.3	42–64	3.2	88–110	78–120
Polyester	1.1–1.4	42–71	2.1–4.5	92–190	60–120
Polyethylene	1.1	23	8.4	20–25	25–29
Polystyrene	0.9	35–40	1.4	56	42–56
Polypropylene	1.2	59	2.8	98	77

9.2.2 Characteristics of composite materials

Compared with traditional materials, composite materials are composed of reinforcements and matrix. Thus they have unique structural characteristics [190].

(1) Heterogeneity, anisotropy and randomness of mechanics performance. Due to the anisotropy of composite materials, the mechanics performance along the fiber direction and perpendicular to the fiber direction are mainly determined by the performance of the matrix material and the bonding capacity between matrix and fibers.
(2) High specific strength and specific modulus. The strength and modulus per unit mass are called specific strength and specific modulus, which are performance indexes to measure the load bearing capacity and rigidity of a material on the premise of equal mass. The specific strength and specific modulus are important indexes for aerospace, aviation and wind turbine applications. Structures made of composite materials can have better performance and lighter mass.
(3) Designability of material. Designers cannot casually alter the performance indexes of metal materials. However, the integral performance of composite materials is not only determined by the characteristics of fibers and matrix but also determined by the fiber content, layup pattern, layup order, etc. This brings great superiorities to the design.
(4) Anti-fatigue performance. Although cracks in composite materials expand and reach the critical dimension faster than in metal materials, there is a process before a fracture appears in composite materials. Therefore, sudden destruction of composite materials because of fatigue will not happen, which makes the fatigue resistance of composite materials better than that of metal materials.
(5) Simple manufacturing process and low cost. Compared with components made of traditional metal materials, less complicated machining equipment and processes are needed in the fabrication of composite components. Additionally, composite materials can be made into thin structures with less material consumed and fewer working hours.
(6) Good thermal stability. Carbon fiber and aramid fiber both have a negative thermal expansion coefficient. Therefore, if combined with matrix materials with a positive thermal expansion coefficient, composite materials with small thermal expansion coefficient can be made. When environmental conditions change, the thermal stress and thermal strain are small.

Furthermore, composite materials also have other excellent characteristics like shock resistance, conductivity, heat resistance, corrosion resistance, etc.

9.2.3 Basic structures of composite materials and analysis methods

The elementary configurations of composite materials are mainly divided into plies and laminates.

9.2.3.1 Plies

The commonly used plies are unidirectional plies and fabric plies. The fibers in unidirectional plies are all aligned in one direction and the fibers in fabric plies are aligned in two directions perpendicular to each other.

The direction of the fiber is usually called the longitudinal direction, which is represented by "1"; the direction perpendicular to the direction of the fiber is called the transverse direction, which is represented by "2"; the direction along the thickness of plies is marked with "3". The axes numbered by 1, 2 and 3 are called the main axes. Plies are inhomogeneous and anisotropic due to the direction of fiber arrangement, though fibers and matrix may be isotropic.

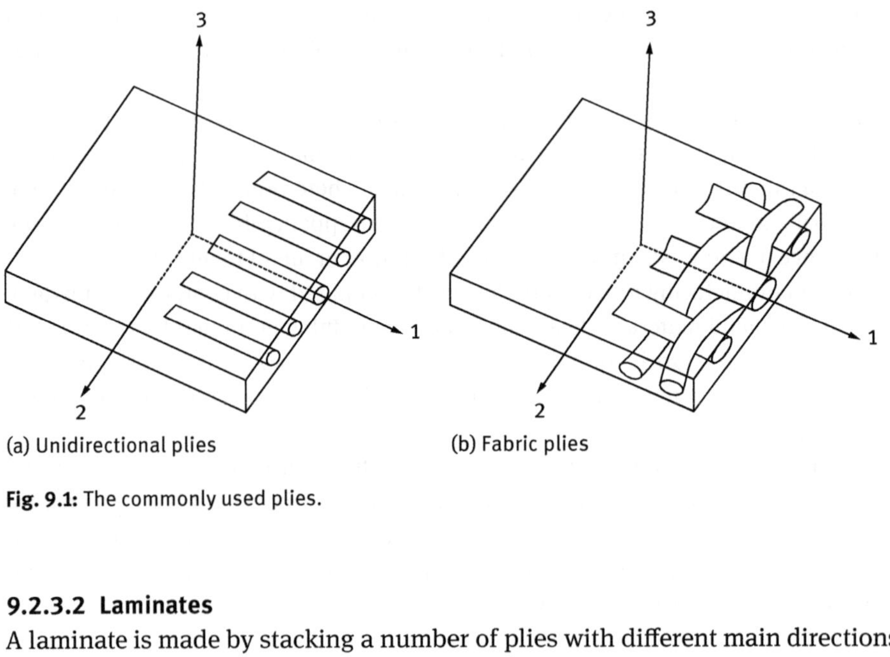

(a) Unidirectional plies (b) Fabric plies

Fig. 9.1: The commonly used plies.

9.2.3.2 Laminates

A laminate is made by stacking a number of plies with different main directions, as shown in Fig. 9.2. Usually, plies of a laminate are bonded together using matrix material which is the same as the matrix material used in each ply. The major purpose of stacking plies is to design a material of the required strength and stiffness (which are related to the main material directions of each ply) to meet the load bearing requirements of the structure. The main material direction of each ply can be determined respectively according to the requirements. The heterogeneity and anisotropy characteristics of

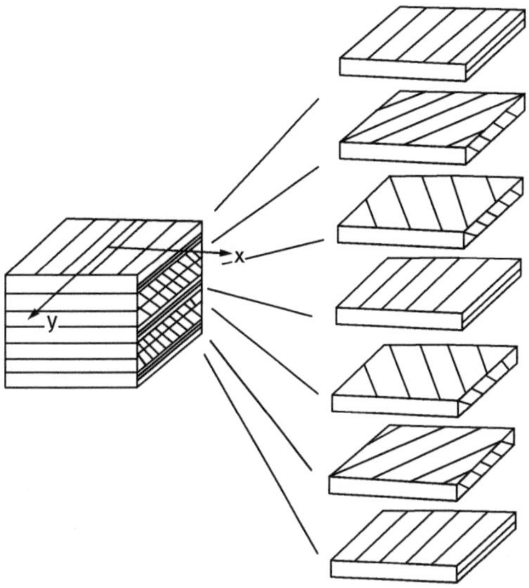

Fig. 9.2: Different plies in a laminate.

laminates are far more complex than those of each ply. So the mechanics analysis and computation of laminates are far more complex than those of each ply.

The methods for analyzing the mechanics of composite materials mainly include micromechanics, macromechanics and structural mechanics, which will be illustrated in the following.

(1) Micromechanics of composite materials: Micromechanics studies the mechanics performance of composite materials through analyzing the interactions between constituents. The fibers and matrix are considered as the basic components in micromechanics of composite materials and are regarded as isotropic and homogeneous materials. The micromechanical performance of composite materials is mainly analyzed based on the geometric shape and layout of fibers, the mechanics performance of fibers and matrix as well as the interactions between fibers and matrix. Although this method is precise, its mathematical derivations are very complex and it is only appropriate to analyze the basic mechanics performance of a ply under a simple stress state.
(2) Macromechanics of composite materials: Composite macromechanics study the average performance of composites in a macro manner with the assumption that the material is uniform. In macromechanics, each ply is regarded as an anisotropic and homogeneous material without considering the inner interactions between fibers and matrix. The stiffness and strength characteristics of a ply are expressed by its average mechanics performance. The mechanics performance of a laminate can be easily analyzed using this method, and the results are of relative high accuracy.

(3) Structural mechanics of composite materials: Through analyzing the average mechanics performance (in a macro manner) of a laminate with the analysis methods for structure mechanics of isotropic and homogeneous materials, a mechanics analysis of the composite structure's components (i.e., laminates, shell, etc.) of various shapes can be made. The structural mechanics of composite materials can be conveniently used to solve the problems of bending, buckling, vibration, fatigue, crack, damage or strength of opening, etc. of laminates and shell structures.

9.2.4 Anisotropy mechanics theory of composite materials

Different to elastic materials in elasticity theory, composite materials are heterogeneous and anisotropic. Composite materials can be regarded as homogeneous when analyzing them from the macroperspective. As the differences between the mechanics performance of fiber and matrix are large, fiber reinforced composite materials should be regarded as anisotropic even from the macroperspective. Therefore, fiber reinforced composite materials can be viewed as homogeneous and anisotropic bodies. So, the equilibrium equations and geometric equations from elastic mechanics can still work in the mechanics of composite materials, but the constitutive equations should be established anew.

(1) Differential equations of motion (or equilibrium):

$$\left.\begin{aligned}
\frac{\partial \sigma_x}{\partial x}+\frac{\partial \tau_{xy}}{\partial y}+\frac{\partial \tau_{xz}}{\partial z}+f_x=0\left(\rho\frac{\partial^2 u}{\partial t^2}\right)\\
\frac{\partial \tau_{yx}}{\partial x}+\frac{\partial \sigma_y}{\partial y}+\frac{\partial \tau_{yz}}{\partial z}+f_y=0\left(\rho\frac{\partial^2 v}{\partial t^2}\right)\\
\frac{\partial \tau_{zx}}{\partial x}+\frac{\partial \sigma_{zy}}{\partial y}+\frac{\partial \sigma_z}{\partial z}+f_z=0\left(\rho\frac{\partial^2 w}{\partial t^2}\right)
\end{aligned}\right\};\tag{9.1}$$

(2) Geometric equations show the relationship between strain and displacement on minor deformation and rotation. Based on elastic mechanics, the equations are shown as follows:

$$\left.\begin{aligned}
\varepsilon_x&=\frac{\partial u}{\partial x}, & \gamma_{xy}&=\frac{\partial v}{\partial x}+\frac{\partial u}{\partial y}\\
\varepsilon_y&=\frac{\partial u}{\partial y}, & \gamma_{yz}&=\frac{\partial w}{\partial y}+\frac{\partial v}{\partial z}\\
\varepsilon_z&=\frac{\partial w}{\partial z}, & \gamma_{zx}&=\frac{\partial w}{\partial x}+\frac{\partial u}{\partial z}
\end{aligned}\right\};\tag{9.2}$$

(3) Constitutive relations: For linear elastic and anisotropic materials, the linear relationships between stress and strain can be obtained through the generalized stress-strain relations (Hooke's law)

$$\begin{Bmatrix} \sigma_x \\ \sigma_y \\ \sigma_z \\ \tau_{yz} \\ \tau_{zx} \\ \tau_{xy} \end{Bmatrix} = \begin{bmatrix} C_{11} & C_{12} & C_{13} & C_{14} & C_{15} & C_{16} \\ C_{21} & C_{22} & C_{23} & C_{24} & C_{25} & C_{26} \\ C_{31} & C_{32} & C_{33} & C_{34} & C_{35} & C_{36} \\ C_{41} & C_{42} & C_{43} & C_{44} & C_{45} & C_{46} \\ C_{51} & C_{52} & C_{53} & C_{54} & C_{55} & C_{56} \\ C_{61} & C_{62} & C_{63} & C_{64} & C_{65} & C_{66} \end{bmatrix} \begin{Bmatrix} \varepsilon_x \\ \varepsilon_y \\ \varepsilon_z \\ \gamma_{yz} \\ \gamma_{zx} \\ \gamma_{xy} \end{Bmatrix}, \tag{9.3}$$

where

$$\begin{Bmatrix} \sigma_x & \sigma_y & \sigma_z & \tau_{yz} & \tau_{zx} & \tau_{xy} \end{Bmatrix}^{\mathrm{T}}$$

is the stress vector;

$$\begin{Bmatrix} \varepsilon_x & \varepsilon_y & \varepsilon_z & \gamma_{yz} & \gamma_{zx} & \gamma_{xy} \end{Bmatrix}^{\mathrm{T}}$$

is the strain vector;

$$\begin{bmatrix} C_{11} & C_{12} & C_{13} & C_{14} & C_{15} & C_{16} \\ C_{21} & C_{22} & C_{23} & C_{24} & C_{25} & C_{26} \\ C_{31} & C_{32} & C_{33} & C_{34} & C_{35} & C_{36} \\ C_{41} & C_{42} & C_{43} & C_{44} & C_{45} & C_{46} \\ C_{51} & C_{52} & C_{53} & C_{54} & C_{55} & C_{56} \\ C_{61} & C_{62} & C_{63} & C_{64} & C_{65} & C_{66} \end{bmatrix}$$

is the stiffness matrix, C_{ij} $(i, j = 1, 3, \ldots, 6)$ is the element in the stiffness matrix called the stiffness coefficient. As fiber reinforced composite materials have been viewed as a homogeneous material in a macroperspective, the stiffness coefficients are constant.

The numbers 1, 2 and 3 are used to represent axes x, y and z respectively. And the stress and strain component symbols are expressed briefly as follows:

$$\begin{Bmatrix} \sigma_x & \sigma_y & \sigma_z & \tau_{yz} & \tau_{zx} & \tau_{xy} \end{Bmatrix}^{\mathrm{T}} \to \begin{Bmatrix} \sigma_1 & \sigma_2 & \sigma_3 & \tau_4 & \tau_5 & \tau_6 \end{Bmatrix}^{\mathrm{T}},$$
$$\begin{Bmatrix} \varepsilon_x & \varepsilon_y & \varepsilon_z & \gamma_{yz} & \gamma_{zx} & \gamma_{xy} \end{Bmatrix}^{\mathrm{T}} \to \begin{Bmatrix} \varepsilon_1 & \varepsilon_2 & \varepsilon_3 & \gamma_4 & \gamma_5 & \gamma_6 \end{Bmatrix}^{\mathrm{T}}.$$

Among homogeneous and anisotropic materials, orthotropic materials are commonly studied in engineering. Materials with three symmetry planes for mechanics performance that are orthotropic to each other are called orthotropic materials. Due to the special performance of orthotropic materials, the number of stiffness coefficients for orthotropic materials is only 9, which is less than that of ordinary anisotropic materials. The stiffness matrix of orthotropic materials is a symmetric matrix. Thus the stress-strain relations for an orthotropic material can be written as

$$\begin{Bmatrix} \sigma_1 \\ \sigma_2 \\ \sigma_3 \\ \sigma_4 \\ \sigma_5 \\ \sigma_6 \end{Bmatrix} = \begin{bmatrix} C_{11} & C_{12} & C_{13} & 0 & 0 & 0 \\ C_{21} & C_{22} & C_{23} & 0 & 0 & 0 \\ C_{31} & C_{32} & C_{33} & 0 & 0 & 0 \\ 0 & 0 & 0 & C_{44} & 0 & 0 \\ 0 & 0 & 0 & 0 & C_{55} & 0 \\ 0 & 0 & 0 & 0 & 0 & C_{66} \end{bmatrix} \begin{Bmatrix} \varepsilon_1 \\ \varepsilon_2 \\ \varepsilon_3 \\ \varepsilon_4 \\ \varepsilon_5 \\ \varepsilon_6 \end{Bmatrix}. \tag{9.4}$$

The strain component can be expressed with the stress component, and then we obtain

$$\begin{Bmatrix} \varepsilon_1 \\ \varepsilon_2 \\ \varepsilon_3 \\ \varepsilon_4 \\ \varepsilon_5 \\ \varepsilon_6 \end{Bmatrix} = \begin{bmatrix} S_{11} & S_{12} & S_{13} & 0 & 0 & 0 \\ S_{21} & S_{22} & S_{23} & 0 & 0 & 0 \\ S_{31} & S_{32} & S_{33} & 0 & 0 & 0 \\ 0 & 0 & 0 & S_{44} & 0 & 0 \\ 0 & 0 & 0 & 0 & S_{55} & 0 \\ 0 & 0 & 0 & 0 & 0 & S_{66} \end{bmatrix} \begin{Bmatrix} \sigma_1 \\ \sigma_2 \\ \sigma_3 \\ \sigma_4 \\ \sigma_5 \\ \sigma_6 \end{Bmatrix}. \tag{9.5}$$

It can be seen from equation (9.5) that the normal stresses only cause linear strain rather than shear strain, and shear stresses only cause shear strain rather than linear strain. Furthermore, shear stresses in one plane do not cause shear strains in another, in other words, there is no coupling effect between shear stresses and strains of different planes.

The stiffness coefficients in equation (9.4) and the flexibility coefficients in equation (9.5) are the inherent characteristics of a material and can be obtained through experiments. The Young's modulus E and Poisson ratio ν of a material along one main direction can be obtained through uniaxial tension tests along that direction. The shear modulus G can be obtained through standard shear tests. And the flexibility matrix of an orthotropic material can be expressed by these engineering constants as follows:

$$[S] = \begin{bmatrix} \frac{1}{E_1} & -\frac{\nu_{21}}{E_2} & -\frac{\nu_{31}}{E_3} & 0 & 0 & 0 \\ -\frac{\nu_{12}}{E_1} & \frac{1}{E_2} & -\frac{\nu_{32}}{E_3} & 0 & 0 & 0 \\ -\frac{\nu_{13}}{E_1} & -\frac{\nu_{23}}{E_2} & -\frac{1}{E_3} & 0 & 0 & 0 \\ 0 & 0 & 0 & \frac{1}{G_{23}} & 0 & 0 \\ 0 & 0 & 0 & 0 & \frac{1}{G_{31}} & 0 \\ 0 & 0 & 0 & 0 & 0 & \frac{1}{G_{12}} \end{bmatrix}. \tag{9.6}$$

The stiffness matrix of an orthotropic material can be obtained through the inverse of $[S]$ in equation (9.6).

9.2.5 Strength criteria of unidirectional plies

Strength of a material refers to the material's capability to withstand destruction when bearing loads. Isotropic materials have the same strength on each direction, in other words, strength is not related to direction. Normally, a material's strength can be represented by ultimate stress. The strength of a composite material is related to direction. For orthotropic materials, there are three main directions, and the strengths on each main direction are different. For fiber reinforced unidirectional plies, the strength along the fiber direction is usually dozens of times that of the strength in the direction transverse to the fiber. Therefore, compared with isotropic materials, the strength of orthotropic materials has the following features.

(1) For isotropic materials, the maximum stress and linear strain in many strength theories are referred to as the principal stress and the principal strain, respectively. But for anisotropic materials, the maximum stress does not necessarily correspond to the dangerous state of the material. So the strength state of anisotropic materials cannot be judged without proper comparison with the actual stress field.
(2) There are three basic strength values for an orthotropic ply when its tensile strength and compressive strength are the same (see in Fig. 9.3).
X: strength along the direction of the fiber (along the main direction 1 of material);
Y: strength perpendicular to the direction of the fiber (along the main direction 2 of material);
S: shear strength within planes 1–2.

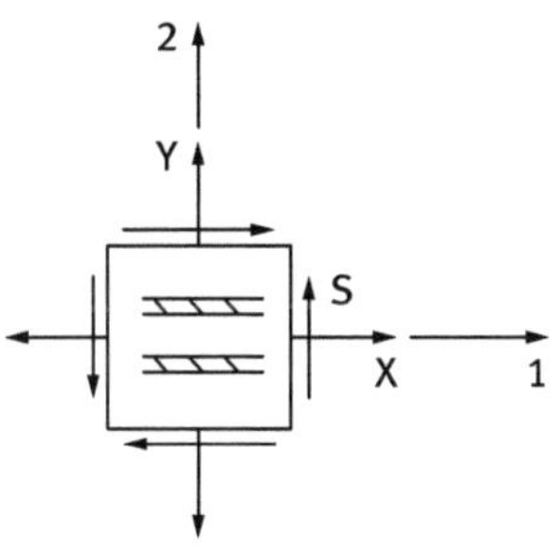

Fig. 9.3: Basic strength values of a fiber reinforced unidirectional ply.

Supposing the basic strengths within planes 1 and 2 of a unidirectional ply are as follows:

$$X = 1500\,\text{MPa}, \quad Y = 50\,\text{MPa}, \quad S = 70\,\text{MPa}.$$

The strength along direction 1 is far higher than that on direction 2. Assuming the stresses caused by loads are $\sigma_1 = 800\,\text{MPa}$, $\sigma_2 = 60\,\text{MPa}$, $\tau_{12} = 40\,\text{MPa}$; $\sigma_1 = 800\,\text{MPa}$, $\sigma_2 = 60\,\text{MPa}$ are the stresses along material direction 1 and 2 respectively. They are not the first principal stress and the second principal normal stresses. Although $\sigma_1 < X$ and $\tau_{12} < S$ shows no danger, $\sigma_2 > Y$ is satisfied which will lead to the destruction of the ply according to some strength theories. The strength of an anisotropic material is a function of the stress direction. However, the strength of an isotropic material is unrelated to the stress direction.
There are five basic strength values for an orthotropic ply when its tensile strength and compressive strength are different (for most composite materials), which is shown as follows
X_t: tensile strength along the fiber direction;
X_c: compressive strength along the fiber direction;
Y_t: tensile strength perpendicular to the fiber direction;
Y_c: compressive strength perpendicular to the fiber direction;
S: shear strength within planes 1–2.

(3) For orthotropic materials, although the tensile strength and the compressive strength in the main directions of the material are different, the shear strength in the main directions of the material remains unchanged regardless of whether the shear stress is positive or negative (we call it direction change here). As is shown in Fig. 9.4, positive shear stress and negative shear stress in the main direction of the material have no influence on the in-plane shear strength. However, if a directional change of shear is not along the main direction of the material, the shear strength will change. As shown in Fig. 9.5, the positive shear stress along the 45° direction (with respect to the main direction of the material) will cause tensile stress along the fiber direction and compressive stress perpendicular to the fiber direction. Negative shear stress will cause compressive stress along the fiber direction and tensile stress perpendicular to the fiber direction. In conclusion, the direction change of shear stress along the 45° direction will let the material reflect different strengths.

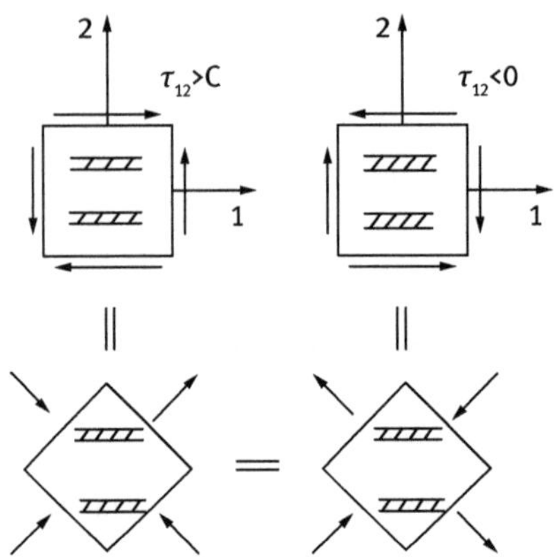

Fig. 9.4: Shear stresses in the main directions.

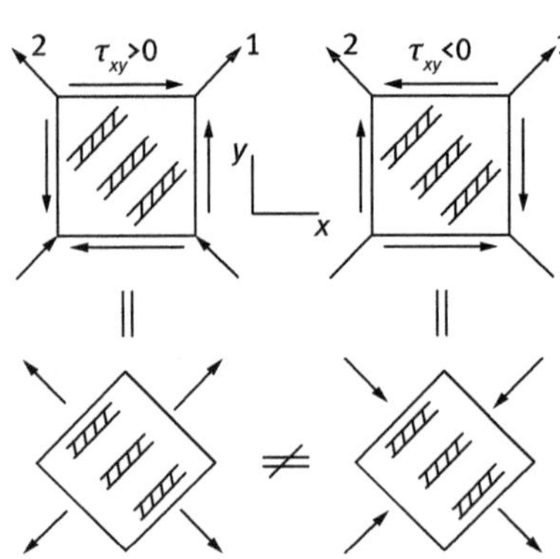

Fig. 9.5: Shear stresses along the 45° direction.

The destruction of laminated composite materials may have the following reasons: mistakes or errors in production; factors not considered (or ignored) in design; irrationality of structural design. Therefore, it is necessary to analyze the strength of laminates. The strength analysis of laminates is more complex than that of the unidirectional plies. The destruction of unidirectional plies mainly has the following two forms: damage to the fiber; damage at the interface of matrix and fiber along the fiber direction. Laminates are made up of unidirectional plies with different directions. The stress on each ply should be analyzed and the state of each ply determined according to the chosen strength criterion. Normally, laminates become ineffective under the external load one ply at a time. In conclusion, the stress analysis of a single ply should be made and the strength of laminates can be predicted by the strength of single ply. Several strength criteria of unidirectional plies will be illustrated next.

(1) Maximum stress criterion: According to the maximum stress criterion, the material will be destroyed when one of the stress components along the main directions of the material reaches the corresponding strength value.
For tensile stress,

$$\left.\begin{aligned} \sigma_1 &= X_t, \\ \sigma_2 &= Y_t, \\ |\tau_{12}| &= S. \end{aligned}\right\} \tag{9.7}$$

For compressive stress, the first two equations above are modified as follows:

$$\left.\begin{aligned} \sigma_1 &= X_c, \\ \sigma_2 &= Y_c, \end{aligned}\right\} \tag{9.8}$$

where σ_1, σ_2, τ_{12} are three stress components along the main directions of the material, which are normally unknown in practical problems. Commonly, the stress components along the non-main directions σ_x, σ_y, τ_{xy} of the material are known. So stress components in the main directions of material can be calculated by coordinates transformation.

(2) Maximum strain criterion: According to the maximum strain criterion, the material will be destroyed if one (or more) of the stress components along the main directions of the material reaches the corresponding strain value.
For tensile strain,

$$\left.\begin{aligned} \varepsilon_1 &= \varepsilon_{X_t}, \\ \varepsilon_2 &= \varepsilon_{Y_t}, \\ |\gamma_{12}| &= \varepsilon_S. \end{aligned}\right\} \tag{9.9}$$

The first two equations above will be revised as follows for compressive strain

$$\left.\begin{aligned} |\varepsilon_1| &= \varepsilon_{X_c}, \\ |\varepsilon_2| &= \varepsilon_{Y_c}, \end{aligned}\right\} \tag{9.10}$$

where

$$\left.\begin{aligned} \varepsilon_{X_t} &= \frac{X_t}{E_1}, & \varepsilon_{X_c} &= \frac{X_c}{E_1}, \\ \varepsilon_{Y_t} &= \frac{Y_t}{E_2}, & \varepsilon_{Y_c} &= \frac{Y_c}{E_2}, \\ \varepsilon_S &= \frac{S}{G_{12}}. \end{aligned}\right\} \tag{9.11}$$

(3) Tsai–Hill strength criterion: The Tsai–Hill strength criterion for orthotropic unidirectional ply is as follows

$$\frac{\sigma_1^2}{X^2} - \frac{\sigma_1\sigma_2}{X^2} + \frac{\sigma_2^2}{Y^2} + \frac{\tau_{12}^2}{S^2} = 1, \tag{9.12}$$

where X and Y respectively represent the strengths in the two main directions of the material, and S represents the shear strength in the main direction of the material. The material will be ineffective if the left-hand side of equation (9.12) is not less than 1.

(4) Hoffman strength criterion: The Tsai–Hill strength criterion is only applicable to unidirectional plies with equal tensile strength and compressive strength ($X_t = X_c = X$, $Y_t = Y_c = Y$) along the main directions of the material. Hoffman revised the Tsai–Hill strength criterion for material ($X_t \neq X_c$, $Y_t \neq Y_c$) and proposed the Hoffman strength criterion.

$$\frac{\sigma_1^2 - \sigma_1\sigma_2}{X_t X_c} + \frac{\sigma_2^2}{Y_t Y_c} + \frac{X_c - X_t}{X_t X_c}\sigma_1 + \frac{Y_c - Y_t}{Y_t Y_c}\sigma_2 + \frac{\tau_{12}^2}{S^2} = 1. \tag{9.13}$$

In expression (9.13), when $X_t = X_c$, $Y_t = Y_c$, it turns into the Tsai–Hill strength criterion.

(5) Tsai–Wu strength criterion: The Tsai–Wu failure criterion was proposed based on tensor theory and considering that composites have different strengths in tension and compression:

$$F_i\sigma_i + F_{ij}\sigma_i\sigma_j + F_{ijk}\sigma_i\sigma_j\sigma_k + \cdots = 1. \tag{9.14}$$

As for unidirectional plies (plane stress state), $i, j, k, \ldots = 1, 2, \ldots, 6$; F_i, F_{ij}, $F_{ijk}, \ldots$ are the strength parameters of the material, which can be determined with basic strength X_t, X_c, Y_t, Y_c through tests. The first two items in the above equation are usually used in engineering design and the strength criterion turns into

$$F_i\sigma_i + F_{ij}\sigma_i\sigma_j = 1. \tag{9.15}$$

And it can be simplified as follows

$$F_1\sigma_1 + F_2\sigma_2 + F_{11}\sigma_1^2 + F_{22}\sigma_2^2 + F_{66}\sigma_6^2 + 2F_{12}\sigma_1\sigma_2 = 1. \tag{9.16}$$

In expression (9.16), the first five strength parameters F_1, F_{12}, F_{11}, F_{22}, F_{66} can be obtained through the uniaxial tension (or compression) tests and pure shear test in the main direction of the material.

$$\left.\begin{aligned}
F_1 &= \frac{1}{X_t} - \frac{1}{X_c}, \quad F_{11} = \frac{1}{X_t X_c}, \\
F_2 &= \frac{1}{Y_t} - \frac{1}{Y_c}, \quad F_{22} = \frac{1}{Y_t Y_c}, \\
F_{66} &= \frac{1}{S^2}, \\
F_{12} &= \frac{1}{2\sigma_0^2}\left[1 - \left(\frac{1}{X_t} - \frac{1}{X_c} + \frac{1}{Y_t} - \frac{1}{Y_c}\right)\sigma_0 - \left(\frac{1}{X_t X_c} + \frac{1}{Y_t Y_c}\right)\sigma_0^2\right].
\end{aligned}\right\} \tag{9.17}$$

9.2.6 Strength analysis of laminates

Laminates are inhomogeneous in the direction of thickness, therefore even under the in-plane loads, the stress state of each ply is different. As a result, damage should firstly start from the ply which firstly reaches the ultimate stress. However, the damage of one ply will not cause the final failure of the whole laminate but the reduction of the laminate's stiffness, which is called stiffness reduction. Following damage to another ply, the stiffness of the laminate further reduces until all the plies have been destroyed, which means the final destruction of a laminate. Therefore, there are two strength indexes of laminates: the strength of first-time ply destruction and the ultimate strength. Due to the complexity of strength analysis of laminates, only ultimate loads are determined in engineering calculations. Strength analysis of laminates is basically carried out in following steps:

(1) setting the ratio between each load;
(2) calculating the stiffness of initial laminate with performance of each ply;
(3) getting the relationship between stress of each ply and external loads;
(4) choosing strength criterion and judging which ply becomes ineffective first and getting the strength of the first destruction of ply;
(5) removing the destroyed ply and calculating the stiffness;
(6) checking whether other plies can continue to bear loads;
(7) applying the loads and repeating steps (3)–(7) until all the plies become ineffective and the limit strength is obtained. Due to the complexity of the strength analysis of laminates, a computer should be utilized.

9.2.7 Structural design principles of composite materials

Unlike metal materials, the designability of composite materials brings great advantages to structural design. The structural design of composite materials is a repetitive process in which various designs (i.e., laminates design, typical structural design and connection design) are integrated. The performance of constituents and the microstructures of composite materials are simultaneously taken into consideration so as to obtain the anticipated characteristics for materials and structures. A number of optimization variables and multilevel design are involved in this work, which is different from the design of traditional materials. The major factors that should be considered in design are design conditions, structure weight, manufacturing process, cost of development and quality control, etc.

Generally, structural design of composite materials includes material design and structural design with each part closely related to the craft plan. The basic steps for the structural design of composite materials are shown in Fig. 9.6.

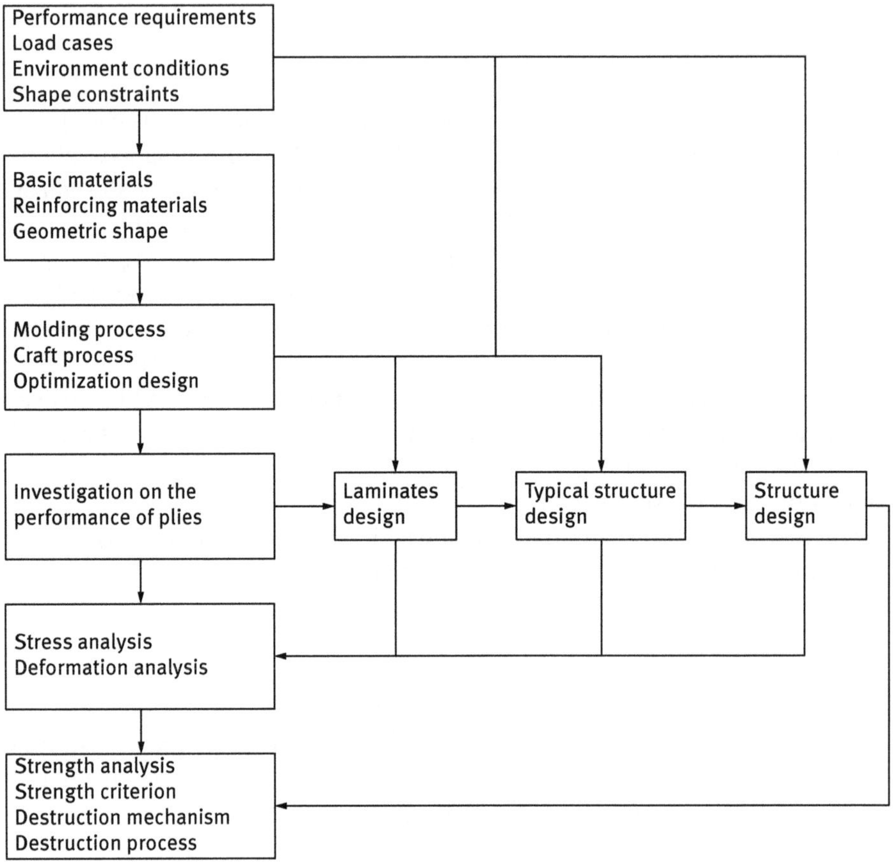

Fig. 9.6: Design steps for structures made of composite materials.

However, material design usually refers to the process in which several components are selected to manufacture composite materials with anticipatory performance. The plies are the basic elements of laminate structures of wind turbine blades. The design of laminates and shells is classified into structural design.

The task of laminate design is to determine the direction of each ply, the number and order of plies. As the mechanics behavior of laminates is complicated, laminates are normally designed into symmetric laminates (with symmetric stacking sequences of plies) with balanced directions of plies. Additionally, due to the complexity brought by numerous directions of plies, the direction of plies is usually limited to 0°, 90° and ±45°. The stacking sequence of plies can be determined according to the following principles:

(1) The direction of each ply should be distributed evenly along the thickness direction of the laminate. The number of ply groups in a laminate should be as high as possible or the number of plies in each group should be as low as possible (normally smaller than 4). In this way, the possibility two kinds of ply groups separating can be reduced.
(2) ±45° plies are usually stacked in pairs so as to decrease the shear stress between ±45° plies and 0° or 90° plies. Furthermore, ±45° plies should be layered on the external surface of laminates so as to improve the compression stability and impact resistance of laminates and the strength of the connection hole.
(3) If a variable thickness laminate is to be designed, plies on the external surface of the laminate should remain the same when altering the internal plies. To avoid shear failure between plies, the width of each stage should be the same and should be no less than 2.5 mm. In order to prevent the ply edge stripping from the stage, one ply is covered on the stage.

9.3 Structural design of wind turbine blades made of composite materials

9.3.1 The geometric shape of the new blade

A new type of wind turbine blade with high performance and light mass has been designed using the improved particle swarm algorithm mentioned in Chapter 8. Further description of the geometric shape of the blade will be made in this section. The detailed layout of the blade's geometry is shown in Tab. 9.3, including the relative locations, the distribution of chords and twists of 25 sections and the airfoils used on each section. The largest chord of this new blade is 3.219 m at spanwise location of 0.20R, where the largest twist of 12.079° is located. The newly-designed wind turbine airfoils adopted on the blade are depicted in Fig. 9.7–9.9. Compared to traditional airfoils, the aerodynamic performance of this new airfoil series had been verified (refer to Chapter 4 for more details). It can be seen that seven standard airfoils and eight transitional airfoils were utilized on this new blade.

Tab. 9.3: Detailed layout of the blade's geometry.

Sections	Spanwise location ($\mu = r/R$)	Chord (m)	Twist (°)	Airfoil
1	0.00	2.115	0.000	Circle
2	0.075	2.115	0.000	Circle
3	0.115	2.326	4.052	Transitional airfoil
4	0.155	2.704	8.013	Transitional airfoil
5	0.20	3.219	12.079	CQU-A400
6	0.24	2.975	10.265	Transitional airfoil
7	0.28	2.756	8.976	CQU-A350
8	0.32	2.557	8.053	CQU-A350
9	0.36	2.376	7.336	Transitional airfoil
10	0.40	2.210	6.709	CQU-A300
11	0.44	2.056	6.131	CQU-A300
12	0.48	1.916	5.575	Transitional airfoil
13	0.52	1.786	5.033	CQU-A250
14	0.56	1.668	4.509	CQU-A250
15	0.60	1.561	4.006	Transitional airfoil
16	0.64	1.462	3.527	CQU-A210
17	0.68	1.375	3.070	CQU-A210
18	0.72	1.299	2.631	Transitional airfoil
19	0.76	1.235	2.207	CQU-A180
20	0.80	1.180	1.796	CQU-A180
21	0.84	1.118	1.397	CQU-A180
22	0.88	1.035	1.012	Transitional airfoil
23	0.92	0.914	0.640	CQU-A150
24	0.96	0.741	0.283	CQU-A150
25	1.00	0.512	0.000	CQU-A150

In order to get a smooth and continuous transition zone at the blade root, appropriate modifications are needed at the root region (geometric curve interpolation). The position of the modified chord, twist and relative thickness along the span of the blade are displayed in Fig. 9.10–9.12.

The three-dimensional coordinates of each blade section (blade element) can be calculated based on the above geometric data of the blade. The three-dimensional model of the blade's surface can be established by 3D modeling software (i.e., Pro/E), which is shown in Fig. 9.13–9.15.

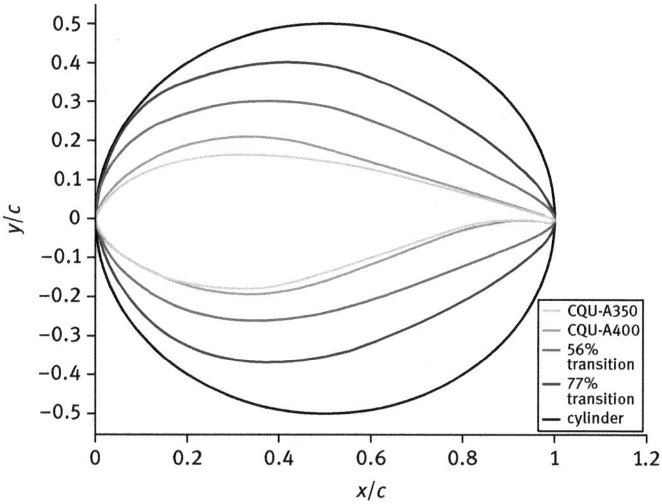

Fig. 9.7: Shapes of the newly-designed airfoils.

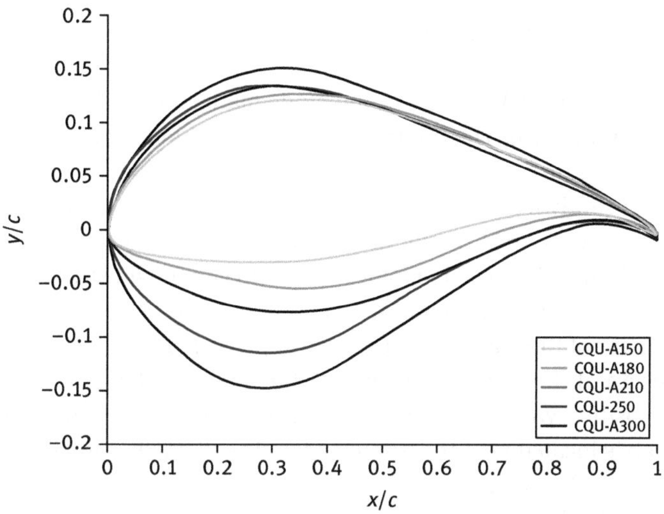

Fig. 9.8: Shapes of the newly-designed airfoils.

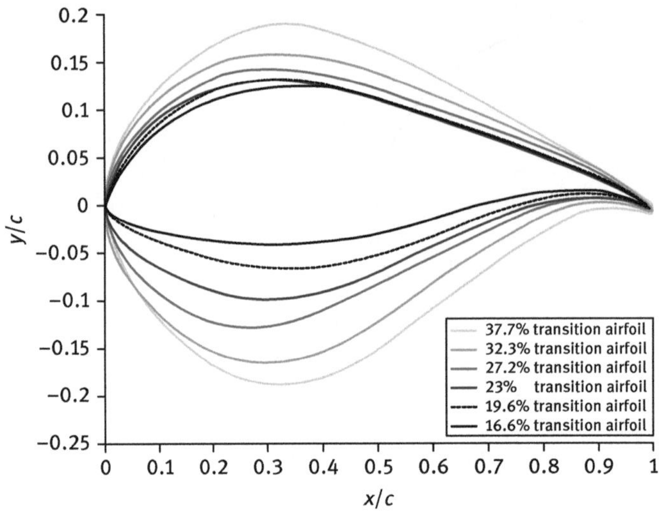

Fig. 9.9: Shapes of the newly-designed airfoils.

9.3.2 Design of internal structure of the blade

Two webs are adopted in a typical 2 MW wind turbine blade made of glass fiber reinforced epoxy resin, shown in Fig. 9.16. The internal space of the blade is divided by webs into three parts called leading edge, cap and trailing edge. The two webs that connect the cap are made of sandwich-style materials, as shown in Fig. 9.17. A typical internal structure of the blade section is shown in Fig. 9.18. The outer surface of the blade is a skin made of gel cloth and biaxial laminates. Gel cloth can provide a smooth external surface and reduce the roughness caused by manufacturing. The cap is the main load bearing component of the blade that bears the flapwise loads.

During the initial design of the webs, the web near the leading edge is usually located at the chordwise position with the maximum relative thickness of each blade section while the other web is located near the trailing edge. The width and thickness of the cap decrease gradually along the span (from root to tip). In this way, the cap and webs constitute the so called box-beam structure with cross sections varying along the blade span. The thicknesses of the laminates should change slowly from cap to the airfoil plane (along the circumferential direction), in case of any suddenly drop in thickness. The cap structure mainly influences the flapwise stiffness of the blade. Reinforcement materials are mixed in the leading edge and trailing edge to enhance the edgewise stiffness. The thickness of the leading edge laminates and the trailing edge laminates gradually decrease along the blade span (from root to tip). Additionally, the suction surface and pressure surface of the blade section (including the leading edge, cap, airfoil plane and trailing edge of the airfoil section) are made up of laminates with symmetric stacking sequence of plies.

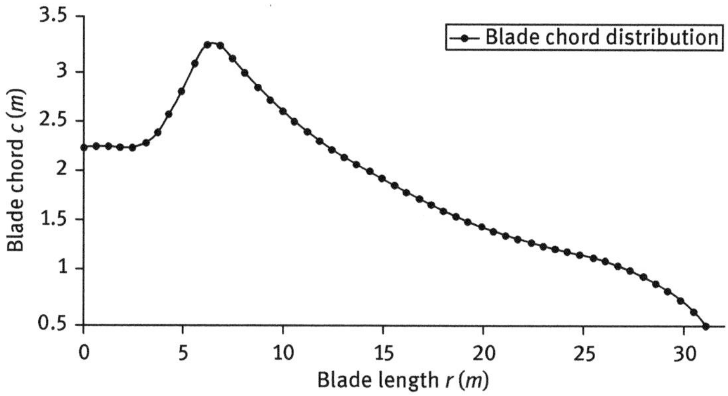

Fig. 9.10: Distribution of blade chord.

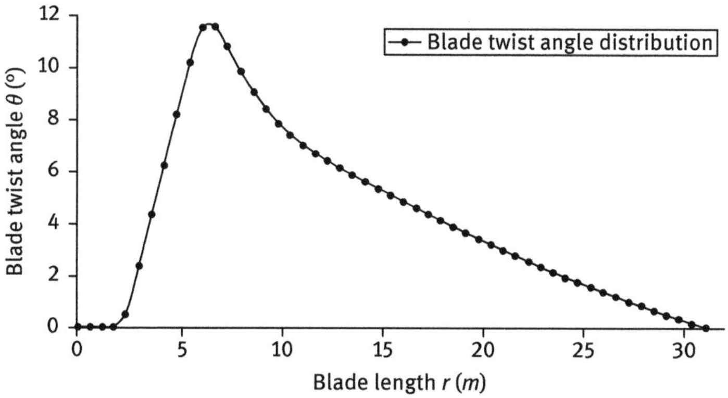

Fig. 9.11: Distribution of blade twist.

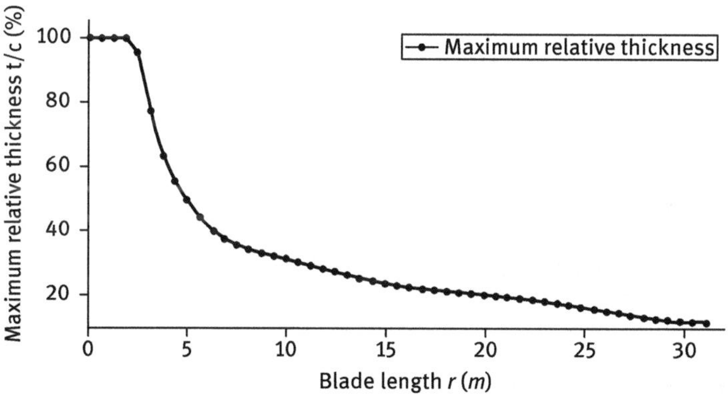

Fig. 9.12: Distribution of maximum relative thickness.

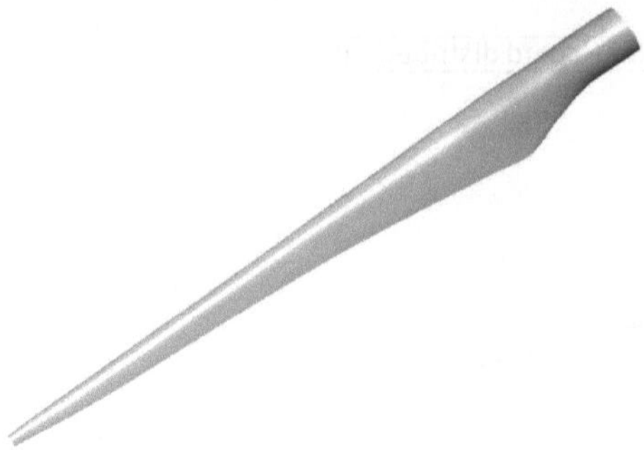

Fig. 9.13: Suction side of blade.

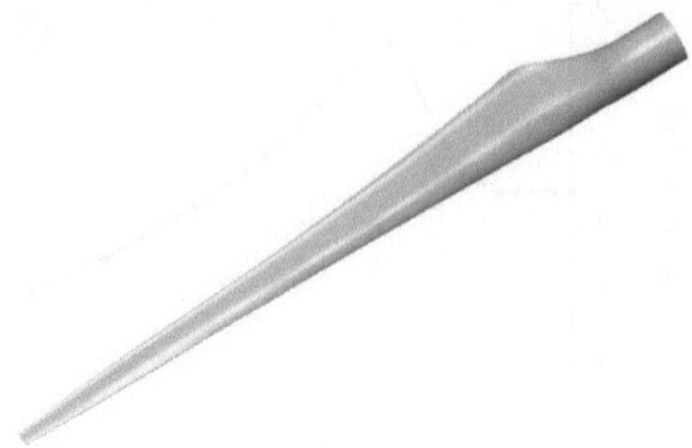

Fig. 9.14: Pressure side of blade.

Fig. 9.15: View of wind turbine rotor.

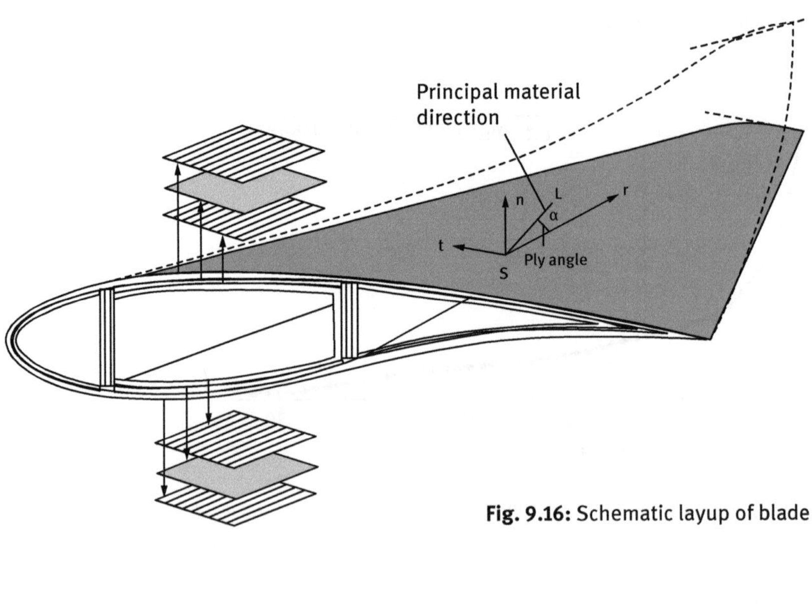

Fig. 9.16: Schematic layup of blade cap.

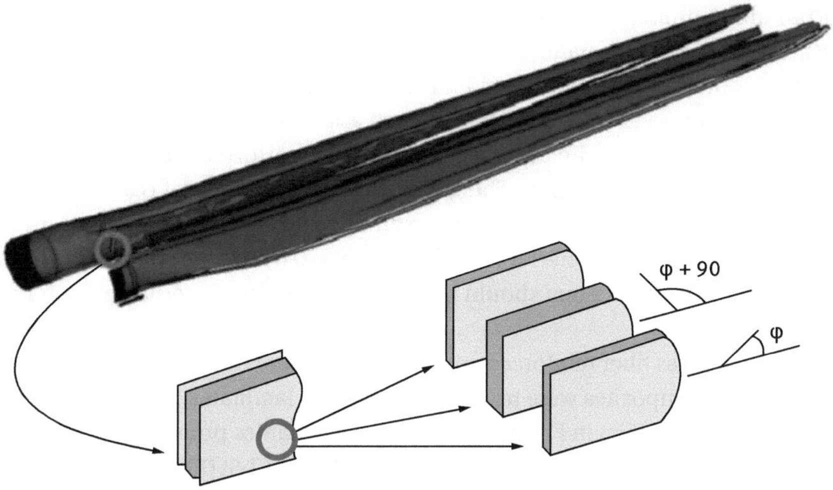

Fig. 9.17: Schematic layup of blade web.

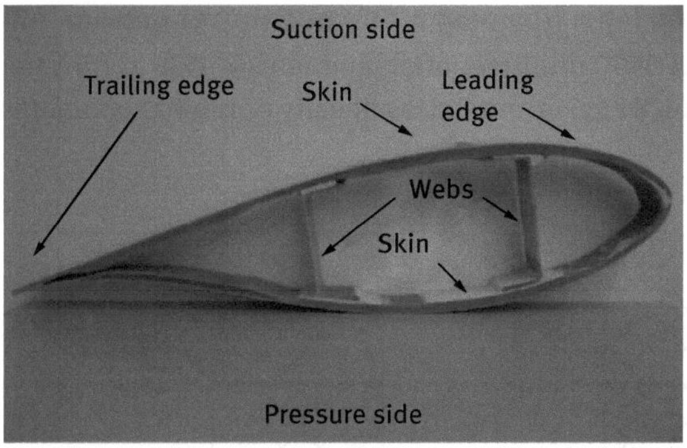

Fig. 9.18: Inner structure of blade section.

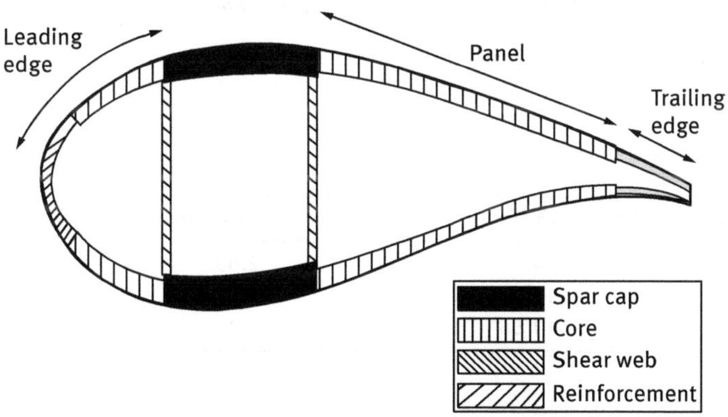

Fig. 9.19: Schematic drawing of blade section material.

The key structural parameters and materials of the blade section are shown in Fig. 9.19. The filling material used in the panel is organic foam. The leading edge and trailing edge of the airfoil are composed of reinforced laminates without filling materials. The cap is composed of thicker reinforced laminates which are mainly made of unidirectional laminates with good anti-tensile and anti-compressive performance. The cap bears most of the flapwise bending loads. The gel cloth with thickness of 0.6 mm should be pasted on the external surface to ensure a smooth and continuous blade surface. What is more, the short fiber material with thickness of 0.4 mm and the bidirectional laminates with thickness of 1.2 mm should be pasted on both the external and internal surface.

In this chapter, glass fiber reinforced epoxy resin laminates are chosen. Three types of glass fiber/epoxy composites were used: unidirectional laminates, biaxial laminates and triaxial laminates (shown in Fig. 9.20). The main mechanics properties of the three laminates chosen in this design are listed Tab. 9.4 [13]. E-LT-5500 is chosen for the unidirectional laminate, and glass fiber reinforced material (EP-3) is chosen for the biaxial laminate. Properties of the triaxial laminate (denoted as SNL Triax in Tab. 9.4), are determined by averaging the test-derived data for the unidirectional and biaxial laminates. Based on the volume fractions (denoted as V_F in Tab. 9.4) indicated in Tab. 9.4, the mass density of the E-LT-5500 unidirectional laminates is 1920 kg/m^3, the mass density of biaxial laminates is 1780 kg/m^3, the mass density of triaxial laminates is 1850 kg/m^3.

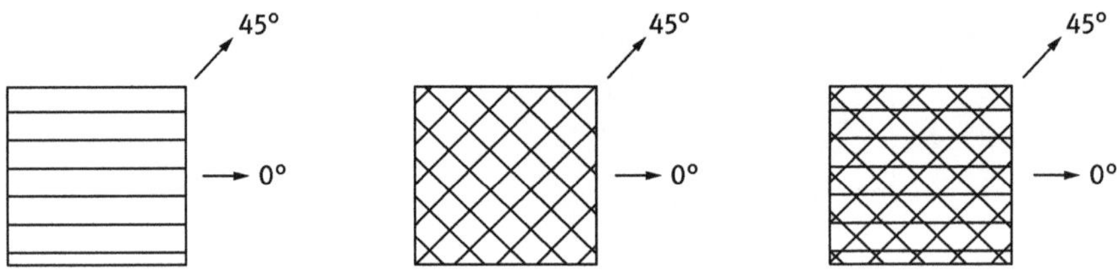

(a) *Undirectional laminate* (b) *Biax laminate* (c) *Triax laminate*

Fig. 9.20: Three types of glass fiber/epoxy composite.

Tab. 9.4: Mechanics properties of laminates.

Materials	V_F (%)	E_1 (GPa)	E_2 (GPa)	Poisson ratio	G_{XY} (GPa)
E-LT-5500 $[0]_2$	54	41.8	14.0	0.28	2.63
EP-3 $[\pm45]_4$	44	13.6	13.3	0.51	11.8
SNL Triax $[\pm45]_2[0]_2$	–	27.7	13.65	0.39	7.2

Tab. 9.5: Mechanics properties of other materials.

Materials	E_1(GPa)	E_2 (GPa)	Poisson ratio	G_{XY} (GPa)	Density (kg/m^3)
Gel cloth	3.44	3.44	0.3	1.38	1235
Reinforcement material	3.5	3.5	0.3	1.4	1100
Foam filling	0.256	0.256	0.3	0.022	200

In addition to the mechanics properties of laminates in Tab. 9.4, the structural characteristics of some other materials are also relevant. Tab. 9.5 lists additional materials used in this design, including coating material, extra reinforced material and foam core material.

To facilitate the initial layout design for the composite wind turbine blade, the laminate materials are named as followed:
(A) unidirectional laminate $[0]_2$, with a thickness of 1.2 mm;
(B) biaxial laminate $[\pm 45]_4$, with a thickness of 2.4 mm;
(C) triaxial laminate $[\pm 45]_2[0]_2$, with a thickness of 1.8 mm;
(D) gel-cloth material, with a thickness of 0.6 mm;
(E) reinforcement material $[0]_2$, with a thickness of 1.2 mm;
(F) foam filling material, the maximum thickness is 30 mm for the leading edge panel and the trailing edge panel of the blade. The maximum thickness is 35 mm for the two webs. The thickness drop along the blade span (from root to tip) is to 1.2 mm.

The thicknesses of laminates at 27 sections along the blade span are given in Tab. 9.6. In the table, the thicknesses of the blade root and four major parts (cap, reinforcement of trailing edge, leading edge panel and trailing edge panel) on each section have also been defined.

Due to the lack of key layout data, based on some literature [192–195] and an equivalent design method often used in composite structural design, the initial layout design for the composite wind turbine blade is determined and listed in Tab. 9.7. The initial layout for the blade design is described below:
(1) The root part of the blade is composed of material C. The maximum thickness of the root part is 96.8 mm, and the thickness gradually reduces along the blade span.
(2) The spar cap is mainly made up of material A. The thickness of the spar cap increase first from the blade root and then decreases to a minimum value at the blade tip. The maximum thickness of the spar cap is 58 mm at the spanwise location of 24 %.
(3) The leading panel and trailing panel of the blade mainly consist of laminate A and foam material F. In order to avoid a discontinuous connection with the spar cap, the thickness of these parts should not have a great thickness drop.
(4) The leading edge and the trailing edge of the blade are composed of laminate A and extra reinforcement E. The thickness of these parts is thinner than the leading panel and trailing panel, and the maximum thickness is about 20 mm at the location

close to blade root. Moreover, the filling material F will be added at the trailing edge of the blade to avoid great deformation caused by aerodynamic loads.

(5) The two webs, which begin at spanwise location of 1 m and terminate at 31 m, are mainly composed of laminate B and foam material F. In the initial design, these two webs (one near to the leading edge and the other near to the trailing edge) are positioned at 24 % of chord (as measured from leading edge to trailing edge) and 54 % of chord respectively. The maximum thickness of the two webs is 35 mm, which appears near the blade root. The thickness gradually decreases along the blade span.

Tab. 9.6: Thickness of laminates along blade.

No.	Span location r/R	Root (mm)	Cap (mm)	LE reinforcement (mm)	TE reinforcement (mm)	LE panel (mm)	TE panel (mm)
1	0.00	98.6	–	–	–	–	–
2	0.025	62.6	–	–	–	–	–
3	0.05	42.6	–	–	–	–	–
4	0.075	22.6	18	4.8	4.8	10	10
5	0.115	–	24	7.2	7.2	15	15
6	0.155	–	36	12.0	12.0	15	15
7	0.20	–	48	18.0	18.0	20	20
8	0.24	–	54	18.0	18.0	30	30
9	0.28	–	48	12.0	12.0	30	30
10	0.32	–	42	12.0	12.0	28	28
11	0.36	–	36	9.6	9.6	25	25
12	0.40	–	30	6.0	6.0	20	20
13	0.44	–	30	6.0	6.0	20	20
14	0.48	–	30	6.0	6.0	20	20
15	0.52	–	27	4.8	4.8	18	18
16	0.56	–	27	4.8	4.8	18	18
17	0.60	–	27	4.8	4.8	18	18
18	0.64	–	27	4.8	4.8	18	18
19	0.68	–	24	4.8	4.8	14	14
20	0.72	–	18	4.8	4.8	12	12
21	0.76	–	18	3.6	3.6	10	10
22	0.80	–	15	3.6	3.6	8	8
23	0.84	–	12	3.6	3.6	6	6
24	0.88	–	7.2	2.4	2.4	4	4
25	0.92	–	6.0	2.4	2.4	2	2
26	0.96	–	2.4	1.2	1.2	2	2
27	1.00	–	1.2	1.2	1.2	1.2	1.2

Note: In the table, LE is short for leading edge, TE is short for trailing edge.

Tab. 9.7: Initial layup of blade.

No.	Span location r/R	Layup					
		Blade root	**Leading edge**	**Leading edge panel**	**Cap**	**Trailing edge panel**	**Trailing edge**
1	0.000	1D + 34C + 2E	–	–	–	–	–
2	0.025	1D + 34C + 2E	–	–	–	–	–
3	0.050	1D + 17C + 2E	–	–	–	–	–
4	0.075	1D + 4C + 15A + 2E	1D + 4C + 5A + 2E	1D + 4C + 4A + F + 2E	1D + 4C + 15A + 2E	1D + 4C + 4A + F + 2E	1D + 4C + 4A + 2E
5	0.115	–	1D + 2C + 6A + 2E	1D + 2C + 6A + F + 2E	1D + 2C + 20A + 2E	1D + 2C + 6A + F + 2E	1D + 2C + 6A + 2E
6	0.155	–	1D + 2C + 10A + 2E	1D + 2C + 10A + F + 2E	1D + 2C + 30A + 2E	1D + 2C + 10A + F + 2E	1D + 2C + 10A + 2E
7	0.20	–	1D + 2C + 12A + 2E	1D + 2C + 12A + F + 2E	1D + 2C + 40A + 2E	1D + 2C + 12A + F + 2E	1D + 1C + 8A
8	0.24	–	1D + 2C + 8A + 2E	1D + 2C + 12A + F + 2E	1D + 2C + 45A + 2E	1D + 2C + 12A + F + 2E	1D + 1C + 8A
9	0.28	–	1D + 2C + 8A + 2E	1D + 2C + 10A + F + 2E	1D + 2C + 40A + 2E	1D + 2C + 10A + F + 2E	1D + 1C + 7A
10	0.32	–	1D + 2C + 7A + 2E	1D + 2C + 10A + F + 2E	1D + 2C + 35A + 2E	1D + 2C + 10A + F + 2E	1D + 1C + 7A
11	0.36	–	1D + 2C + 7A + 2E	1D + 2C + 8A + F + 2E	1D + 2C + 30A + 2E	1D + 2C + 8A + F + 2E	1D + 1C + 7A
12	0.40	–	1D + 2C + 6A + 2E	1D + 2C + 5A + F + 2E	1D + 2C + 25A + 2E	1D + 2C + 5A + F + 2E	1D + 1C + 6A
13	0.44	–	1D + 2C + 4A + 2E	1D + 2C + 5A + F + 2E	1D + 2C + 25A + 2E	1D + 2C + 5A + F + 2E	1D + 1C + 4A
14	0.48	–	1D + 2C + 4A + 2E	1D + 2C + 5A + F + 2E	1D + 2C + 25A + 2E	1D + 2C + 5A + F + 2E	1D + 1C + 4A
15	0.52	–	1D + 2C + 4A + 2E	1D + 2C + 4A + F + 2E	1D + 2C + 23A + 2E	1D + 2C + 4A + F + 2E	1D + 1C + 4A
16	0.56	–	1D + 2C + 4A + 2E	1D + 2C + 4A + F + 2E	1D + 2C + 23A + 2E	1D + 2C + 4A + F + 2E	1D + 1C + 3A
17	0.60	–	1D + 2C + 4A + 2E	1D + 2C + 4A + F + 2E	1D + 2C + 22A + 2E	1D + 2C + 4A + F + 2E	1D + 1C + 3A
18	0.64	–	1D + 2C + 4A + 2E	1D + 2C + 4A + F + 2E	1D + 2C + 22A + 2E	1D + 2C + 4A + F + 2E	1D + 1C + 3A
19	0.68	–	1D + 2C + 3A + 2E	1D + 2C + 4A + F + 2E	1D + 2C + 20A + 2E	1D + 2C + 3A + F + 2E	1D + 1C + 2A
20	0.72	–	1D + 2C + 3A + 2E	1D + 2C + 4A + F + 2E	1D + 2C + 15A + 2E	1D + 2C + 3A + F + 2E	1D + 1C + 2A
21	0.76	–	1D + 2C + 2A + 2E	1D + 2C + 3A + F + 2E	1D + 2C + 15A + 2E	1D + 2C + 3A + F + 2E	1D + 1C + 2A
22	0.80	–	1D + 2C + 2A + 2E	1D + 2C + 3A + F + 2E	1D + 2C + 13A + 2E	1D + 2C + 3A + F + 2E	1D + 1C + 2A
23	0.84	–	1D + 2C + 2A + 2E	1D + 2C + 3A + F + 2E	1D + 2C + 10A + 2E	1D + 2C + 2A + F + 2E	1D + 1C + 2A
24	0.88	–	1D + 2C + 2A + 2E	1D + 2C + 2A + F + 2E	1D + 2C + 6A + 2E	1D + 2C + 2A + F + 2E	1D + 1C + 1A
25	0.92	–	1D + 2C + 1A + 2E	1D + 2C + 2A + F + 2E	1D + 2C + 5A + 2E	1D + 2C + 2A + F + 2E	1D + 1C + 1A
26	0.96	–	1D + 2C + 1A + 2E	1D + 2C + 1A + 2E	1D + 2C + 2A + 2E	1D + 2C + 1A + 2E	1D + 1C + 1A
27	1.00	–	1D + 2C + 1A + 2E	1D + 2C + 1A + 2E	1D + 2C + 2A + 2E	1D + 2C + 1A + 2E	1D + 1C + 1A

In addition, the entire internal and external blade surfaces have triaxial material with thickness of 1.2 mm. Extra reinforcement material with thickness of 1.2 mm is also pasted onto the internal blade surface. The external surface is covered by 0.6 mm of gel coat. The extra reinforcement and gel coat are included to produce a more realistic blade design weight. After finishing the initial layout, the thickness distribution of all parts of the blade is shown in Fig. 9.21. Generally, the thickness of the blade is declining along the blade spanwise.

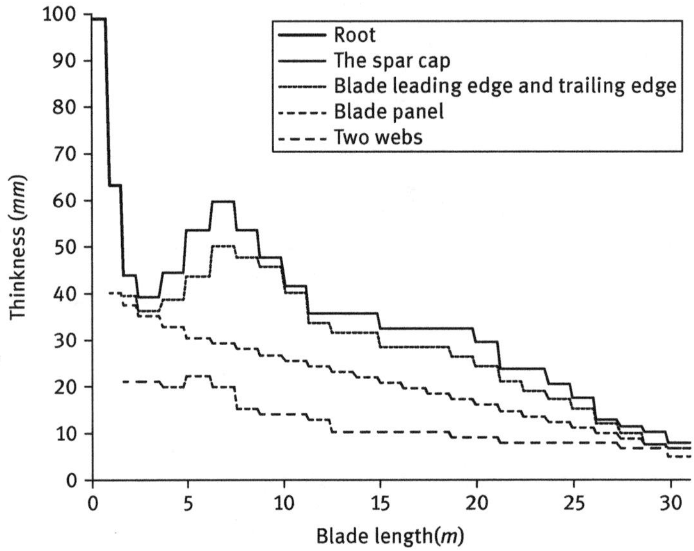

Fig. 9.21: Thickness distribution of each part of the blade.

Mass and stiffness of the blade are two important parameters in structural design. In order to verify the rationality of the two parameters in the initial blade structure, the stiffness and mass of the blade sections were calculated using the software PreComp developed by American Renewable Resources Lab. Fig. 9.22–9.24 respectively show the distribution of blade flapwise stiffness, edgewise stiffness and torsional stiffness along blade span. It is found that the edgewise stiffness (Fig. 9.23) reaches its maximum value at the spanwise position of $0.2R$, which is away from the root of the blade. The flapwise and torsional stiffness reach their maximum values at the root of the blade. Generally, the stiffness declines along the blade span from root to tip. The mass distribution along the blade span is shown in Fig. 9.25, with the trend of decreasing from blade root to tip. One reason for the large mass and stiffness at the blade root is that it bears the largest bending moment. The second reason is that the effective bearing area of the root has been decreased due to the connection between the blade root and the hub, which calls for thicker laminates to improve stiffness and strength.

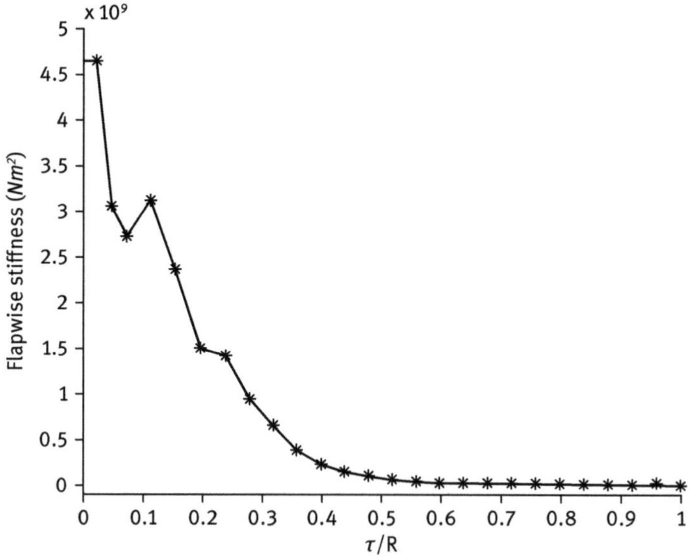

Fig. 9.22: Distribution of flapwise stiffness.

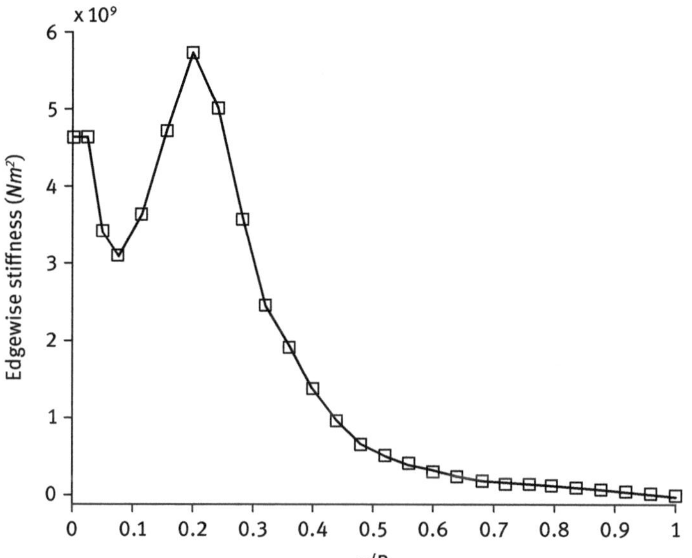

Fig. 9.23: Distribution of edgewise stiffness.

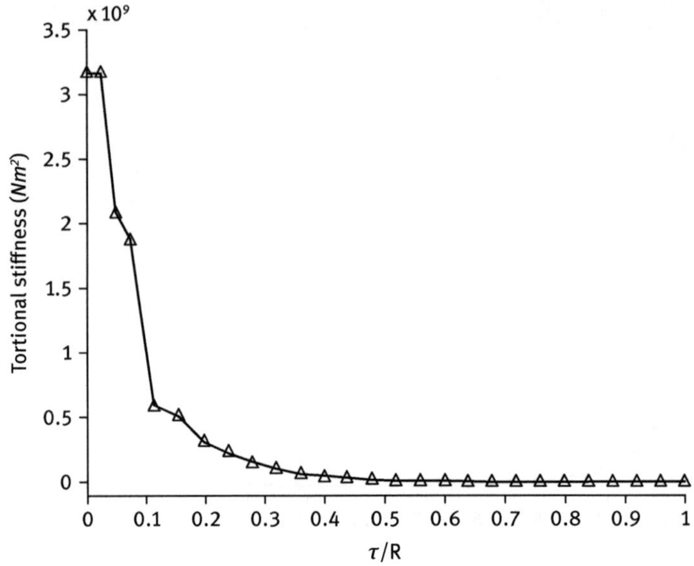

Fig. 9.24: Distribution of torsional stiffness.

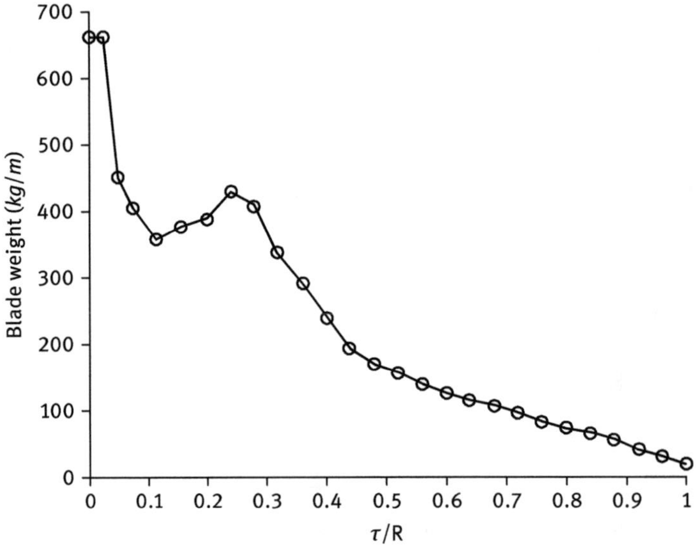

Fig. 9.25: Distribution of blade mass.

9.4 Parametric finite element modeling of composite wind turbine blades

For the traditional method to build the finite element model of a wind turbine blade, a geometrical model was created based on cross section profiles of shell and spar using CAD software. Then the wire frame model, which is the fundamental geometrical model, was transferred to finite element software such as ANSYS, ABAQUS and so on [196–198]. However, the pre-processing operations of this method for building a blade finite element model are very complicated and time consuming. Moreover, it is inconvenient for post-processing and optimization using ANSYS software. Therefore, based on the parametric geometric representations of wind turbine blade shapes illustrated in Chapter 7 and the initial structural design of a composite wind turbine blade in Section 9.3, the parametric finite element modeling of a composite material wind turbine blade has been studied. A program for building a parametric finite element model of a wind turbine blade that combined MATLAB and APDL language of ANSYS software is developed. Based on this study, a more complete blade finite element model can be quickly obtained and even the airfoil and blade aerodynamic-structural integrated design can be achieved, which make the parallel design and analysis of composite wind turbine blades possible.

9.4.1 The integrated representation of three-dimensional blade shapes

Through fitting the airfoil series with different relative thickness, chord and twist distribution along the blade span, the three-dimensional blade shape parametric representation model was obtained. Based on the 3D integrated expression proposed in Chapter 7, the coordinates of blade shape can be obtained

$$
\begin{cases}
x = \left\{\left[a\left(\rho(\theta) + \dfrac{1}{\rho(\theta)}\right)\cos\theta - X_M\right]\cos\beta(u)\right.\\
\qquad \left. + a\left(\rho(\theta) - \dfrac{1}{\rho(\theta)}\right)\sin\theta\sin\beta(u)\right\}c(u),\\
y = \left\{\left[-a\left(\rho(\theta) + \dfrac{1}{\rho(\theta)}\right)\cos\theta - X_M\right]\sin\beta(u)\right.\\
\qquad \left. + a\left(\rho(\theta) - \dfrac{1}{\rho(\theta)}\right)\sin\theta\cos\beta(u)\right\}c(u),\\
z = u\cdot R,
\end{cases}
\tag{9.18}
$$

where a is the $\frac{1}{4}$ airfoil chord; X_M is the chordwise position of airfoil aerodynamic center, usually $X_M = 0.25$; u is the spanwise position, $u \in [0, 1]$; $c(u)$ is the chord distribution function; $\beta(u)$ is the twist distribution function; $\rho(\theta)$ is the airfoil shape function (integrated expression), $\rho(\theta) = C_0 + C_1\theta + C_2\theta^2 + \cdots + C_k\theta^k + \cdots$. Different airfoils can be obtained through controlling the coefficients of the airfoil shape function.

The integrated expression of a three-dimensional curved blade surface shown in equation (9.18) can be used to represent any kind of geometric blade shapes.

9.4.2 Parametric representation of chord and twist of wind turbine blades

Based on the data for geometric shape such as airfoil contour, and chord and twist distribution shown in Tab. 9.3, the distribution of chord and twist can be obtained by fitting. Since the curvature of the transition zone (from root to the $0.2R$ of the blade) is large, the fitting of the whole blade is not proper. Therefore, the segmented function is used to make the fitting expression. The distributions of blade chord and twist are shown as follows:

$$c(u) = \begin{cases} 2.215 & (0 \le r/R \le 0.075), \\ 0.81205 + 54.29161u - 779.37204u^2 & \\ \quad + 4782.26405u^3 - 9706.73532u^4 & (0.075 \le r/R \le 0.20), \\ 7.25536 - 33.45450u + 100.42422u^2 - 173.711u^3 & \\ \quad + 153.7301u^4 - 53.74936u^5 & (0.20 \le r/R \le 1.0); \end{cases} \tag{9.19}$$

$$\beta(u) = \begin{cases} 0.0 & (0 \le r/R \le 0.075), \\ 3.0679 - 140.0002u + 1604.1728u^2 & \\ \quad - 2161.65501u^3 - 6266.86334u^4 & (0.075 \le r/R \le 0.20), \\ 42.5896 - 259.14597u + 769.7534u^2 - 1178.56972u^3 & \\ \quad + 883.57322u^4 - 258.32742u^5 & (0.20 \le r/R \le 1.0). \end{cases} \tag{9.20}$$

Fig. 9.26 and 9.27 are the fitting curve for chord and twist of the blade, respectively. Using a segmented function, we can perfectly fit the local area with high curvature. Thus, the geometric parameters of the blade in equation (9.18) can all be expressed by the shape functions. Given the airfoil series, the three-dimensional coordinate data can be calculated.

9.4.3 Parametric finite element modeling of wind turbine blades

In the traditional modeling method, a three-dimensional model is established utilizing CAD software and then the model is imported into the finite element software ANSYS for pre-processing. However, this is time consuming and also inconvenient for post-processing and optimization. Combining MATLAB and APDL language, without the aid of any three-dimensional modeling software, finite element modeling can be directly established in software like ANSYS. The basic process is as follows. Based on the

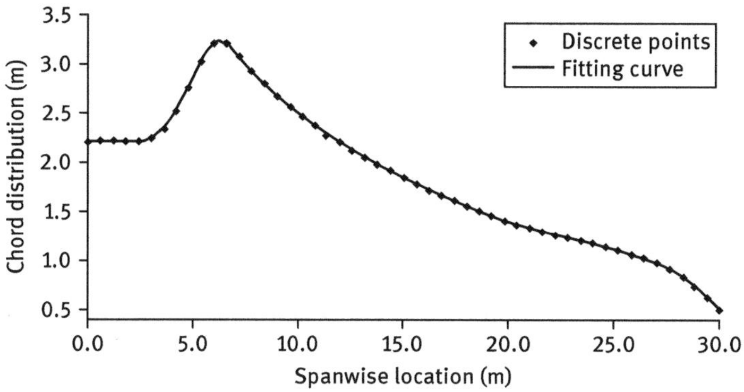

Fig. 9.26: Fitting curve of blade chord.

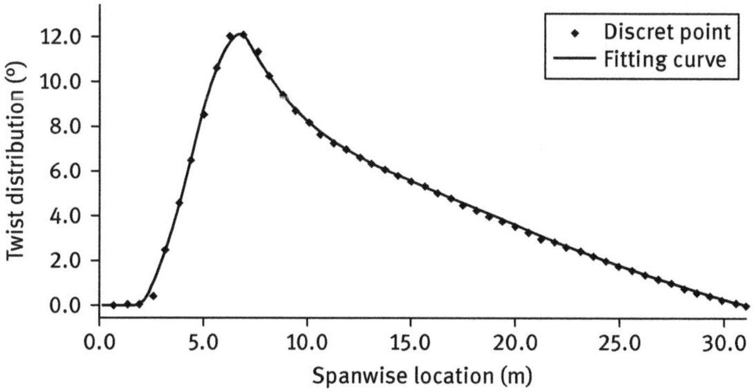

Fig. 9.27: Fitting curve of blade twist.

integrated expression for the 3D shape of the blade, coordinate data of the blade surface was calculated and saved in a specific output format through MATLAB. Then a data transfer interface was set up using APDL language macro files with MATLAB. Then APDL language is used to build a parametric model. The specific processes in ANSYS finite element software are as follows:

(1) Writing macro file for data input: Write a macro file, named `bladedata.mac`, using `*create` of APDL language. The macro file can transfer the 3D geometric data of the blade calculated by MATLAB into ANSYS software. The command to build the macro file is

```
*create,bladedata,mac
Finish
/clear
*dim, star, , 140*27, 3
*vread, star(1,1), bladeout, dat, , jik, 3, 140*27
(f10.6, f12.7, f10.3)
*end
```

(2) Establishing complex 3D shape of the blade: B-splines of airfoil sections of the whole blade were established in finite element software ANSYS using the BSPLIN command and *DO loop sentence (APDL language). Since most of the blade is hollow, the curve model should be built with the use of the ASKIN command, as shown in Fig. 9.28–9.30.
(3) Establishing webs and trailing edge filling of the blade: As for large and medium-sized wind turbine blades, the cap should be used to bear most of the bending load. The cap adopts the rectangular beam structure [199], with its chordwise position shown in Fig. 9.31. For initial configuration on every blade section, the two webs locate at 0.24 and 0.54 chordwise positions respectively.
The location of the cap can be taken as an optimization variable so as to improve the structural properties. Therefore, the chordwise location of the cap was also taken as a design variable. The method is as follows: since the airfoil coordinates of each section are known, the maximum and minimum values of key points on each section were recorded by the *get command; the chord of each section was calculated; the key points of the cap on each section were built to generate a plane. Boolean operators are used to get the box-like beam of each blade cross section and the 3D shape of the blade was obtained finally, which is shown in Fig. 9.32.

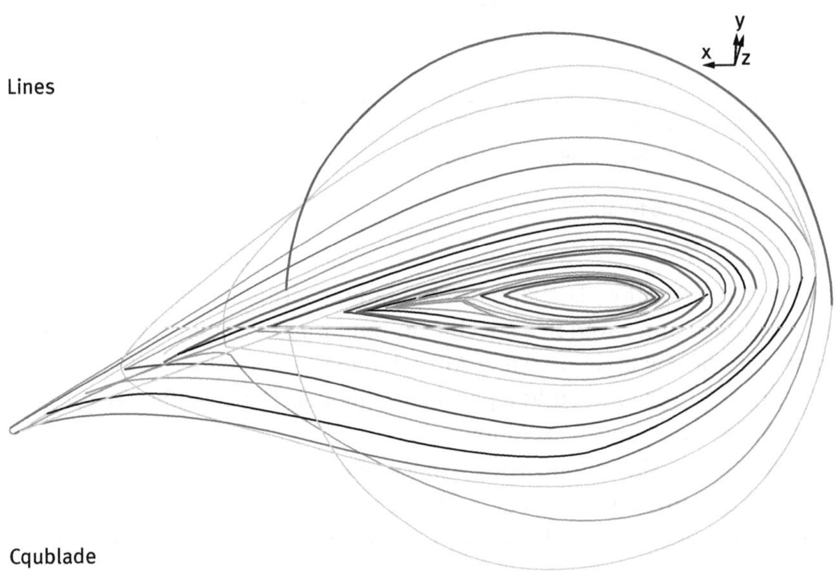

Fig. 9.28: Shape expression of the blade.

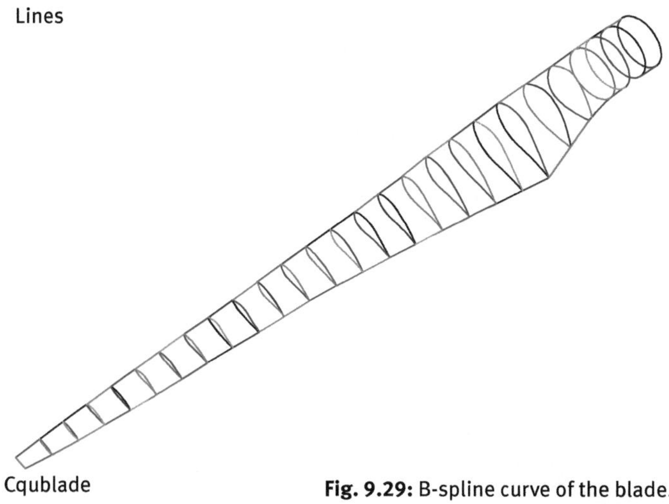

Fig. 9.29: B-spline curve of the blade.

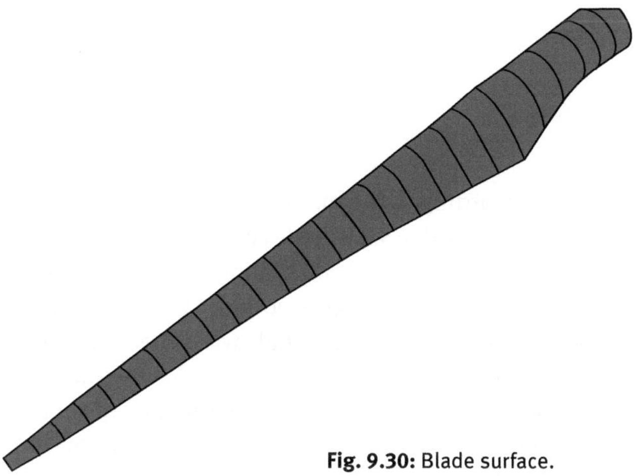

Fig. 9.30: Blade surface.

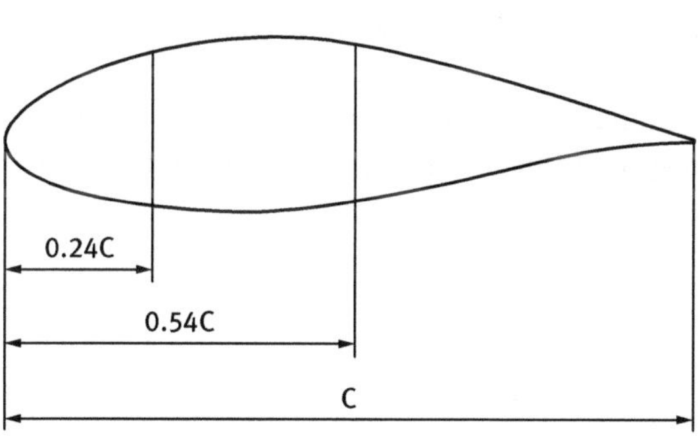

Fig. 9.31: Initial position of the cap.

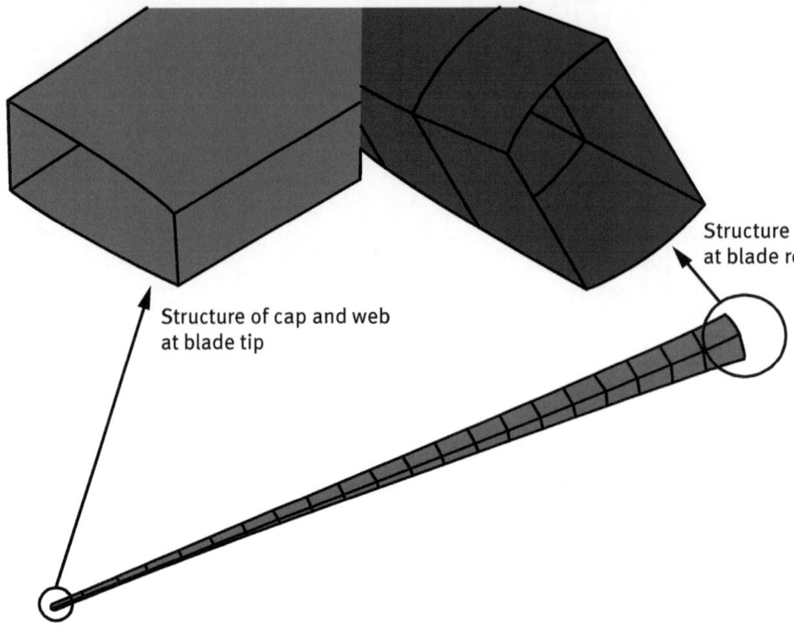

Fig. 9.32: Schematic drawing of cap and webs.

(4) Attributing material properties: According to the material attributes of the wind turbine blade and the laminate data in Section 9.3, shell181 cell was used on the surface blade to obtain the aerodynamic shape. The airfoil trailing edge was simulated with ssolid185 cell. Then the web and airfoil rib were added. The finite element model of a wind turbine blade made of composite materials was generated as long as the material, thickness and layering angle had been defined. As the composite structure of a wind turbine blade is complex, the variation in the thickness of the laminates along the blade span is in a stepped distribution, as shown in Fig. 9.21. The more sections it is divided into, the more continuous the thickness of laminates along the span will be. The finite element model of the whole wind turbine blade is shown in Fig. 9.33. The detailed structure of a typical blade section is shown in Fig. 9.34. In order to illustrate the laminate in detail, one part of the blade is chosen and the stacking sequence is displayed. As shown in Fig. 9.35–9.37, the laminate is composed of more than 100 plies.

Fig. 9.33: Finite element model of the blade.

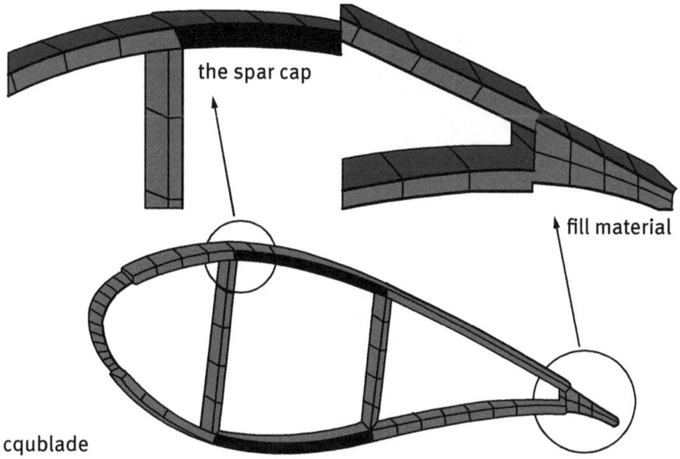

Fig. 9.34: Details of typical blade section.

In order to understand the internal laminates of the blade, some typical blade sections are chosen for demonstration. As shown in Fig. 9.38–9.42, the laminate is getting thinner along the span. And the blade section gradually changes from a circle to an airfoil with large relative thickness and then to an airfoil with small relative thickness.

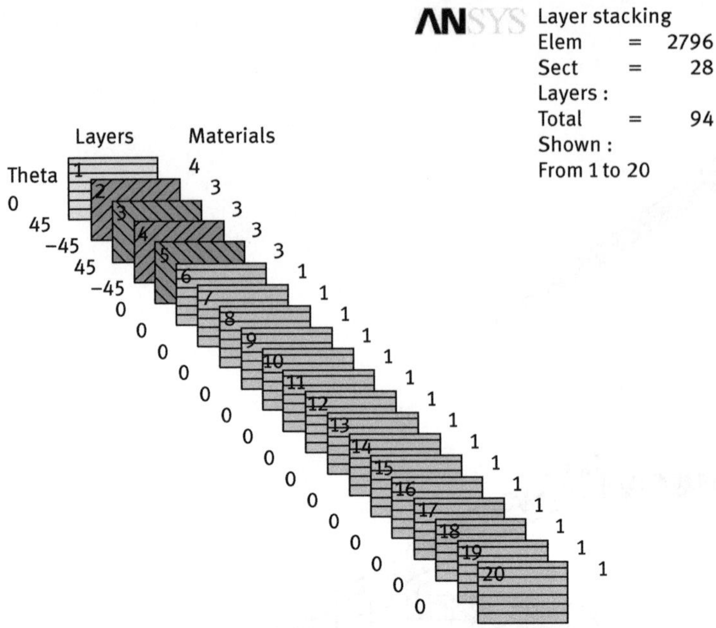

Fig. 9.35: Plies of one part of blade (1–20).

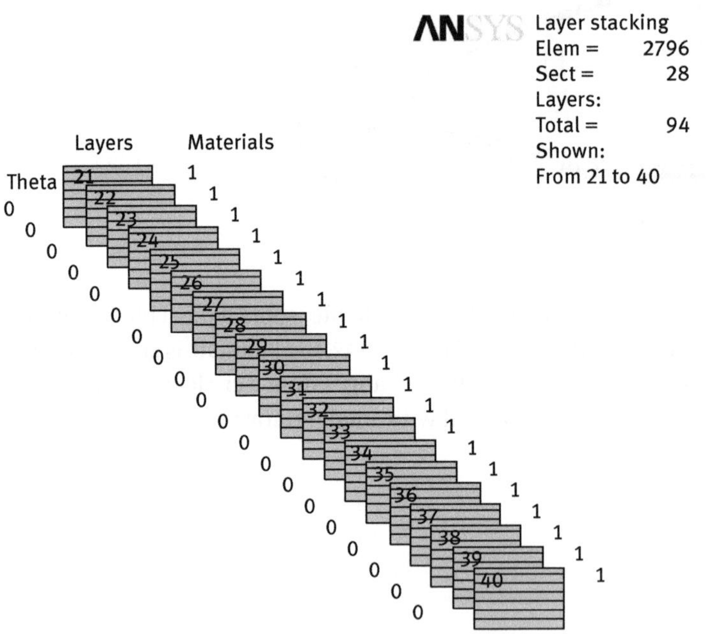

Fig. 9.36: Plies of one part of blade (21–40).

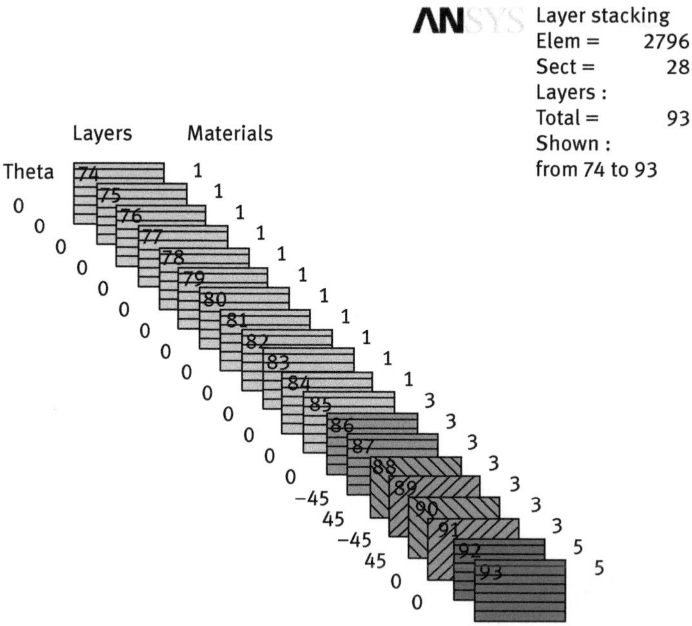

Fig. 9.37: Plies of one part of blade (74–93).

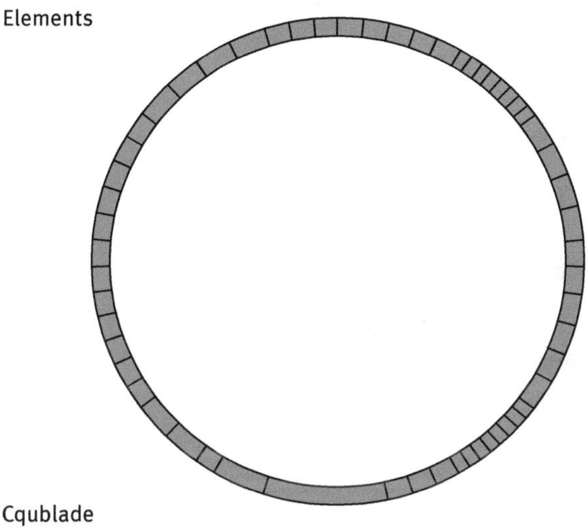

Fig. 9.38: Blade section at the blade root.

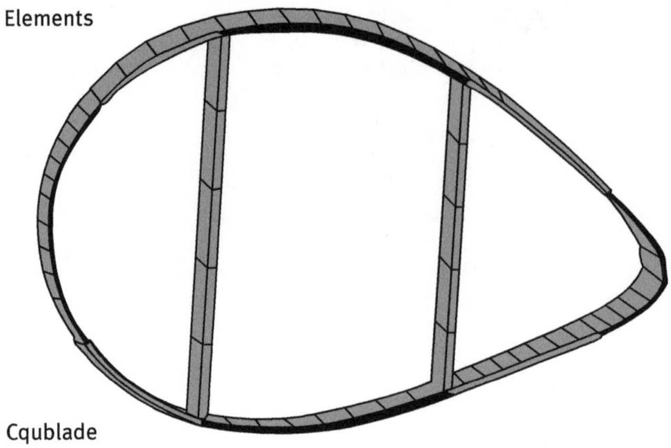

Fig. 9.39: Blade section at 3.5 m span location.

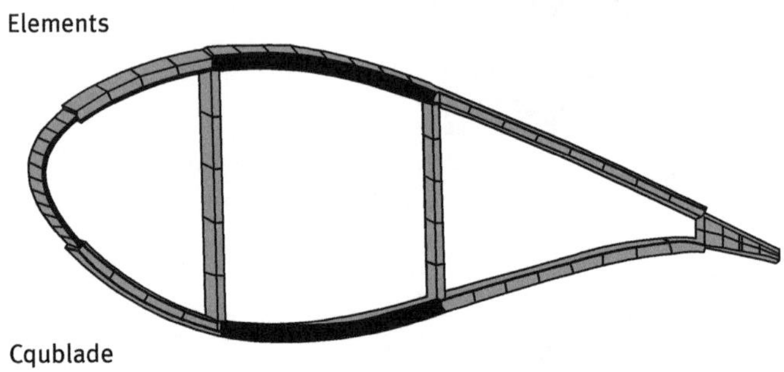

Fig. 9.40: Blade section at 7.5 m span location.

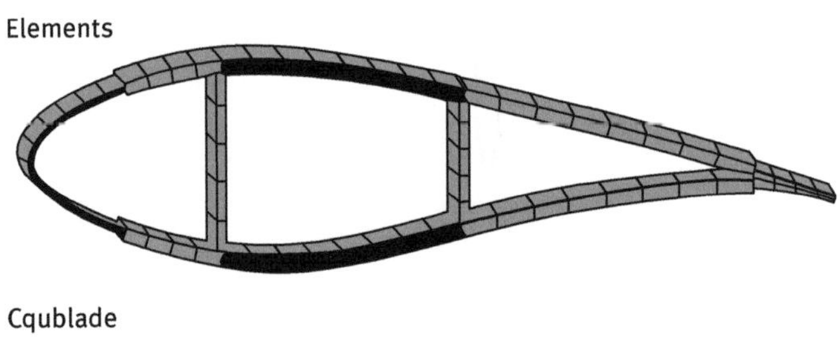

Fig. 9.41: Blade section at 15.5 m span location.

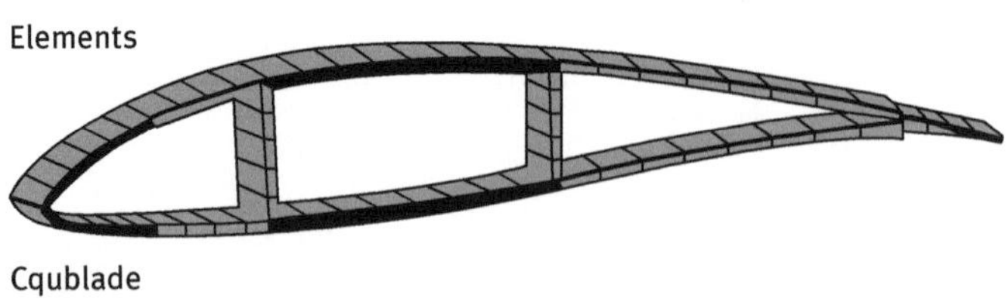

Fig. 9.42: Blade section at 26.5 m span location.

9.5 A new fluid-structure interaction method for blade design

The finite element parametric model for the blade design is established based on the 3D parametric representation of blade shape. Due to the coupling between the structure and aerodynamic load, the blade will deflect and vibrate, which will have a significant effect on the structural properties of the wind turbine blade. Based on the modified Blade Element Momentum theory (refer to Chapter 6), a new one-way fluid-structure interaction method is introduced, which is depicted in Fig. 9.43. The force of each blade element is not applied to the finite element model as concentrated force. Instead, a program specific for interpolating the pressure distribution on the composite material blade was designed.

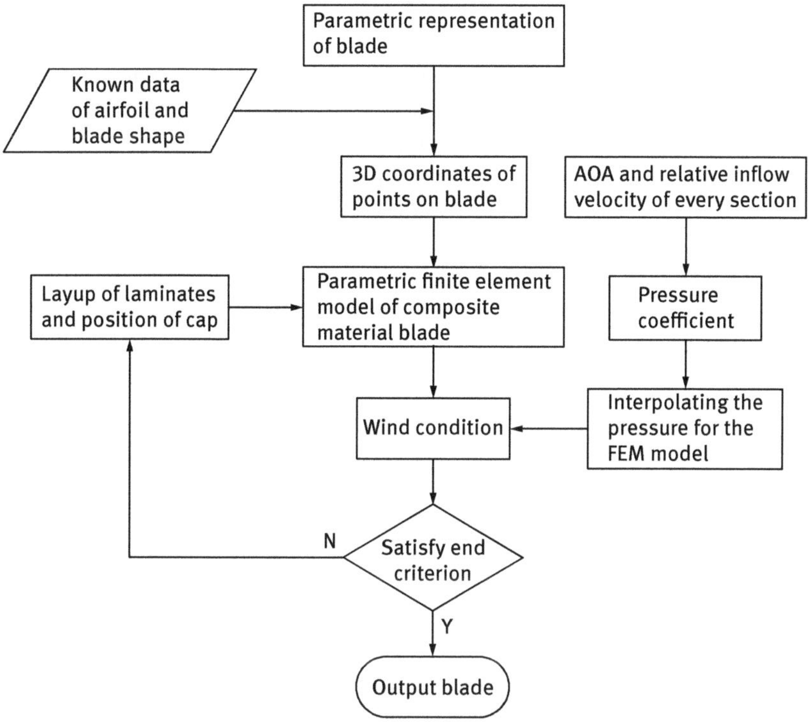

Fig. 9.43: The flowchart of the fluid-structure interaction method.

9.5.1 The operating conditions of wind turbines

The wind turbine's operating conditions must first be defined in order to calculate the force on each blade section. Based on data from a typical 2 MW wind turbine under normal operating conditions, the rated wind speed is set as 12.5 m/s, the rotor reaches a maximum speed of 20 rpm and the cut-out wind speed is set as 25 m/s. Therefore, the aerodynamic load of the blade for the newly-designed 2 MW rotor was calculated considering the rated operating conditions in this section.

9.5.2 The local angle of attack and pressure distribution

After determining the operating conditions of the wind turbine, the local angle of attack for each blade section is computed by the modified BEM theory, which considers the axial and tangential induction factor, tip-loss factor, and blade shape parameters. The aerodynamic performance (such as lift coefficient and drag coefficient) of an airfoil can be obtained by wind tunnel experimental prediction methods when the initial conditions are determined (such as Reynolds number, Mach number, and angle of attack). The aerodynamic performance before stall of an airfoil can be predicted accurately by RFOIL software. Therefore, RFOIL software is chosen to calculate the aerodynamic characteristics for airfoils for different angles of attack. Although the three-dimensional rate effects cannot be considered by using this method, compared with the traditional fluid calculation method (such as using FLUENT software), it can save calculation time greatly, which is essential to optimize the structure of the wind turbine blade.

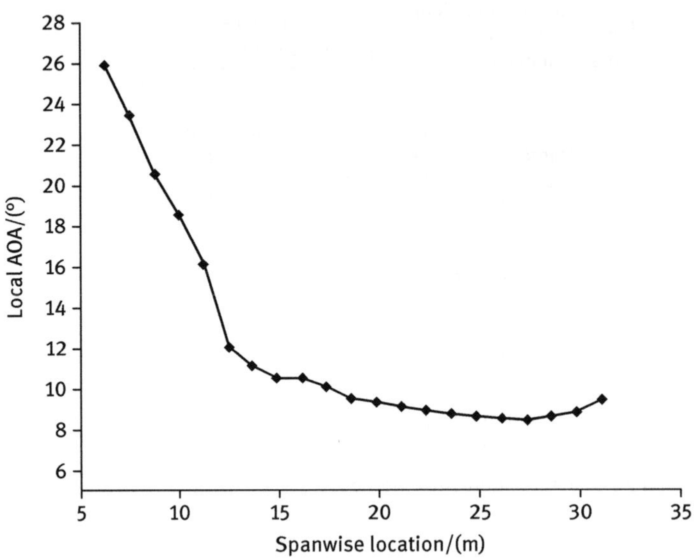

Fig. 9.44: The local angle of attack.

The aerodynamic characteristics of the airfoil were calculated by RFOIL and then taken into a modified BEM method. Then the tangential and axial induction factors were obtained through iterations. The local AOA of each airfoil section was obtained according to the shape parameters of the blade. The spanwise variation of AOA for the newly-designed 2 MW rotor is shown in Fig. 9.44.

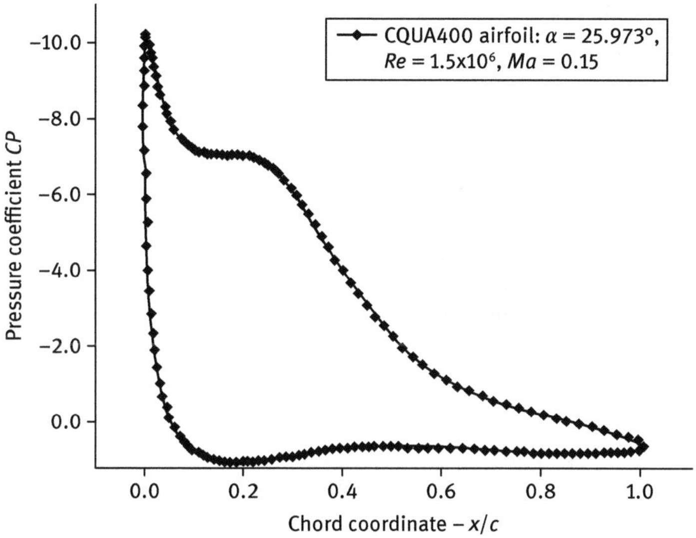

Fig. 9.45: Pressure distribution at $r/R = 0.20$.

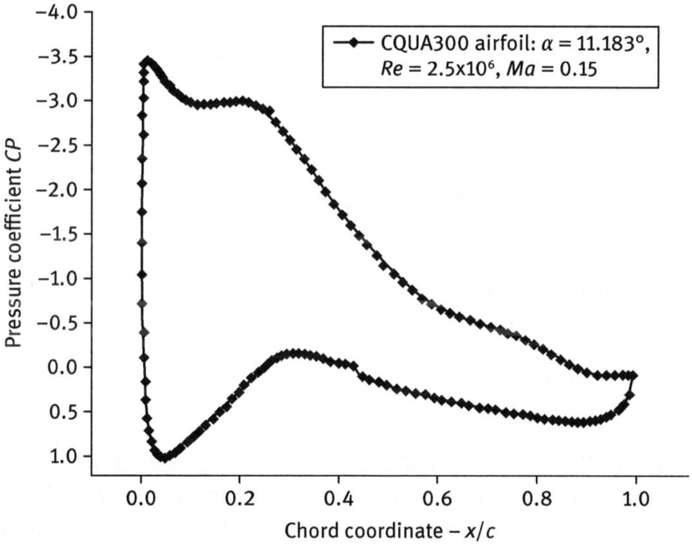

Fig. 9.46: Pressure distribution at $r/R = 0.40$.

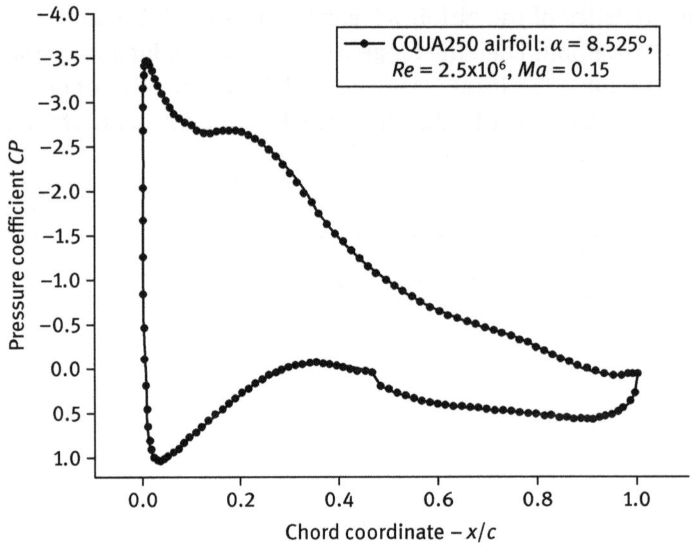

Fig. 9.47: Pressure distribution at $r/R = 0.56$.

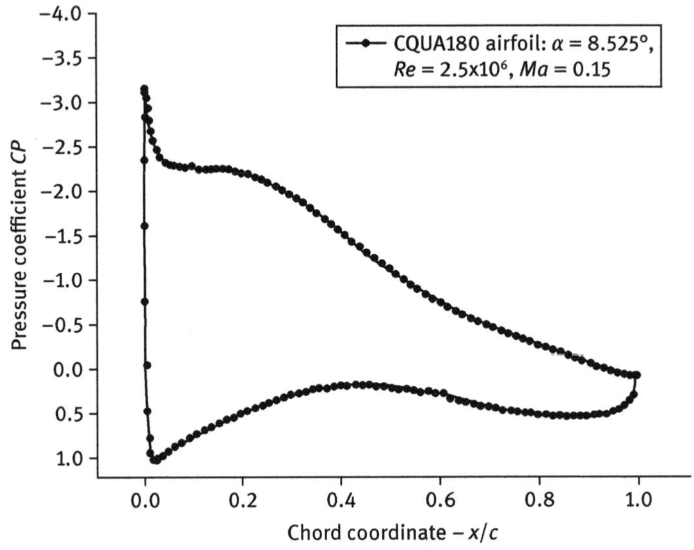

Fig. 9.48: Pressure distribution at $r/R = 0.80$.

Since the local AOA is known, the pressure distribution of each airfoil section can be obtained under the design condition. The pressure coefficients on four typical airfoil sections are depicted in Fig. 9.45–9.48 with Reynolds number of $Re = 3.0 \times 10^6$ and Mach number of $Ma = 0.15$.

9.5.3 The interpolation of aerodynamic forces

However, aerodynamic load on the blade surface is not the pressure coefficient calculated by RFOIL but the pressure. According to the definition of the pressure coefficient as shown in equation (9.21), the pressure p can be computed through equation (9.22):

$$C_\mathrm{p} = \frac{p - p_\infty}{\frac{1}{2}\rho U^2}, \tag{9.21}$$

$$p = \frac{1}{2}\rho U^2 C_\mathrm{p} + p_\infty, \tag{9.22}$$

where ρ is the air density of $1.205\,\mathrm{kg/m^3}$; C_p is the pressure coefficient; p_∞ is the standard atmospheric pressure; U is the relative velocity which is the synthesis speed of the blade rate speed and the wind speed, and it must be calculated. The relative velocity, shown in Fig. 9.49, can be calculated based on the modified BEM theory as mentioned before. As can be seen from Fig. 9.49, the relative velocity varies linearly along the span and reaches the maximum value of 66.021 m/s, which shows that the loads at the blade tip are larger than those at the blade root.

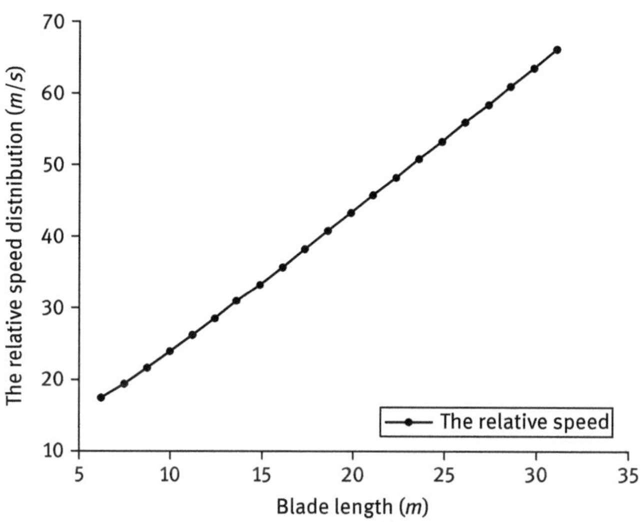

Fig. 9.49: The distribution of relative velocity.

In order to simplify the calculation, the entire wind turbine blade is divided into 24 sections. A varying aerodynamic load will be imposed on the corresponding sections. ANSYS software can apply the pressure load in the form of a function on the finite element model. Therefore, the pressure distribution for each blade section can be translated into a function with a high order polynomial to be applied to the finite element model. The pressure distribution of the upper and lower surfaces is fitted respectively because of the large curvature of the pressure distribution for the upper and lower surfaces. The pressure of the blade section can be fitted by a 10-order fitting function. The compared results between the fitting pressure distribution and the discrete pressure distribution of four typical cross sections are shown in Fig. 9.50–9.53. It can be seen that the pressure with the high order polynomial fitting can substitute the discrete pressure.

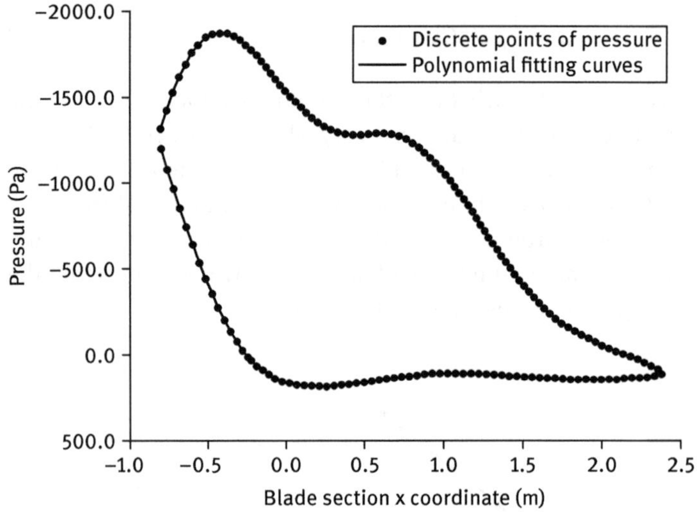

Fig. 9.50: Pressure distribution at $r/R = 0.20$.

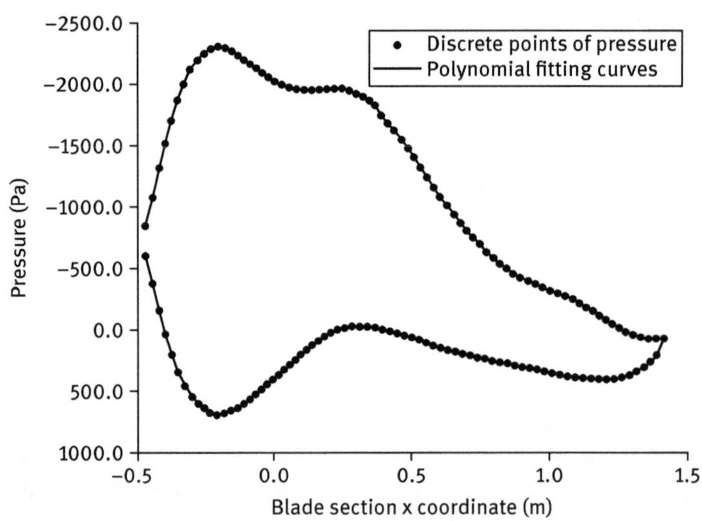

Fig. 9.51: Pressure distribution at $r/R = 0.40$.

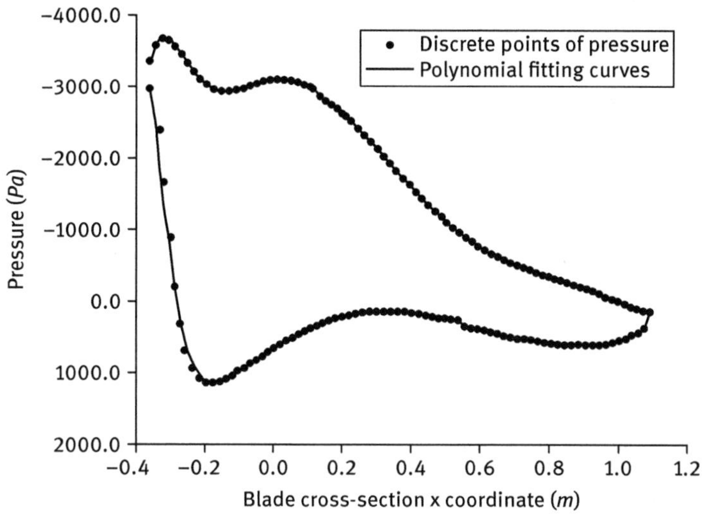

Fig. 9.52: Pressure distribution at $r/R = 0.56$.

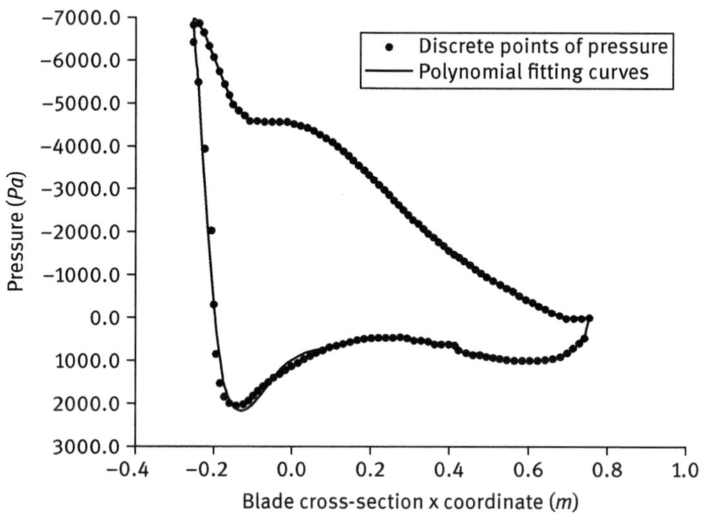

Fig. 9.53: Pressure distribution at $r/R = 0.80$.

An aerodynamic interpolation procedure that applied the pressure load to the structural finite element model was developed. The pressure at a certain point on the blade surface was interpolated using the pressure data on these 24 sections. The typical cross section of the finite element model which needs to be interpolated is depicted in Fig. 9.54. And the pressure distribution on the whole blade after interpolating the aerodynamic load is show in Fig. 9.55 and 9.56.

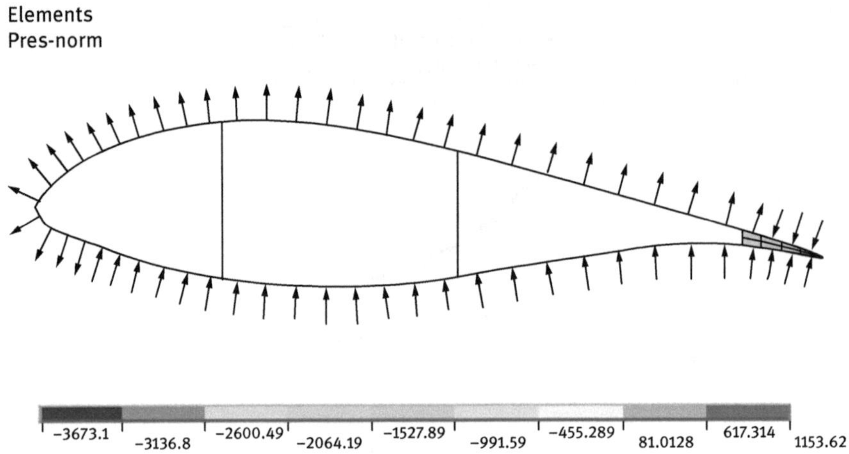

Fig. 9.54: Typical cross section of the finite element model.

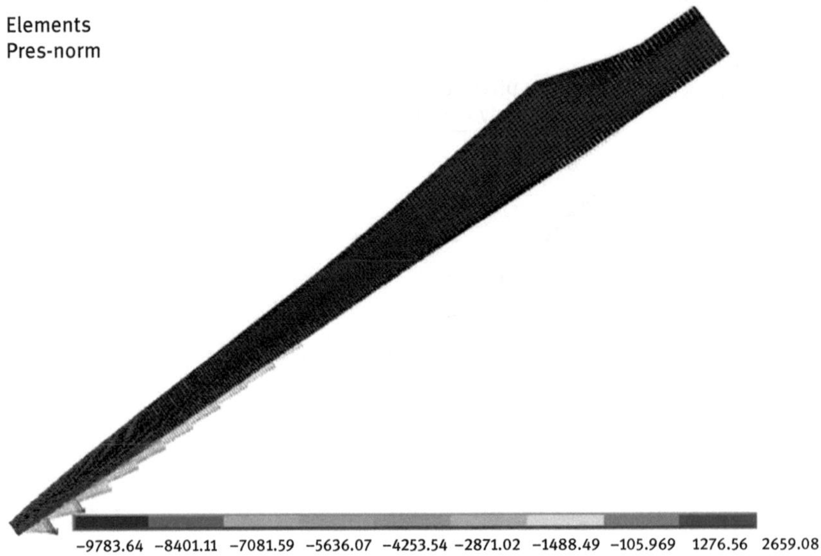

Fig. 9.55: Pressure on the pressure side.

In this way, a program that interpolated the aerodynamic load on the composite wind turbine blade is achieved. This program has two advantages. First, compared with the pressure computed by the CFD method, the aerodynamic pressure loading in this method can shorten the calculation time. Second, the highly coupled aeroelastic problem of the structural and aerodynamic models can be solved.

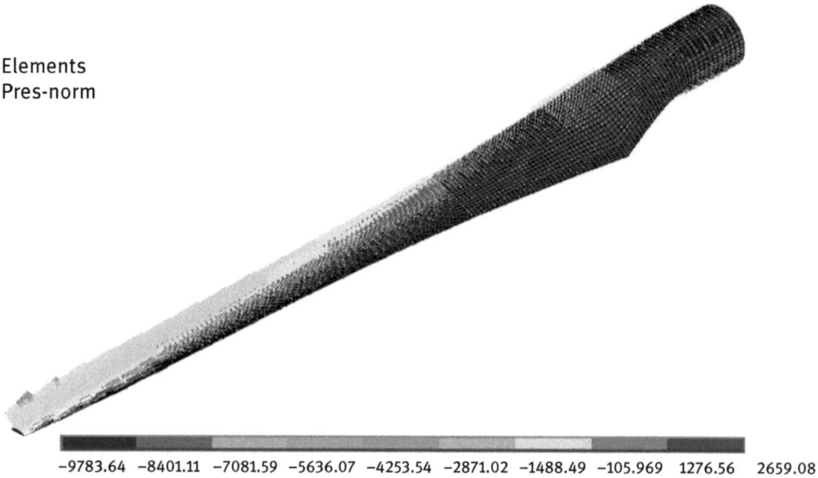

Fig. 9.56: Pressure on the suction side.

9.6 Study of the structural optimization of the wind turbine blade made of composite materials

The finite element method has been widely used in many complicated applications. So, based on the finite element method and parametric modeling method, a relatively complete model was established for the wind turbine blade made of composite materials. The inner structure of a blade section usually consists of the airfoil leading edge, airfoil panel, the spar cap, the trailing edge and the shear webs. A procedure combining finite element analysis and particle swarm algorithm to optimize composite structures of the wind turbine blade is developed, with the optimization goal of minimizing the weight. The design variables are thickness of the laminates and the location of the blade spar. The Tasi–Wu failure criterion and the maximum deflection of blade tip are selected as constraints for this optimization.

Building the parametric finite element model of the wind turbine blade with the combination of MATLAB and APDL language of ANSYS, optimization for the newly-designed 2 MW composite wind turbine blade is performed. The comparison between the optimization result and the original blade is shown.

9.6.1 The optimization model

9.6.1.1 Objective function

Generally speaking, the cost of the materials for the blade will fall as the mass of the blade is reduced without considering the manufacturing costs of the blade. Moreover, the lighter mass can also be beneficial to improving the fatigue life of the blade on the premise of the requirements for strength and stiffness of the blade being met. Therefore, the minimum mass for the wind turbine blade is taken as the objective function:

$$F(X) = \min(\text{mass}_{\text{blade}}). \tag{9.23}$$

9.6.1.2 Design variables

In light of the complexity of the structure of the wind turbine blade in which the maximum number of the layouts reaches more than 100 layers, it is difficult to proceed to an optimization design for the blade if the thickness of every layout of each blade section is considered as a design variable. Thus, the thickness of laminates as described before (refer to Section 9.4.2) is taken as the design variables, such as the thickness of unidirectional laminate, the thickness of biaxial laminate, and the thickness of triaxial laminate. Considering the true blade, the thickness of the blade panel, the thickness of the webs and the reduction factor can also be taken as design variables. In addition, in our second design scheme, the position of the spar cap of the blade section is also taken into account. Therefore, two design schemes are presented in this section. In the first one the position of the spar cap is fixed. In the second, the position of the spar cap is varied. The variables of the two design schemes are defined as follows:

$$X_1 = \begin{bmatrix} x_1 & x_2 & x_3 & x_4 & x_5 & x_6 & x_7 \end{bmatrix}^{\mathrm{T}}, \tag{9.24}$$

$$X_2 = \begin{bmatrix} x_1 & x_2 & x_3 & x_4 & x_5 & x_6 & x_7 & x_8 & x_9 \end{bmatrix}^{\mathrm{T}}. \tag{9.25}$$

According to the initial structural design of the wind turbine blade, the constraints of the design variables are shown in Tab. 9.8 (c is the chord length of each blade section).

Tab. 9.8: Design variables constraint range.

Design variables (mm)	Minimum	Maximum
Thickness of unidirectional laminate x_1	1.0	1.4
Thickness of biaxial laminate x_2	0.5	0.8
Thickness of triaxial laminate x_3	0.5	0.8
Thickness of coating x_4	1.0	1.4
Thickness of panel x_5	18.0	24.0
Thickness of two webs x_6	30.0	40.0
Thickness of reduced factor of sandwich x_7	1.0	1.4
Location of the leading edge web x_8	$0.15c$	$0.25c$
Location of the trailing edge web x_9	$0.50c$	$0.60c$

9.6.1.3 Design constraints

In the process of optimizing the design of the blade, it is not only necessary to meet the requirements for structural strength, but also to prevent the collision of the blade tip with the tower. Because the blade is made of composite materials that exhibit anisotropy, the strength constraint of maximum stress is not appropriate. The Tasi–Wu failure criterion is selected in this case to perform failure verification. For the outer plies of the blade surface, when all elements are free from the likelihood of failure, the optimized analysis procedure is completed. In addition, the tilt angle of the rotor is generally 6 on the condition of normal operation. The maximum allowable distance between the blade tip and tower is set to 3.22 m:

$$d_{\mathrm{tip}_{\max}} \leq 3.2. \tag{9.26}$$

W Tsai–Wu failure criterion:

$$F_1\sigma_1 + F_2\sigma_2 + F_{11}\sigma_1^2 + F_{22}\sigma_2^2 + F_{66}\sigma_6^2 + 2F_{12}\sigma_1\sigma_2, \leq 1 \tag{9.27}$$

where σ_1, σ_2 and σ_6 are the three in-plane stress components in the local orthotropic ply axis; and F_1, F_2, F_{11}, F_{22}, F_{66} are the strength parameters of the composite material. The stress components in this inequality are then computed by the FE program.

Finally, the weight load is imposed, and boundary conditions consisting of fixing all six degrees of freedom of nodes are placed at the root.

9.6.2 Optimization algorithm combined with finite element method

For the utilization of intelligence algorithm in composite structural design, a genetic algorithm is widely used [200–204] in the simple composite plate and shell structures. The application of a particle swarm optimization algorithm (PSO) in some studies in recent years [205–208] has also achieved good results. However, the shape of the wind turbine blade and the thickness of composite laminates vary along spanwise locations. What is more, the internal structure is more complex than those of simple composite plate and shell structures, which brings great difficulties to optimizations. This is the reason why the particle swarm algorithm is rarely used, together with the finite element method, in the optimization of composite wind turbine blades. In view of this, the PSO algorithm is chosen as the structural optimal algorithm of the wind turbine blade. A method combining the PSO algorithm and a finite element program to optimize the composite structure of the wind turbine blade with variable thickness is presented. The optimized design flow chart is shown in Fig. 9.57. This flow chart mainly comprises three parts: the parametric modeling of the blade, the PSO algorithm and aerodynamic loading applied to the blade element.

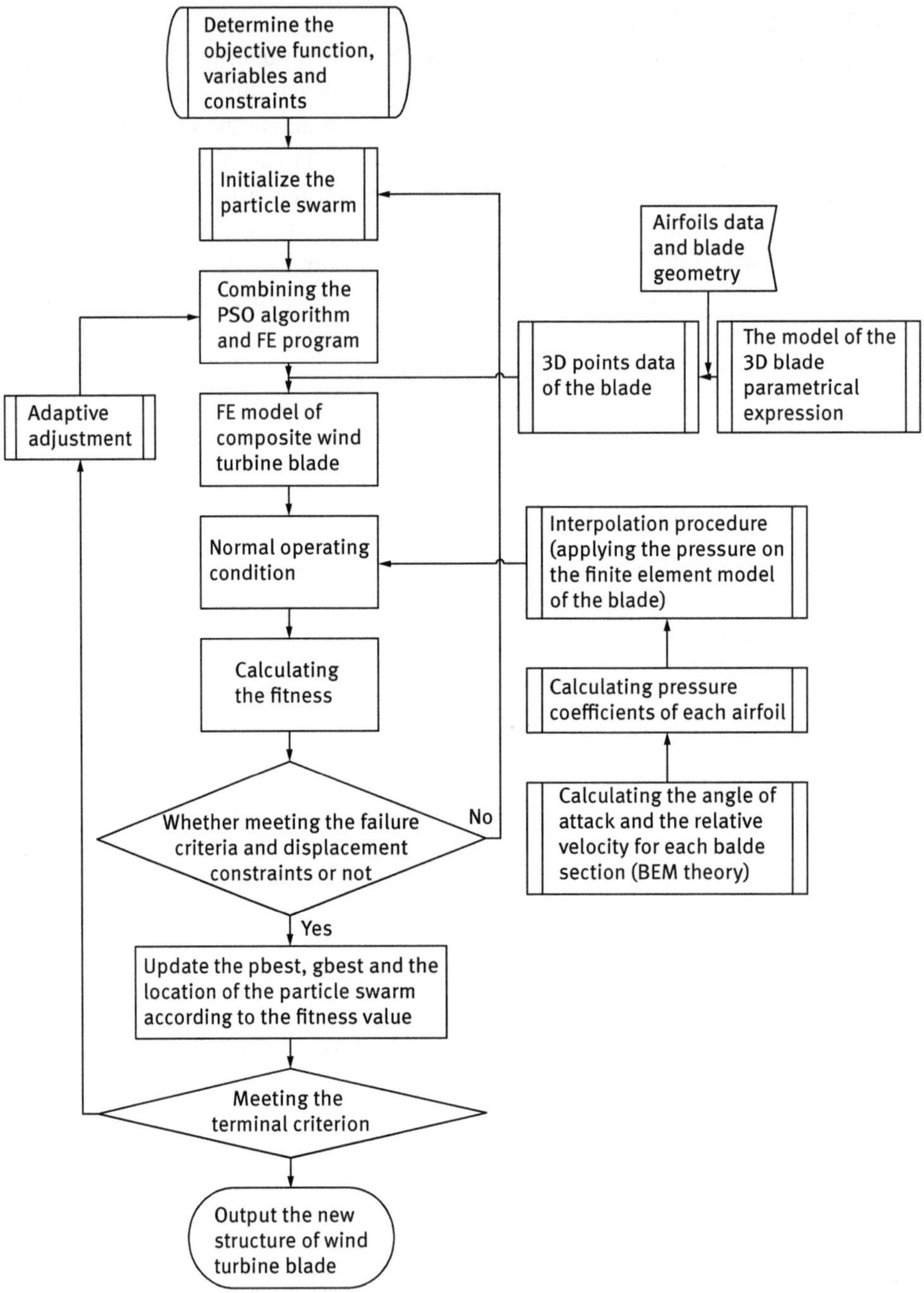

Fig. 9.57: The optimization flowchart combining PSO algorithm with finite element method.

9.7 Optimization results

Based on the two optimized design schemes, a procedure combining finite element analysis and PSO algorithm to optimize composite structures of wind turbine blades is developed. The basic parameters of the improved PSO algorithm are: learning factor $C_1 = C_2 = 0.5$, variable dimension $n = 7$ (the position of the spar cap is fixed) or $n = 9$ (the position of the spar cap is varied), population size of 30, the maximum number of generations is set as 120, and the inertia factor is taken as 0.85. Under the condition of the requirements of the failure criterion and the displacement constraint, the generation result of the two optimal schemes is shown in Fig. 9.58 and 9.59. It can be seen that the optimal results converge when the number of iteration step reaches 120. The manufacturing factors of composite laminate are taken into account; the optimized results of the design variables have been rounded up (as shown in Tab. 9.9).

The partial comparison results for the structural parameters between the two optimization schemes and the original blade are shown in Fig. 9.60–9.63. Tab. 9.10 shows the comparison results for mass and displacement and so on. From the figures and Tab. 9.10, the mass of the original blade is 6.859 t, which is closed to the weight of a typical 2 MW wind turbine blade (the mass of a typical 2 MW blade is about 6.528 t [194]. This can be taken as proof of the rationality of the model. After optimization, the mass of the two design schemes is 6.305 t and 6.096 t, respectively. This means that the mass of the blade can be reduced greatly by the two design schemes. The mass reduction of scheme II is particularly evident with a value of about 11.518 %. And the blade tip deflection is smaller than with scheme I. The main reason is that the location of the spar

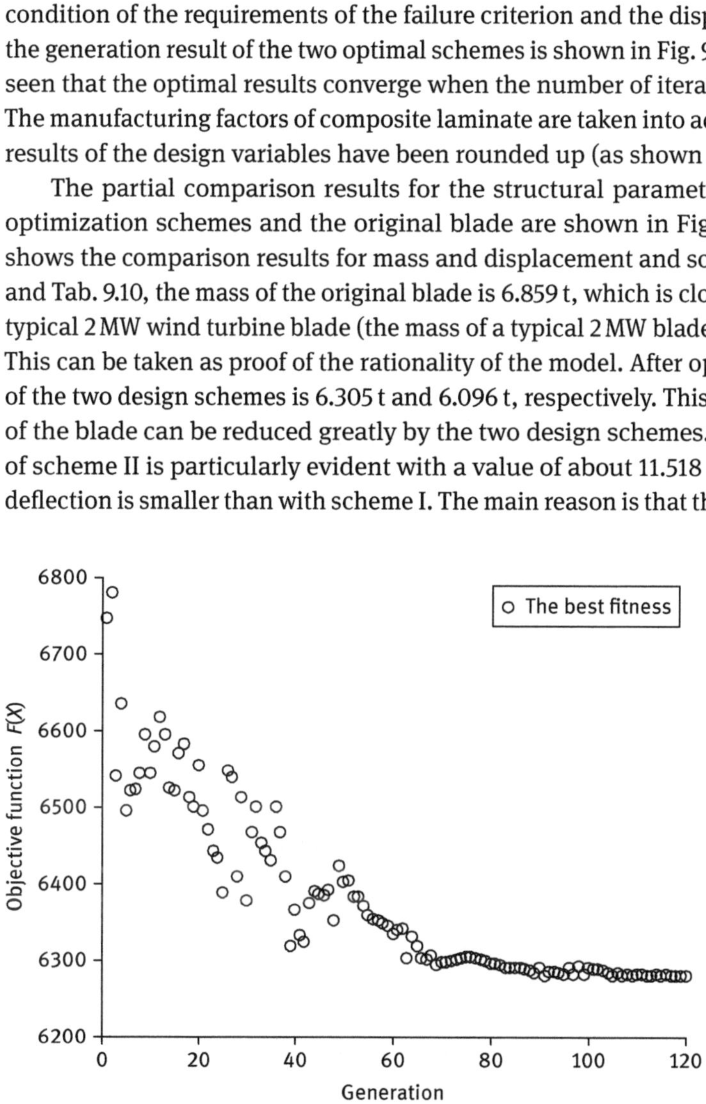

Fig. 9.58: The iterative course of optimization scheme I.

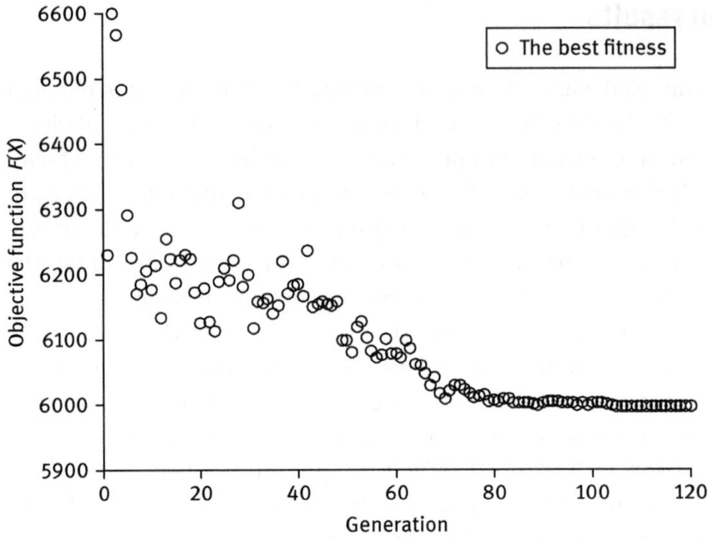

Fig. 9.59: The iterative course of optimization scheme II.

Tab. 9.9: Results for the design variables.

Design variables (mm)	Initial	Scheme I	Scheme II
Thickness of unidirectional laminate x_1	1.2	1.06	1.06
Thickness of biaxial laminate x_2	0.6	0.62	0.51
Thickness of triaxial laminate x_3	0.6	0.56	0.50
Thickness of coating x_4	0.6	0.47	0.52
Thickness of panel x_5	20	20.72	16.08
Thickness of two webs x_6	30.0	31.87	31.83
Thickness of reduced factor of sandwich x_7	1.0	1.15	1.06
Location of the leading edge web x_8	$0.24c$	$0.24c$	$0.196c$
Location of the trailing edge web x_9	$0.54c$	$0.54c$	$0.524c$

Tab. 9.10: Optimization results comparison.

	Initial blade	Optimization scheme I	Optimization scheme II
Mass (t)	6.859	6.305	6.069
Tip displacement (m)	2.896	3.214	3.167
Maximum tensile strain	0.00301	0.00332	0.00328
Maximum compressive strain	0.00320	0.00352	0.00347

cap is taken as a variable for scheme II (changed from 0.24–0.54c to 0.196–0.524c). As a result, the spar cap becomes wider, which is beneficial to controlling tip displacement. In addition, the thickness of the spar cap and the panel are both more or less reduced, which can contribute to a mass saving for the wind turbine blade.

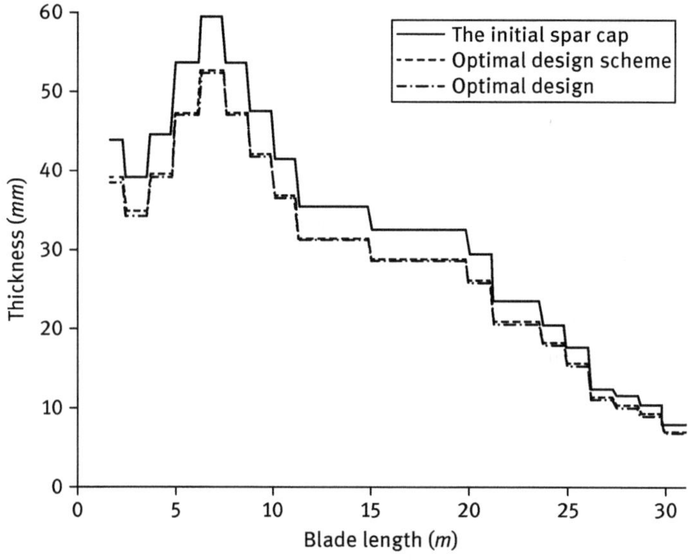

Fig. 9.60: The blade spar thickness.

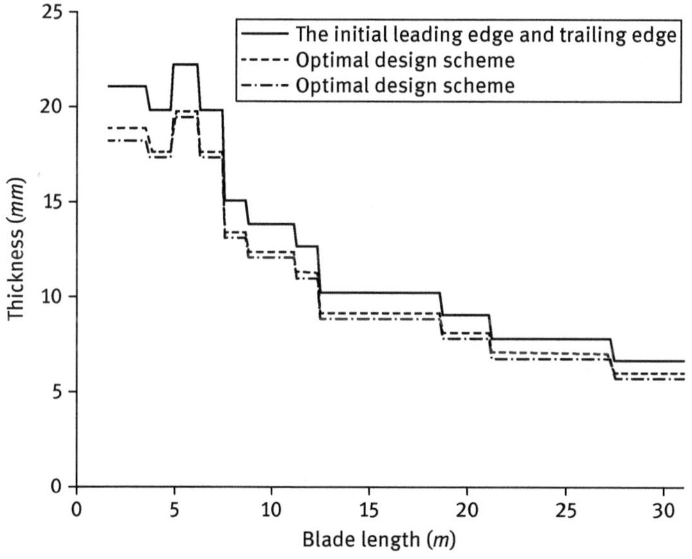

Fig. 9.61: The leading edge and trailing edge thickness.

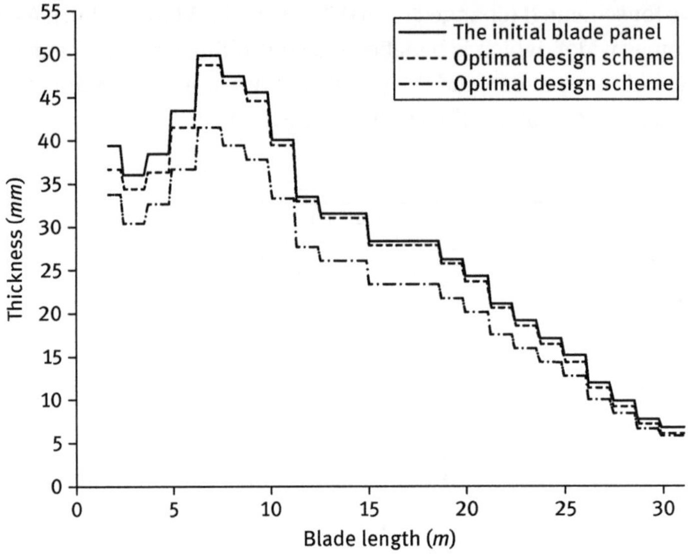

Fig. 9.62: The blade panel thickness.

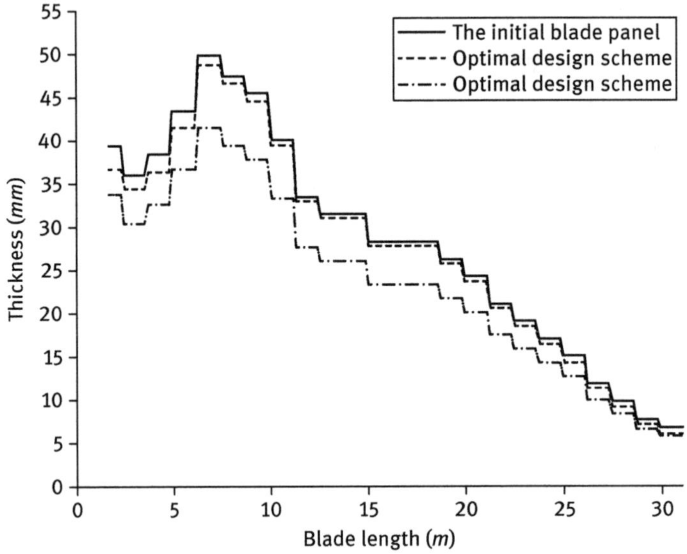

Fig. 9.63: The shear web thickness.

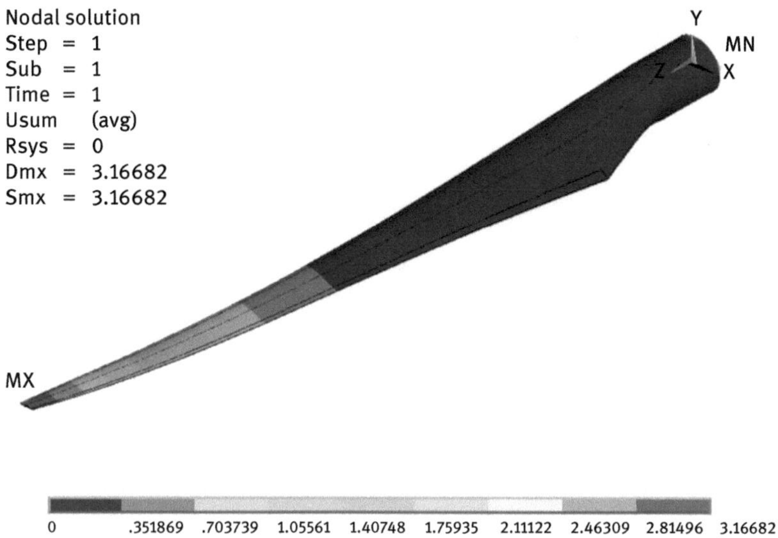

Fig. 9.64: The deflection of the blade.

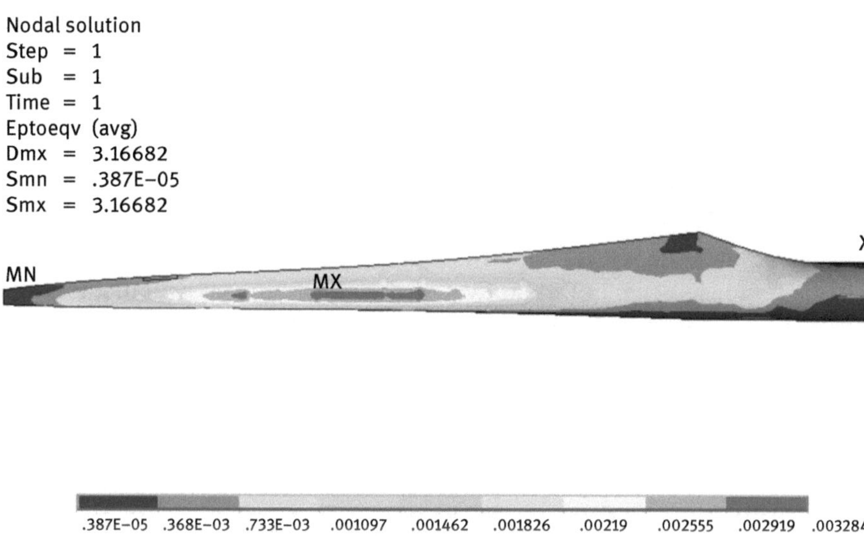

Fig. 9.65: Maximum tensile strain of the blade.

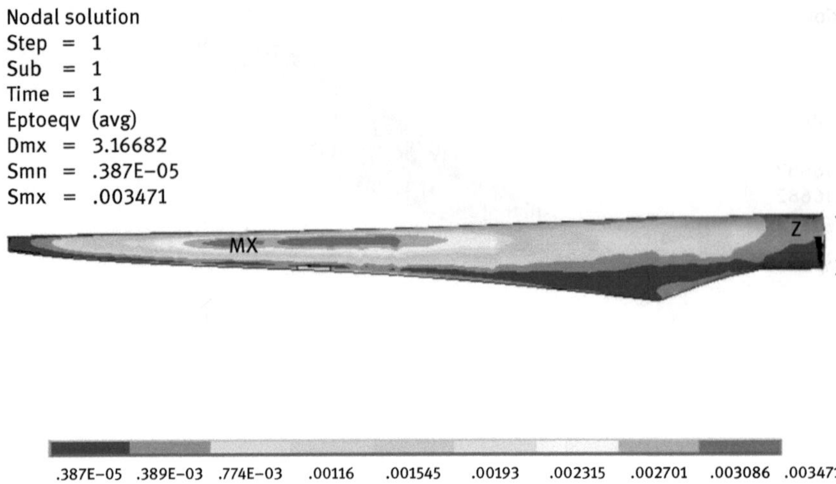

Fig. 9.66: Maximum compressive strain of the blade.

Furthermore, the strain of the blade is analyzed. The results of the strain analysis for scheme II under normal conditions are shown in Fig. 9.64–9.66. It is concluded that the main bending strain of the blade is located from 40 % to 80 % of the spar cap, but not near to the root. There are two reasons to explain this. One reason is that the normal operation of the wind turbine rated operating condition, which has less impact on the strength of the root than the limit wind load condition, is considered. The other reason is that the application of the aerodynamic load on the blade elements is different. The general approach of loading is to translate the pressure into concentrated load and then apply it on the finite element model. However, the pressure is directly applied on the blade finite elements in this study. Compared with the general method, the approach to aerodynamic load taken in this study is more accurate. Therefore, the results of the strain analysis on the rated operating condition are credible.

9.8 Chapter conclusions

After the basics of composite materials were introduced briefly, the initial laminate plies for a 2 MW wind turbine blade were designed based on the equivalent design method. A parametric element model of the composite wind turbine blade was built. Based on the modified BEM theory, a novel method of fluid-structure interaction was presented, which is the key issue for loading the aerodynamic pressure of the structural optimization. An optimized model of the wind turbine blade was established. Coupling an FE program and a PSO algorithm leads to a powerful tool suitable for composite wind turbine optimization. The procedure proposed not only allows thickness variation, but also permits the spar cap location to vary over the structure (scheme II). Results obtained in the two optimization schemes reported in this paper show that this procedure leads to significant weight savings. The mass reduction of scheme II is largest, with a value of about 11.518 %, which showed the superiority of scheme II. Then the strain of the blade was analyzed, and the location of the main bending strain was identified. There is still more work to do in the future. For example, the power coefficient should be considered if the chord lengths and twist angles are changed when optimizing the minimum mass and COE (cost of energy). In addition, with increasing power of wind turbines, the length of the blade becomes longer. As a result, it is a great challenge to the structural strength of the blade. Some new composite materials must be selected to replace the original materials. Therefore, another step will focus on the influence of materials selection on the structural properties of the wind turbine blade.

10 Analysis of the aeroelastic coupling of wind turbine blades

10.1 Introduction

The wind turbine blades are fundamental parts of a wind turbine, which is a rigid-flexible coupling multibody system. The aeroelastic stability analysis of blades is of vital importance to this multibody system. The vibration stability of the wind turbine system is determined by the coupled vibrations between the edgewise vibrations of blades and the transmission system, between the flapwise vibrations of the wind turbine blades and the tower, between the vibrations of the wind turbine blades and the pitching system, etc. If the rigidity of the tower and drive train are considered to be large enough, the dynamic characteristics of the wind turbine system can be basically reflected by those of the blades.

As the rated power of the wind turbine increases, the tower's height and the rotor diameter also increase. Consequently, the blades' rigidity become increasingly weaker and the flexibility become increasingly larger. The alternating loads on blades, including wind loads, inertia force and elastic force, would incur deformation and vibration of the blades. In turn, deformation and vibration will affect the incoming flow. In order to estimate the load and power characteristics more accurately when designing a wind turbine, it is of vital importance to consider the blades' vibrations in the loads.

Based on principles from structural dynamics and aerodynamics, the dynamic equations for wind turbine blades are established in this chapter. The characteristics of the blades' first mode flapwise vibrations, first mode edgewise vibrations and second mode flapwise vibrations are considered. The displacement, velocity and acceleration of the blades caused by the above-mentioned vibrations are studied, as well as their influence on the inflow velocity and the blade loads.

10.2 The structural kinetic model of wind turbine blades

According to the Hamilton principle for the holonomic-nonconservative dynamical systems, there exists

$$\int_{t_1}^{t_2} (\delta U - \delta T - \delta W)\,\mathrm{d}t = 0, \tag{10.1}$$

where U stands for system elastic energy; T for system kinetic energy; W for the sum of virtual work done by the generalized external force.

DOI 10.1515/9783110344387-011

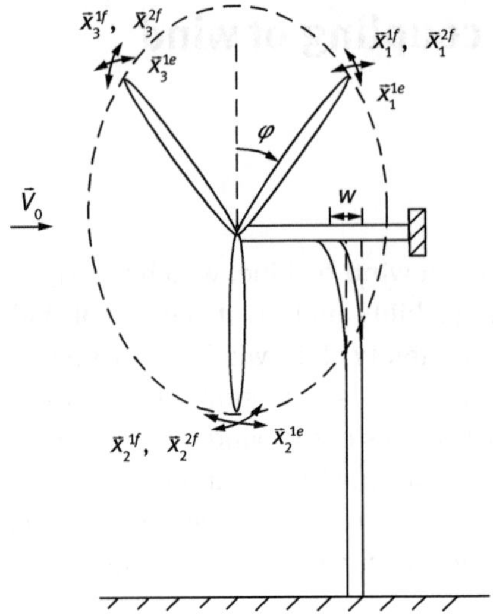

Fig. 10.1: Schematic diagram of the wind turbine's degrees of freedom.

Conducting variational operation on the above formula and according to the virtual work principle, the kinetic equation for the wind turbine blades is established:

$$M\ddot{x} + C\dot{x} + Kx = F_g, \tag{10.2}$$

where M stands for mass matrix; C for damping matrix; K for stiffness matrix; F_g for generalized force associated with external loads; x for a number of unknowns associated with the DOF. For simplicity, 11 DOF in total are used to describe a three-bladed wind turbine, as shown in Fig. 10.1. For each wind turbine blade, the first three eigenmodes (the first and second flapwise and the first edgewise modes) are counted. The other two unknowns are the displacement in the axial direction of the rotor root and the azimuth displacement of one blade. In summary, the unknowns can be written as

$$x = \left(w, \varphi, x_1^{1f}, x_1^{1e}, x_1^{2f}, x_2^{1f}, x_2^{1e}, x_2^{2f}, x_3^{1f}, x_3^{1e}, x_3^{2f}\right), \tag{10.3}$$

where w is the axial displacement of the whole rotor; φ is the azimuth displacement of the blades; and x_i^j is the deflection coefficient of the j-th mode on the i-th blade. After knowing the loads and the deformations, the vibration velocities and accelerations of each mode are calculated from equation (10.2).

Since only the first three eigenmodes are considered, the total deformation of a blade is calculated from a linear combination of the three major modes [184]:

$$\vec{u} = x^{1f} \cdot \vec{u}^{1f}(x) + x^{1e} \cdot \vec{u}^{1e}(x) + x^{2f} \cdot \vec{u}^{2f}(x). \tag{10.4}$$

Once the displacement is known, the velocity and the acceleration of the blade are found as follows:

$$\dot{\vec{u}} = \dot{x}^{1f} \cdot \vec{u}^{1f}(x) + \dot{x}^{1e} \cdot \vec{u}^{1e}(x) + \dot{x}^{2f} \cdot \vec{u}^{2f}(x), \tag{10.5}$$

$$\ddot{\vec{u}} = \ddot{x}^{1f} \cdot \vec{u}^{1f}(x) + \ddot{x}^{1e} \cdot \vec{u}^{1e}(x) + \ddot{x}^{2f} \cdot \vec{u}^{2f}(x). \tag{10.6}$$

The velocity $\dot{\vec{u}}$ is the vibration velocity of the blade and will contribute to the relative velocity as

$$\vec{V}_{\mathrm{rel}} = \vec{V}_0 - \vec{V}_{\mathrm{rot}} + \vec{W} - \dot{\vec{u}}, \tag{10.7}$$

where $\vec{V}_0$ is the wind speed, $\vec{V}_{\mathrm{rot}}$ is the rotational speed and $\vec{W}$ is the induced velocity. In equation (10.7), $\vec{V}_{\mathrm{rot}} + \dot{\vec{u}}$ is the blade velocity, corresponding to the relative wind speed because to the blade motion has a negative sign.

10.3 The coordinate transformation

Fig. 10.2 is the coordinate system for the dynamic analysis of wind turbine blades. In this figure, coordinate system $R_0(x_0, y_0, z_0)$ is the inertial system, $R_{\mathrm{t}}(x_{\mathrm{t}}, y_{\mathrm{t}}, z_{\mathrm{t}})$ is the tower coordinate system, $R_{\mathrm{c}}(x_{\mathrm{c}}, y_{\mathrm{c}}, z_{\mathrm{c}})$ is the nacelle coordinate system, $R_{\mathrm{r}}(x_{\mathrm{r}}, y_{\mathrm{r}}, z_{\mathrm{r}})$ is the rotor coordinate system, $R_{\mathrm{f}}(x_{\mathrm{f}}, y_{\mathrm{f}}, z_{\mathrm{f}})$ is the flapwise vibration coordinate system, $R_{\mathrm{e}}(x_{\mathrm{e}}, y_{\mathrm{e}}, z_{\mathrm{e}})$ is the edgewise vibration coordinate system.

Here, the transition from nacelle coordinate system to tower coordinate system is

$$\begin{bmatrix} x_{\mathrm{t}} \\ y_{\mathrm{t}} \\ z_{\mathrm{t}} \end{bmatrix} = \begin{bmatrix} \cos\beta & \sin\beta & 0 \\ -\sin\beta & \cos\beta & 0 \\ 0 & 0 & 1 \end{bmatrix} \begin{bmatrix} x_{\mathrm{c}} \\ y_{\mathrm{c}} \\ z_{\mathrm{c}} \end{bmatrix}, \tag{10.8}$$

where β is the yaw angle. The nacelle coordinate system and the rotor coordinate system are transformed through

$$\begin{bmatrix} x_{\mathrm{c}} \\ y_{\mathrm{c}} \\ z_{\mathrm{c}} \end{bmatrix} = \begin{bmatrix} 1 & 0 & 0 \\ 0 & \cos\psi & \sin\psi \\ 0 & -\sin\psi & \cos\psi \end{bmatrix} \begin{bmatrix} x_{\mathrm{r}} \\ y_{\mathrm{r}} \\ z_{\mathrm{r}} \end{bmatrix}, \tag{10.9}$$

where ψ is the azimuth angle of the blade.

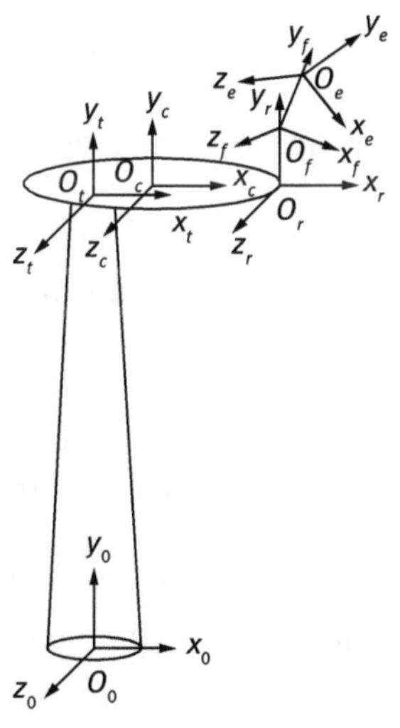

Fig. 10.2: Coordinate system of the wind turbine.

10.4 The wind load model

The wind model is indispensable when calculating the load on a wind turbine. Wind can be classified with the normal wind model or the extreme wind model.

10.4.1 The normal wind model

Normal wind comes from the average statistical nature of natural wind. When calculating the loads of wind, the wind profile model can be used:

$$\frac{\bar{V}(z)}{\bar{V}(z_S)} = \left(\frac{z}{z_S}\right)^a, \tag{10.10}$$

where $\bar{V}(z)$ is the average wind velocity at height Z (relative to the ground); $\bar{V}(z_S)$ is the average wind velocity at z_S (relative to the ground and usually refers to the hub height); a is the wind profile index (usually the constant a takes the value 2). So the unperturbed inflow velocity distribution at one point on the rotor plane can be expressed as

$$V(r, \psi) = V_h \left(1 + \frac{r \cos \psi}{H_h}\right)^a, \tag{10.11}$$

where r is the radial distance between the point to the hub center; ψ is the azimuth angle of the point on the rotor plane; V_h is the inflow velocity at hub height H_h.

10.4.2 The extreme wind model

The wind turbine should also be designed to withstand extreme wind, such as various extreme gusts. Extreme conditions can happen when the wind turbine is working, shut down or parked. Only a brief introduction is given here. Extreme wind speed can be classified into the following models:

(1) Extreme wind speed model (EWM). This extreme wind speed occurs every 1–50 years.

(2) Extreme operating gust (EOG). The wind speed first drops to the smallest value from the beginning and then rises abruptly until it reaches the maximum value. After that it drops abruptly to the smallest value and finally rises up to the initial value. The gust's amplitude and duration change along with the repetition period.

(3) Extreme direction change (EDC). The continuous change of wind direction follows the shape of the cosine curve with its amplitude and duration changing along with the repetition period.

(4) Extreme turbulence model (ETM). The extreme turbulence model uses the normal wind profile model and turbulence with its amplitude and duration changing along with the repetition period.

(5) Extreme coherent gust with direction change (ECD). The wind speed and direction all change instantly, just like in the EDC and the ETM models.
(6) Extreme wind shear (EWS). Both the transient (positive and negative) vertical shear and the transient horizontal shear can be considered.

10.5 Results validation

In order to validate the implementation of the structural model, the code is validated against the aeroelastic software FLEX which is used by most wind turbine companies [209]. FLEX is the wind turbine load calculation software developed by the Department of Wind Energy, Technical University of Denmark. It is applicable for constant or variable speed controlled, pitch or stall controlled wind turbines and can be used to calculate the nonlinear loads and responses considering multiple degrees of freedom of the blades.

The validation was conducted on a three blades of a 2 MW wind turbine, with blade length of 30.56 m. The shape parameters of the blades are provided in Tab. 10.1. In order to run two codes at identical conditions, only 11 DOF are activated in the FLEX code, and the classic Glauert tip-loss correction is used in both codes. The input to the aeroelastic computations, the mass and stiffness distributions of the 2 MW blades are listed in Tab. 10.2. The results obtained from the two codes are plotted in Fig. 10.3. The comparison of the rotor torque is shown in Fig. 10.4. As seen from these two figures, the total thrust and the shaft torque have almost the same amplitude and variation period. For the first flapwise deflection (Fig. 10.5), the new code predicts the same dominant frequency but with smaller amplitude, which means that the blade would deflect less in the flapwise direction. For the first edgewise deflection (Fig. 10.6), the frequency and the amplitude predicted by the new code are very similar to those from the FLEX code, but the mean edgewise positions are different. Since the eigenmodes in the FLEX code are defined on the plane passing through the blade tip, whereas the new code defines the eigenmodes depending locally on the airfoil cross section, small differences are expected. From the comparison, we can conclude that the newly-developed aeroelastic code is capable of correctly predicting the structural response of a wind turbine.

10.6 Case analysis

Based on the above aeroelastic model, a structural dynamics analysis was conducted on the blades of the 2 MW wind turbine in Section 10.5. The input wind speed is at 14 m/s (rated wind speed), the tip speed ratio is 5 and the rotational speed of rotor is 2.3 rad/s. Then the flapwise and edgewise displacement, velocity and acceleration of the blade can be obtained.

Tab. 10.1: Shape parameters of the 2 MW rotor blades.

Rotor radius (m)	Chord (m)	Twist (°)	Relative thickness (%)
1.46	2.0	0	100
2.96	2.0	0	100
6.46	3.3	8.0	30.58
9.46	3.0	7.0	24.10
12.46	2.7	6.0	21.13
15.46	2.4	5.0	18.70
18.46	2.1	4.0	16.81
21.46	1.8	3.0	15.46
24.46	1.5	2.0	14.38
27.46	1.2	1.0	13.30
30.56	0.02	0.0	12.2

Tab. 10.2: Structural parameters of the 2 MW rotor blades.

r (m)	EI_1 (MNm²)	EI_2 (MNm²)	*m* (kg/m)
1.46	14480.00	14480.00	3460.00
2.75	1901.60	1963.60	442.64
6.46	566.57	1400.30	337.19
9.46	235.68	851.64	275.64
12.46	119.16	525.11	229.02
15.46	59.83	329.37	191.40
18.46	29.76	207.20	166.92
21.46	13.80	120.99	159.47
24.46	5.39	59.94	84.52
27.46	1.51	24.54	47.88
30.56	0.16	7.22	19.99

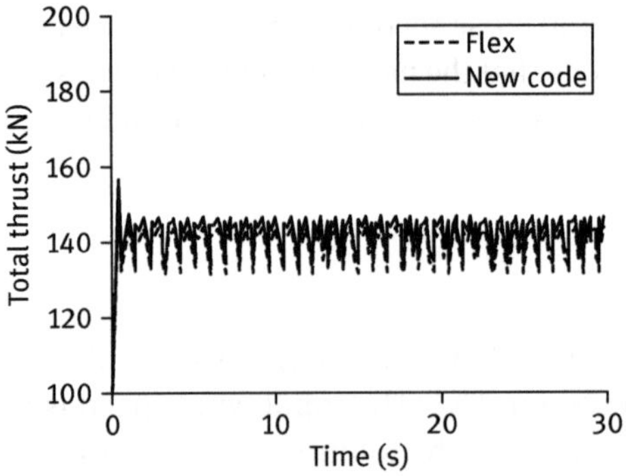

Fig. 10.3: Comparison of thrust on the rotor.

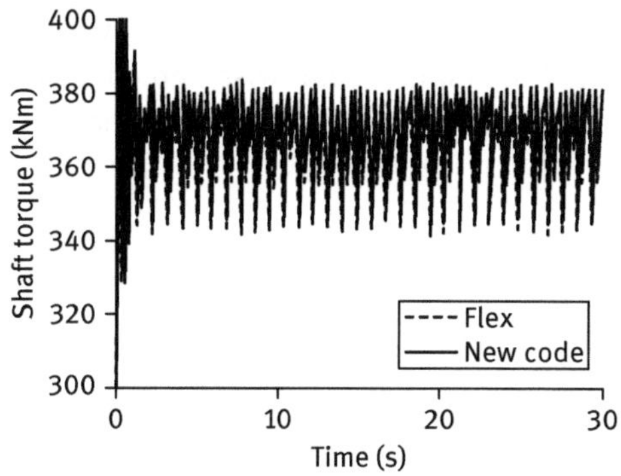

Fig. 10.4: Comparison of torque on the rotor.

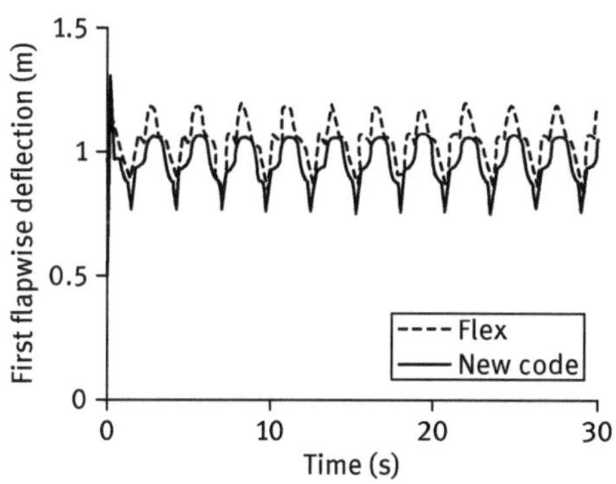

Fig. 10.5: The first flapwise deflection at the tip.

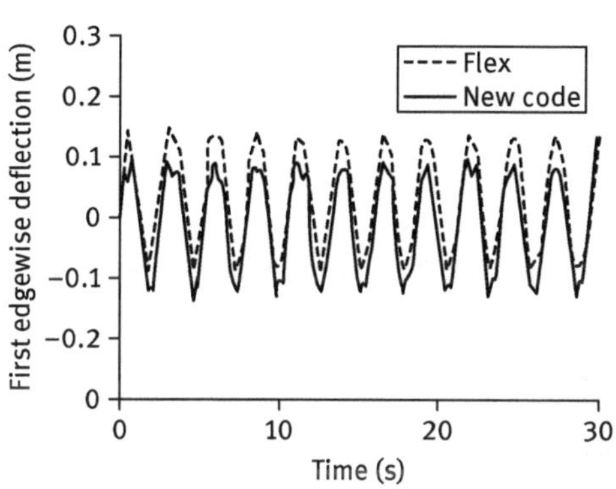

Fig. 10.6: The first edgewise deflection at the tip.

The first mode vibration flapwise and edgewise displacement of the blade tip are shown in Fig. 10.7. It can be seen that when the blades are working at their rated wind speed, the average flapwise deformation at the blade tip reaches about 1.2 m, but the edgewise deformation is smaller. The first mode flapwise and edgewise vibration velocity of the blade tip are shown in Fig. 10.8. It could be found that when the blades start working, the vibration speed is relatively higher. The flapwise vibration speed then turns into periodical change with the maximum speed reaching 1 m/s. It can be seen that the influence of blade vibration on relative inflow velocity is quite obvious. However, the edgewise vibration speed is relatively smaller. The first mode flapwise and edgewise vibration accelerated velocity of the blade tip is shown in Fig. 10.9 and is relatively stable.

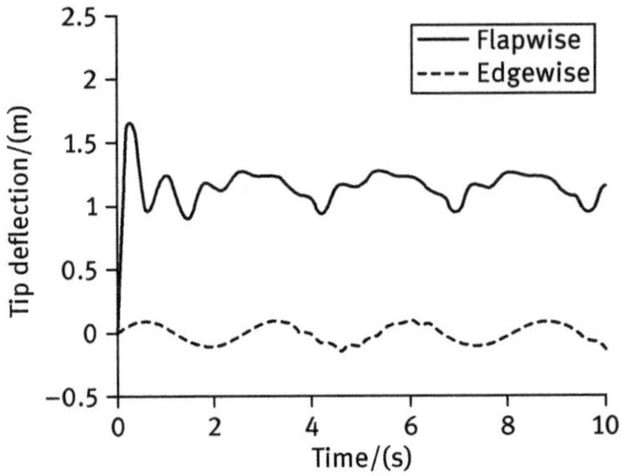

Fig. 10.7: The first mode vibration displacement.

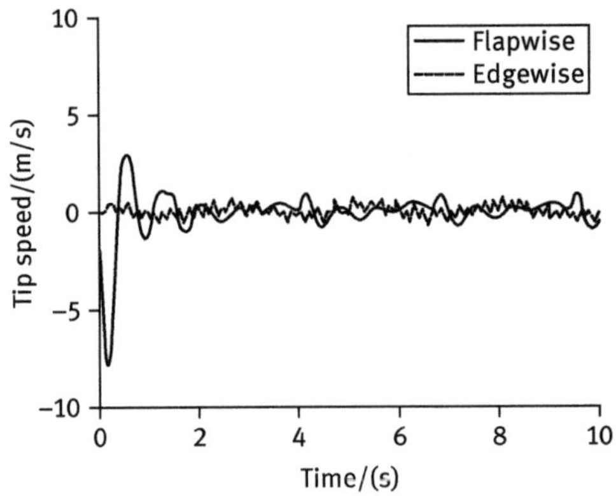

Fig. 10.8: The first mode vibration velocity.

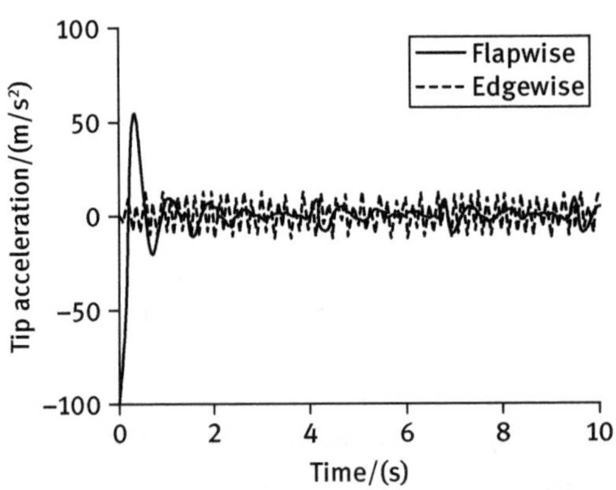

Fig. 10.9: The first mode vibration accelerated velocity.

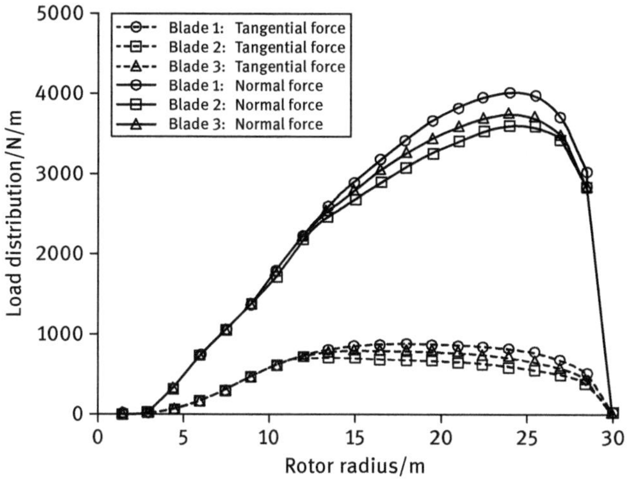

Fig. 10.10: Distribution of loads along the blade span.

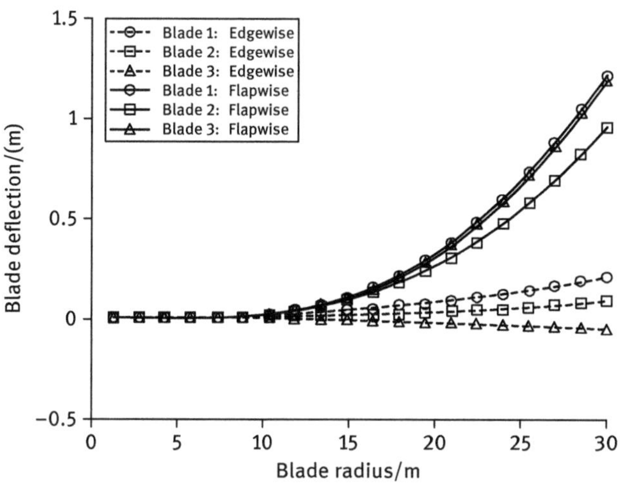

Fig. 10.11: Deformation along blade span.

The load and the deformation at different blade span locations are different. Fig. 10.10 is the normal force and tangential force on each blade at a certain time. The normal force and tangential force in the zone with radial position from 15 m to 25 m is relatively larger. Such sections are the main part contributing to the power output. Fig. 10.11 is the deformation of each blade at the same time. It could be found that the deformation of the blades increases gradually along the blade span with the maximum flapwise deformation reaching 1.25 m at the tip. The edgewise deformation is much smaller because of the weaker edgewise loads and the larger edgewise stiffness.

The power characteristics of the wind turbine are provided in Fig. 10.12. It can be seen that the rotor reaches its maximum power at rated wind speed 14 m/s. When the wind speed exceeds 14 m/s, the pitch control could maintain the output power of 2 MW. The relationship between the wind turbine's power coefficient and wind speed are depicted in Fig. 10.13. It can be seen that, during the normal working wind speed 7–14 m/s, the power coefficient of the rotor is relatively high and reaches the maximum value 0.463 at wind speed of 9 m/s.

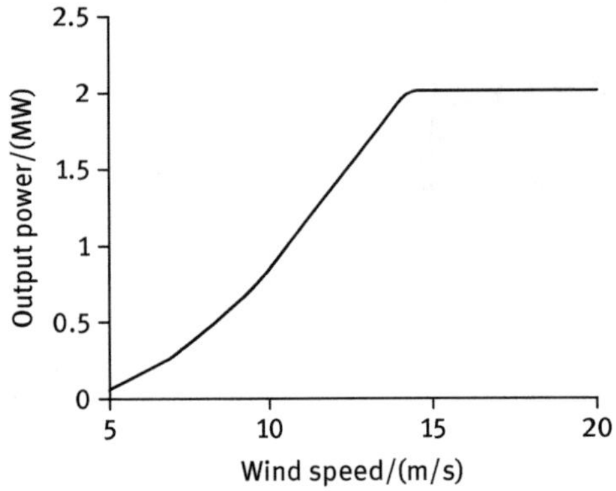

Fig. 10.12: Power characteristics.

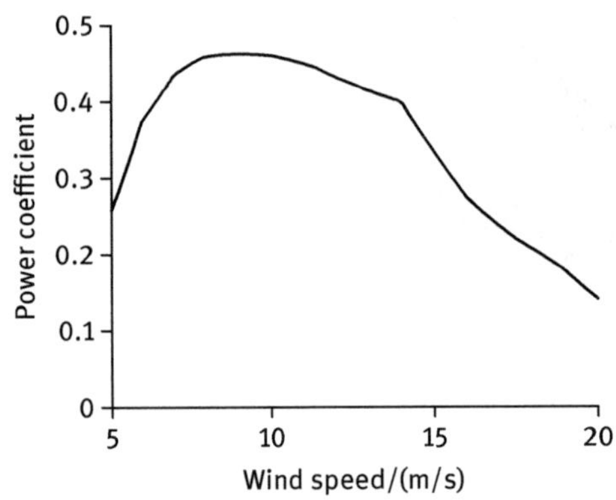

Fig. 10.13: Power coefficient.

10.7 Chapter conclusions

Based on principles from structural dynamics, coupling the vibration of blades into the wind turbine blade aerodynamic model, the dynamic equations for the wind turbine blades were established in this chapter. The first mode flapwise vibrations, first mode edgewise vibrations and second mode flapwise vibrations of blades were studied. The displacement, velocity and acceleration of the blades caused by the above-mentioned vibrations were studied. A comparison between this model and the widely used commercial software FLEX was conducted and the reliability of this new model was validated. Finally, calculations and analysis were done on a 2 MW wind turbine and the distribution of loads along the blade span and the power characteristics of the wind turbine were shown.

11 Aeroelastic stability analysis of two-dimensional airfoil sections for wind turbine blades

11.1 Introduction

The static aeroelastic basic equations suitable for the 2D airfoil sections of a wind turbine blade can be used to analyze the interactions between elastic deformation and aerodynamic force [210], such as torsional divergence and load redistribution. With increasing rated power of wind turbines, the rotor diameter has exceeded a magnitude of 100 m. The stiffness of the blade is becoming increasingly weaker and the flexibility of blade greater. The aerodynamic, gravitational and inertial forces on a wind turbine blade will cause deformation or vibration, which will seriously affect the normal operation of the wind turbine. Therefore, the static aeroelastic stability must be considered when designing the wind turbine blade [92]. J. C. Newman and P. A. Newman [211] adopted the finite element method to study the nonlinear static aeroelastic problem of aircraft wings. Joseph P. Hepp [212], from Duke University, utilized fluid-solid coupling method to study the static aeroelastic performance of the wing on ARW-2 (the Second Aeroelastic Research Wing) aircraft. A comparison between the calculated pressure distribution and experimental data was made and the reasons for the errors were analyzed. Martin Puterbaughd [213], on the basis of the theory of elasticity, simplified the 2D airfoil section into a beam element. Then the deformation and load redistribution of the typical airfoil were studied. The relationship between the elastic modulus and thickness of the airfoil was derived. Hoogedoorn Eelco et al. [214], by combining XFOIL and MATLAB, studied the elastic deformation of a 2D airfoil section for a wind turbine blade and drew the conclusion that the elasticity of a 2D airfoil section can increase the lift coefficient and lift-to-drag ratio. Li Deyuan [215] used the improved B–L (Beddoes–Leishman) dynamic stall model when analyzing the unsteady aerodynamic characteristics of an airfoil section. His results show that this model can successfully simulate the dynamic characteristics of wind turbine airfoils. The studies above mostly focused on the long straight wing of aircraft. However, the wind turbine blade airfoil differs from the wing airfoil in terms of airfoil camber and thickness, which makes the model more complex.

Therefore, the coupling mechanism between lift of airfoil and elastic torsional deformation is studied in this chapter. An aeroelastic feedback model for an airfoil section of a wind turbine blade is established. The critical condition of torsional divergence and the iteration model for redistributing aerodynamic force are derived. The effects of torsion stiffness, distance from stiffness center to aerodynamic center and different airfoils on the divergence speed, additional elastic angle and lift force are determined. Based on the classical airfoil section aeroelastic model, the quasi-steady and steady aerodynamic and structural coupling models are established. Then the flutter stability

DOI 10.1515/9783110344387-012

of airfoil sections from a 5 MW wind turbine blade is analyzed. The influences of the chordwise position of elastic center and gravity center on the flutter speed are analyzed thoroughly. The influence of different aerodynamic models on the flutter speed is also studied. Secondly, adopting the improved B–L dynamic stall aerodynamic model [216], the nonlinear aeroelastic stability analysis of a 2D airfoil section with three degrees of freedom is studied. Comparisons are made between the performance of linear and nonlinear aerodynamic models. The influence of airfoil camber on blade aeroelastic limit cycle oscillation is analyzed. The influence of structural damping on the nonlinear aeroelastic system response is investigated.

11.2 Static aeroelastic stability analysis of 2D airfoil section for wind turbine blades

11.2.1 Static aeroelastic model of wind turbine airfoil section

The wind turbine blade 2D airfoil section is shown in Fig. 11.1, with an initial angle of α_0 and inflow speed V. The airfoil section will reach equilibrium at the new angle of attack α under the action of aerodynamic force and elastic force.

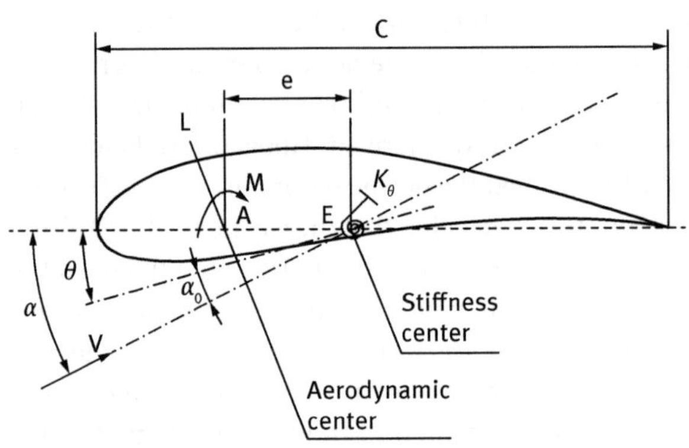

Fig. 11.1: Typical wind turbine blade 2D airfoil section.

The aerodynamic force on the airfoil can be divided into the lift force L and the moment M around the stiffness center (denoted by E) [217, 218]:

$$L = C_L q S, \tag{11.1}$$

$$M = M_A + Le = C_{MA} q S c + Le, \tag{11.2}$$

where

$$C_L = C_{L,0} + \frac{\partial C_L}{\partial \alpha}(\alpha_0 + \theta)$$

is the lift coefficient; $C_{L,0}$ is the lift coefficient when the angle of attack is zero; $\partial C_L/\partial\alpha$ is the airfoil section lifting line slope rate; $q = \frac{1}{2}\rho V^2$ is the dynamic pressure; ρ is air density; $S = C \times 1$ is the section area of unit length; M is the moment around the stiffness center; M_A is the moment around the aerodynamic center (positive with the rising of leading edge); C_{MA} is the moment coefficient around the aerodynamic center; e is the distance from aeroelastic center to rigid center (positive pointing trailing edge).

Normally, for wind turbine airfoils, C_{MA} and $\partial C_L/\partial\alpha$ are both functions of airfoil shape, and $C_{L,0}$, M_A are not zero. The aerodynamic moment around stiffness center is

$$M = C_{MA} qSc + \left[C_{L,0} + \frac{\partial C_L}{\partial\alpha}(\alpha_0 + \theta)\right] eqS. \tag{11.3}$$

According to the equilibrium condition of the aerodynamic moment and the elastic moment, we obtain that

$$K_\theta \theta = C_{MA} qSc + \left[C_{L,0} + \frac{\partial C_L}{\partial\alpha}(\alpha_0 + \theta)\right] eqS, \tag{11.4}$$

where K_θ is the torsional spring constant. Equation (11.4) can be written as:

$$\left(K_\theta - eqS\frac{\partial C_L}{\partial\alpha}\right)\theta = \left(C_{L,0} + \frac{\partial C_L}{\partial\alpha}\alpha_0\right) eqS + C_{MA} qSc. \tag{11.5}$$

Through the equation (11.5), the relationship between the lift coefficients, aerodynamic center, dynamic pressure and airfoil torsional angle can be obtained:

$$\theta = \frac{eqS\left(\frac{\partial C_L}{\partial\alpha}\alpha_0 + C_{L,0}\right) + qScC_{MA}}{K_\theta - \frac{\partial C_L}{\partial\alpha} eqS}. \tag{11.6}$$

From equation (11.6), it can be seen that the denominator can be 0 when the dynamic pressure q reaches a certain value. The elastic torsion angle θ tends to infinity at this condition. This phenomenon is known as torsional divergence. It is clear that the spring has been destroyed before θ reaches infinity. Thus, the divergence condition of an airfoil section is

$$K_\theta - \frac{\partial C_L}{\partial\alpha} eqS = 0. \tag{11.7}$$

Thus the dynamic pressure at divergence is:

$$q_D = \frac{K_\theta}{\frac{\partial C_L}{\partial\alpha} eS}. \tag{11.8}$$

With the dynamic pressure $q = \frac{1}{2}\rho V^2$, the divergence speed of the airfoil section can be obtained:

$$V_D = \sqrt{\frac{2K_\theta}{\rho\frac{\partial C_L}{\partial\alpha} eS}}. \tag{11.9}$$

From equation (11.8), it can be found that when the e is negative (the stiffness center is on the left of the aerodynamic center) the divergence pressure q_D is negative, which has no physical meaning. It implies that the 2D airfoil section is aeroelastic stable under any dynamic pressure. In the case of supersonic inflow, as the aerodynamic center moves towards the trialing edge, the risk of torsional divergence is greatly reduced. It can also be seen from equation (11.8) that the divergence dynamic pressure q_{D} has no relationship to the initial angle of attack α_0. This phenomenon can also be analyzed through letting the right part (right of the equal sign) in equation (11.5) equal zero. When $\alpha_0 = 0$, we can obtain through the equations (11.1)–(11.5) that

$$\left(K_\theta - \frac{\partial C_{\mathrm{L}}}{\partial \alpha} eqS\right)\theta = 0. \tag{11.10}$$

This is a homogeneous equation with the elastic torsion angle θ as the variable. In addition to one solution of $\theta = 0$, this equation can also have any nonzero solutions if the coefficient of θ is zero. So we obtain

$$K_\theta - \frac{\partial C_{\mathrm{L}}}{\partial \alpha} eqS = 0. \tag{11.11}$$

Through equation (11.11), the dynamic pressure q can be solved which is the divergent dynamic pressure represented in equation (11.8). Taking the elastic torsion angle θ in (11.6) into the expression of lift L (11.1), the coupling relationship between the lift and the elastic torsion angle can be obtained:

$$L = C_{\mathrm{L}} qS = \left\{ C_{\mathrm{L},0} + \frac{\partial C_{\mathrm{L}}}{\partial \alpha}\left[\alpha_0 + \frac{eqS\left(\frac{\partial C_{\mathrm{L}}}{\partial \alpha}\alpha_0 + C_{\mathrm{L},0}\right) + qScC_{\mathrm{MA}}}{K_\theta - \frac{\partial C_{\mathrm{L}}}{\partial \alpha} eqS}\right]\right\} qS. \tag{11.12}$$

Equation (11.12) is the iterative model for the aerodynamic load redistribution. When the divergence condition in equation (11.7) is established, the lift force tends to infinity. And when the dynamic pressure q is less than the divergence dynamic pressure q_{D}, the denominator of equation (11.12) will be greater than zero. The lift will have a definite value and will change with dynamic pressure q and the angle of attack α_0.

11.2.2 Analysis of the aeroelastic feedback system for a typical airfoil

According to the above mentioned model, the aeroelastic feedback system of three typical airfoils NACA 0018, NACA 64418 and Risø-A1-18 [106] was analyzed. The lift coefficient and lift moment coefficient of the three airfoils (with the angle of attack range from 0 to 20 degrees) are shown in Fig. 11.2 and 11.3.

Because the relationship between lift and torsional angle is derived from the linear elastic theory, the angle of attack range from 0 to 6 degrees was chosen to get $C_{\mathrm{L},0}$, $\partial C_{\mathrm{L}}/\partial \alpha$ using XFOIL software. And the lifting moment coefficient C_{MA} at an angle of attack of 6° was calculated. The calculation results are shown in Tab. 11.1.

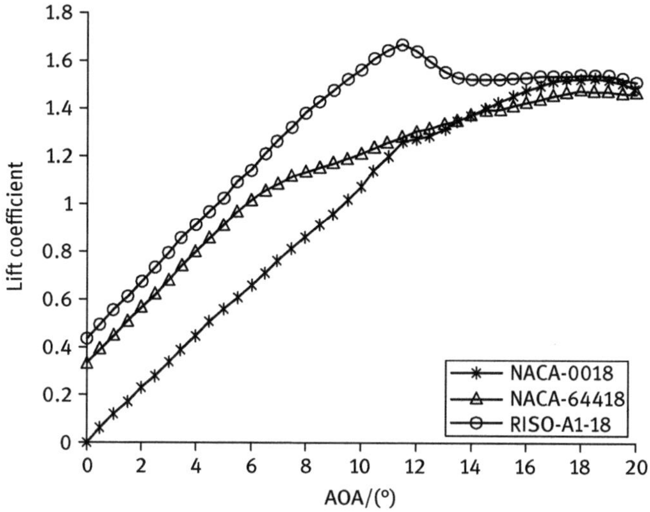

Fig. 11.2: Variation of lift coefficient.

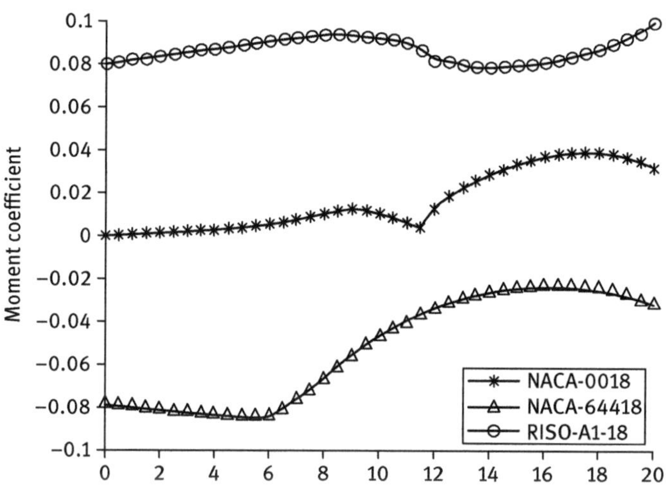

Fig. 11.3: Variation of lift moment coefficient.

Tab. 11.1: The parameters of three airfoils.

Airfoil name	$C_{L,0}$	$\partial C_L/\partial\alpha$	C_{MA}
NACA 0018	0	6.2660	0.0049
NACA 64418	0.3289	6.5929	−0.0839
Risø-A1-18	0.4314	6.8166	0.09

The other related parameters are as follows: chord length $c = 1$ m, air density $\rho = 1.225$ kg/m^2, inflow velocity 25 m/s. For the sake of simplicity, the aerodynamic center is taken to be $0.25c$, stiffness center is $0.35c$. So the distance from the aerodynamic center to the stiffness center is about $0.1c$ [219]. The influence of torsional rigidity on the airfoil is shown in Fig. 11.4–11.6, with torsional stiffness varying from 10^4 to 10^5 Nm^2. It can be seen that the torsional stiffness has a great effect on the static elastic performance

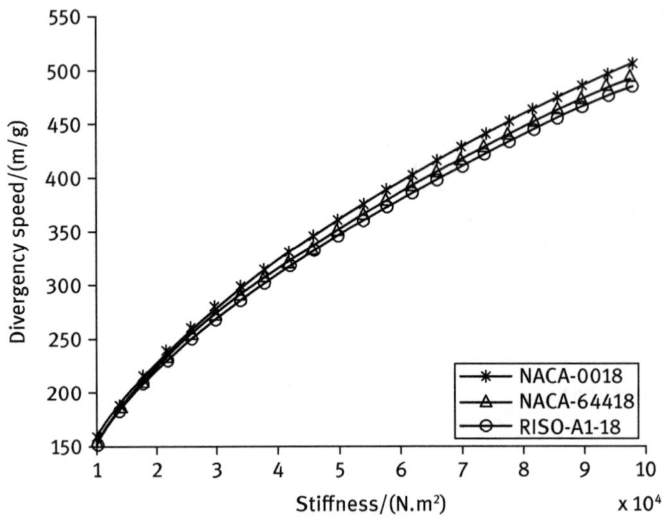

Fig. 11.4: Effect of torsion stiffness on divergence speed.

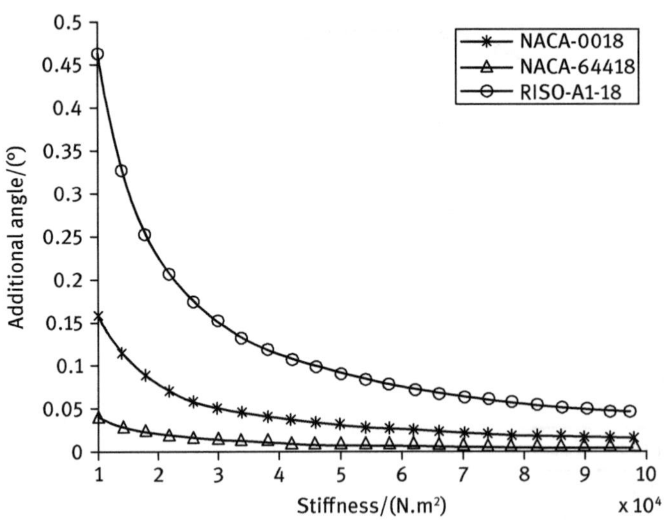

Fig. 11.5: Effect of torsion stiffness on elastic torsion angle.

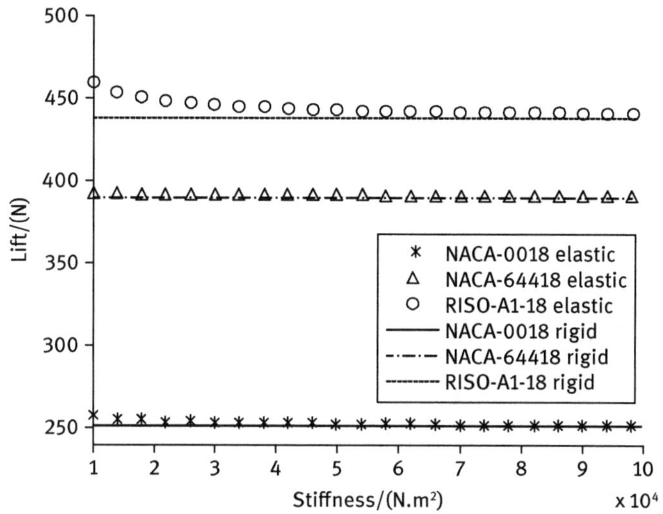

Fig. 11.6: Effect of torsion stiffness on lift force.

of the airfoil section. With increasing torsional stiffness, the divergence speed increases rapidly, the elastic torsion angle decreases sharply and the lift distribution tends to be stable.

The effects of e (distance from the stiffness center to the aerodynamic center) on the airfoil are shown in Fig. 11.7–11.9. It can be seen that with decreasing distance, the divergence speed increases rapidly, the elastic torsion angle decreases and the lift distribution tends to be stable. Therefore, increasing the distance e can also provide lower divergence speeds.

With the torsional stiffness fixed at 10^4 Nm^2, the aeroelastic performance of different airfoils was studied. According to Fig. 11.10, the elastic torsion angle changes linearly with angle of attack. Risø-A1-18 has the largest elastic torsion angle. When the angle of attack is 0, the additional angle of attack is about 0.28 degrees which can increase the lift coefficient and lift-to-drag ratio. The NACA 0018 airfoil has the second largest additional angle. However, the elastic torsion angle of the NACA 64418 airfoil is negative, which reduces the aerodynamic performance of the wind airfoils. Due to the influence of the elastic torsion angle, the lift force distribution has the same trend, as shown in Fig. 11.11. Therefore, when the torsional stiffness is smaller, different airfoils have different static aeroelastic sensitivity. So it is of vital importance to design suitable airfoils for large wind turbine blades.

With the torsional stiffness fixed at 10^5 Nm^2, the aeroelastic performance of different airfoils was studied. As can be seen from Fig. 11.12 and 11.13, the airfoils have relative smaller elastic torsion angles when stiffness is large. The change of aeroelastic lift distribution is smaller compared with stiffness of 10^5 Nm^2, which shows that the torsional stiffness has a great influence on the static aeroelastic characteristics of an airfoil section.

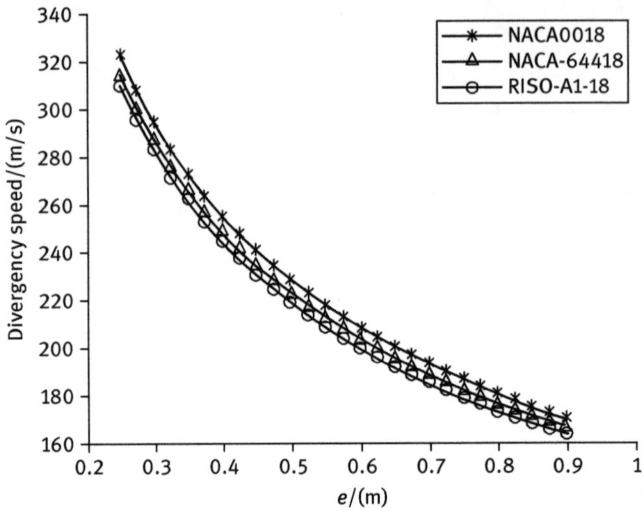

Fig. 11.7: Effect of distance from stiffness center to aerodynamic center on divergence speed.

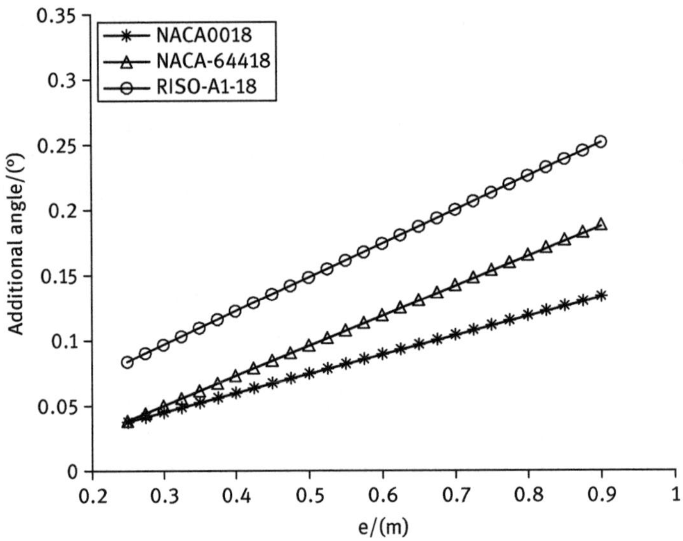

Fig. 11.8: Effect of torsion stiffness on additional torsional angle.

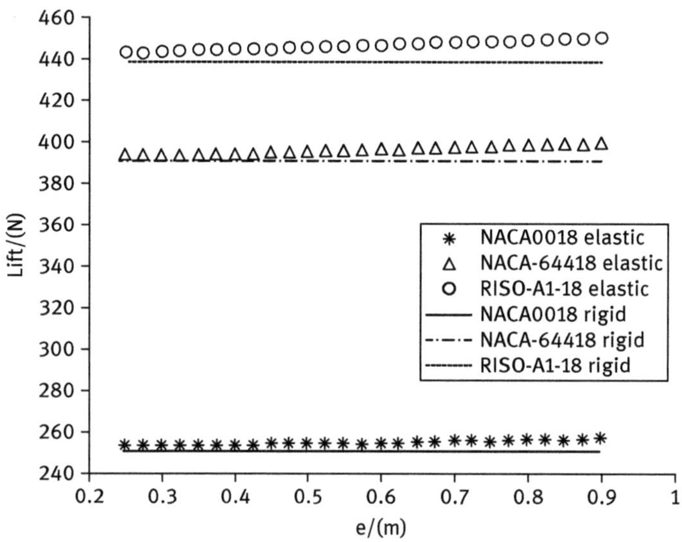

Fig. 11.9: Effect of torsion stiffness on lift force.

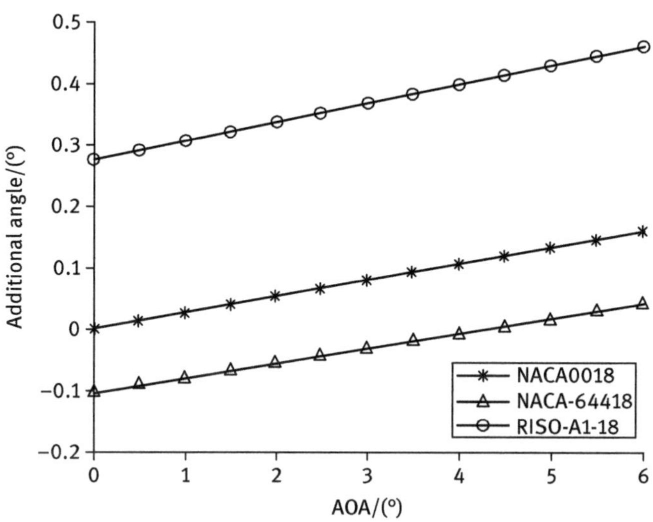

Fig. 11.10: Additional torsion angle of different airfoils (twist stiffness is 10^4 Nm^2).

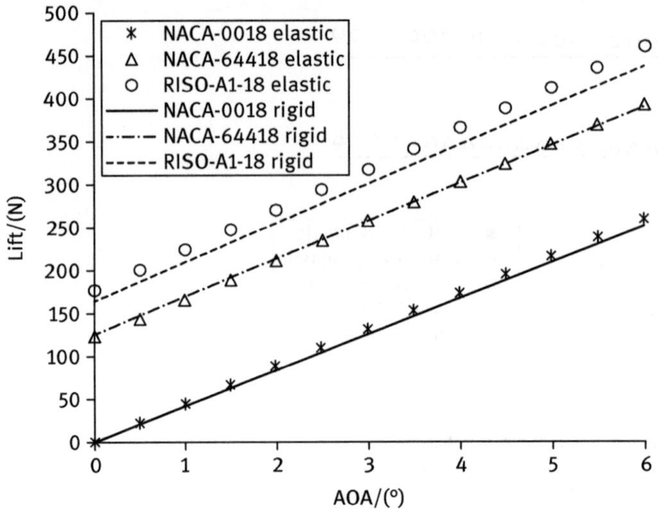

Fig. 11.11: Lift of different airfoils (twist stiffness is 10^4 Nm^2).

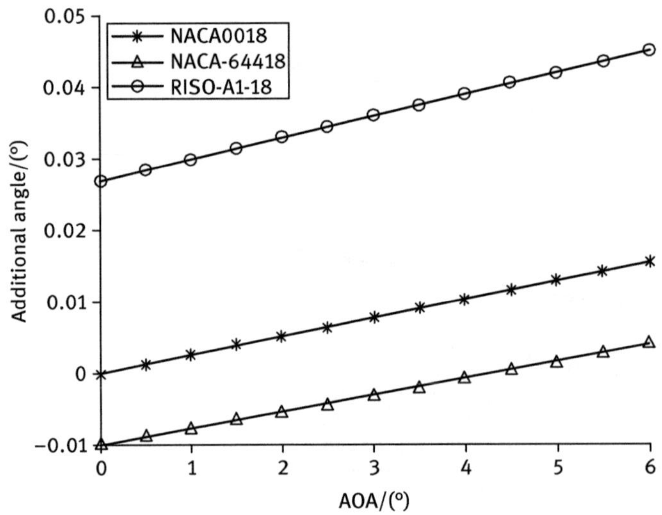

Fig. 11.12: Additional torsion angle of different airfoils (twist stiffness is 10^5 Nm^2).

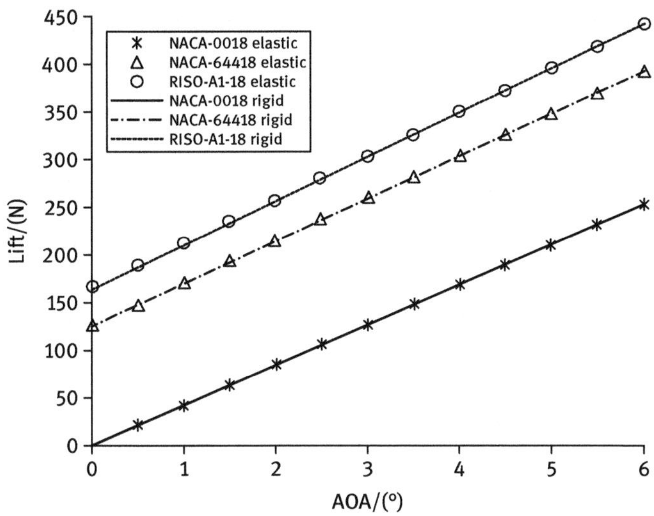

Fig. 11.13: Lift of different airfoils (twist stiffness is 10^5 Nm^2).

11.3 Classic flutter problem

11.3.1 Structural dynamic model

The mechanical model for an airfoil section of unit length is shown in the Fig. 11.14. It has two degrees of freedom, one is the vertical displacement of the rigidity center h and the other is the rotational angle of airfoil section θ. The airfoil section is supported by a tension spring and a torsion spring with respective stiffness of K_h and K_θ. x_f is the chordwise position of rigidity center; x_c is the chordwise position of the center of mass. The hypothesis is adopted that the airfoil sections are rigid without deformation and the elastic, aerodynamic, mass centers of the blade sections are located in the same

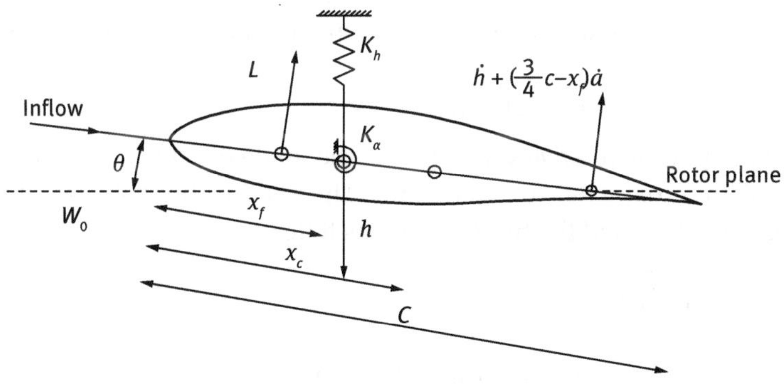

Fig. 11.14: Airfoil model of two degrees of freedom.

line. As it is difficult to predict the structural damping, the structure damping term is not considered here.

The Lagrange equation is used to derive the vibration equation of the elastic blade [210]:

$$\begin{bmatrix} m & S \\ S & I_\alpha \end{bmatrix} \begin{Bmatrix} \ddot{h} \\ \ddot{\theta} \end{Bmatrix} + \begin{pmatrix} K_h & 0 \\ 0 & K_h \end{pmatrix} \begin{Bmatrix} h \\ \theta \end{Bmatrix} = \begin{Bmatrix} Q_h \\ Q_\theta \end{Bmatrix}, \tag{11.13}$$

where $m = \int_0^c \bar{\rho}\,\mathrm{d}x$ is the blade mass per unit length; $\bar{\rho}$ is the chordwise density of the airfoil section; $S = m(x_\mathrm{c} - x_\mathrm{f})$ is the inertia about the Y_f axis (the origin of Y_f is the rigidity center, the direction of Y_f is perpendicular to chord) per unit length; $I_\alpha = \int_0^c \bar{\rho}(x - x_\mathrm{f})^2\,\mathrm{d}x$ is mass moment of inertia about the Y_f axis per unit length; Q_h is the aerodynamic force caused by the vibration of the airfoil section; Q_θ is the aerodynamic moment caused by the vibration of the airfoil section.

11.3.2 The aerodynamic model

Quasi-steady aerodynamic force and unsteady aerodynamic force were analyzed in the aerodynamic elasticity studies here, in other words, the flutter analysis. The flutter speed of the airfoil under these loads was calculated and the differences were compared. The effects of the chordwise position of the elastic center and gravity center on the flutter speed were studied.

11.3.2.1 The quasi-steady aerodynamic force

The quasi-steady aerodynamic was proposed by Grossman. According to the theory of thin airfoil, the quasi-steady assumption was introduced. The free vortex shed from the trailing edge of the airfoil can be neglected. The attached vorticity distribution on the thin airfoil can keep the flow attached and satisfy the Kutta condition. Due to the linearization hypothesis, flutter is only related to the additional aerodynamic force caused by vibration. The aerodynamic force of the wind turbine blade is approximately expressed by the aerodynamic force on the 2D plane. The airfoil lift force [92] is expressed as

$$L = \tfrac{1}{2}\rho c W^2 C_\mathrm{L}(\alpha), \tag{11.14}$$

where ρ is the air density, W is relative velocity, α is the angle of attack, C_L is the lift coefficient of airfoil (approximately equal to $2\pi\alpha$). The relationship between the relative velocity, angle of attack and the twist angle is as follows:

$$W = \sqrt{W_0^2 + \dot{h}^2} \tag{11.15}$$

and

$$\alpha = \arctan\left(\frac{W_0 \sin\theta + \dot{h} + \left(\frac{3}{4}c - x_\mathrm{f}\right)\dot{\alpha}}{W_0 \cos\theta}\right), \tag{11.16}$$

where W_0 is the steady relative inflow velocity. Through taking the equation (11.15) into equation (11.14) and making linear treatment $\theta = \dot{h} = \dot{\theta} = 0$, the expression for airfoil quasi-steady aerodynamic force was obtained:

$$L = \rho W_0^2 c\pi \left(\theta + \frac{\dot{h}}{W_0} + \left(\frac{3}{4}c - x_f \right) \frac{\dot{\theta}}{W_0} \right). \tag{11.17}$$

Then the aerodynamic moment about rigidity center is obtained:

$$M = \rho W_0^2 ec^2 \pi \left(\theta + \frac{\dot{h}}{W_0} + \left(\frac{3}{4}c - x_f \right) \frac{\dot{\theta}}{W_0} \right) - \frac{1}{16} \rho W_0 c^3 \pi \dot{\theta}, \tag{11.18}$$

where E is the distance between the elastic center and the aerodynamic center.

11.3.2.2 The unsteady aerodynamics

Due to the relative low wind speed for wind turbines operating conditions, the Theodorsen unsteady aerodynamic model for incompressible viscous flow was introduced. The model considers the effect of free vortex in the airfoil wake. Due to the complex derivation, the final expression of lift was directly given. When an airfoil section is fluttering harmonically with frequency w, the aerodynamic lift L and moment $M_{x,f}$ around rigidity center per unit length are

$$L = \rho\pi b^2 W_0 \theta + \pi\rho W_0 c C(k) \left[W_0 \theta + \dot{h} + \left(\frac{3}{4}c - x_f \right) \dot{\theta} \right] + \rho\pi b^2 \left[\ddot{h} - \left(x_f - \frac{c}{2} \right) \ddot{\theta} \right], \tag{11.19}$$

$$M_{x,f} = -\left(\frac{3}{4}c - x_f \right) \rho\pi b^2 W_0 \dot{\theta} + \pi\rho W_0 ec^2 C(k) \left[W_0 \theta + \dot{h} + \left(\frac{3}{4}c - x_f \right) \dot{\theta} \right] + \left(x_f - \frac{c}{2} \right) \rho\pi b^2 \left[\ddot{h} - \left(x_f - \frac{c}{2} \right) \ddot{\theta} \right] - \frac{\rho\pi b^4}{8} \ddot{\theta}, \tag{11.20}$$

where $C(k)$ is the Theodorsen function, $k = \omega b / W_0$ is the equivalent frequency (ω is the frequency), b is half of the chord length. For convenience, the approximate expression of the Theodorsen function is given here [218]:

$$C(k) = 1 - \frac{0.165}{1 - \frac{0.0455}{k} j} - \frac{0.335}{1 - \frac{0.3}{k} j}, \tag{11.21}$$

$$C(k) = 1 - \frac{0.165}{1 - \frac{0.041}{k} j} - \frac{0.335}{1 - \frac{0.32}{k} j}. \tag{11.22}$$

Equation (11.21) is more appropriate when $k \leq 0.5$ and equation (11.22) is more appropriate when the $k > 0.5$.

11.3.3 The aerodynamic-structural coupling calculation model

11.3.3.1 The quasi-steady aerodynamic-structural coupling model

Taking equations (11.17) and (11.18) into equation (11.13), we obtain that

$$\begin{bmatrix} m & S \\ S & I_\theta \end{bmatrix} \begin{Bmatrix} \ddot{h} \\ \ddot{\theta} \end{Bmatrix} + \rho W_0 c\pi \begin{bmatrix} 1 & \left(\frac{3}{4}c - x_\mathrm{f}\right) + \frac{c}{4} \\ -ec & \left(\frac{c}{2} - x_\mathrm{f}\right)^2 + \left(\frac{3}{4}c - x_\mathrm{f}\right)\frac{c}{4} \end{bmatrix} \begin{Bmatrix} \dot{h} \\ \dot{\theta} \end{Bmatrix} + \begin{bmatrix} K_h & 0 \\ 0 & K_\theta \end{bmatrix} \begin{Bmatrix} h \\ \theta \end{Bmatrix}$$
$$+ \rho W_0^2 c\pi \begin{bmatrix} 0 & 1 \\ 0 & -ec \end{bmatrix} \begin{Bmatrix} h \\ \theta \end{Bmatrix} = \begin{Bmatrix} 0 \\ 0 \end{Bmatrix}. \tag{11.23}$$

Equation (11.23) can also be written as

$$A\ddot{q} + (B + \rho W_0 c\pi C)\dot{q} + (D + \rho W_0^2 c\pi E)q = 0. \tag{11.24}$$

According to vibration theory, differential equations of motion (11.24) can be transformed into the state equation:

$$\dot{z} = Qz, \tag{11.25}$$

where

$$Q = \begin{bmatrix} -A^{-1}(B + \rho W_0 c\pi C) & -A^{-1}(D + \rho W_0^2 c\pi E) \\ I & 0 \end{bmatrix},$$

$$z = \begin{bmatrix} \dot{x} & x \end{bmatrix}^\mathrm{T},$$

$$x = \begin{bmatrix} h & \theta \end{bmatrix}^\mathrm{T},$$

$$A = \begin{bmatrix} m & S \\ S & I_\theta \end{bmatrix},$$

$$B = \begin{bmatrix} 0 & 0 \\ 0 & 0 \end{bmatrix},$$

$$C = \begin{bmatrix} 1 & \left(\frac{3}{4}c - x_\mathrm{f}\right) + \frac{c}{4} \\ -ec & \left(\frac{c}{2} - x_\mathrm{f}\right)^2 + \left(\frac{3}{4}c - x_\mathrm{f}\right)\frac{c}{4} \end{bmatrix},$$

$$D = \begin{bmatrix} K_h & 0 \\ 0 & K_\theta \end{bmatrix},$$

$$E = \begin{bmatrix} 0 & 1 \\ 0 & -ec \end{bmatrix}.$$

11.3.3.2 The unsteady aerodynamic and structural coupling model

When the wind turbine blade is under a simple harmonic vibration, the displacement and rotation of vibration can be expressed as

$$\begin{cases} h = h_0 \mathrm{e}^{\mathrm{j}wt}, \\ \theta = \theta_0 \mathrm{e}^{\mathrm{j}wt}. \end{cases} \tag{11.26}$$

Taking equation (11.26) into the equations (11.19) and (11.20) and after simplification, we obtain

$$L = \left\{ \rho\pi b^2 W_0 \mathrm{j} w\theta_0 + \pi\rho W_0 c C(k)\left(W_0\theta_0 + \mathrm{j}wh_0 + \left(\frac{3}{4}c - x_\mathrm{f}\right)\mathrm{j}w\theta_0 \right) + \rho\pi b^2\left(-w^2 h_0 + \left(x_\mathrm{f} - \frac{c}{2}\right) w^2\theta_0\right)\right\}\mathrm{e}^{\mathrm{j}wt}, \tag{11.27}$$

$$M_{x,\mathrm{f}} = \left\{ -\left(\frac{3}{4}c - x_\mathrm{f}\right)\rho\pi b^2 W_0 \mathrm{j}w\theta_0 + \pi\rho W_0 ec^2 C(k)\left(W_0\theta_0 + \mathrm{j}wh_0 + \left(\frac{3}{4}c - x_\mathrm{f}\right)\mathrm{j}w\theta_0\right) + \left(x_\mathrm{f} - \frac{c}{2}\right)\rho\pi b^2\left(-w^2 h_0 + \left(x_\mathrm{f} - \frac{c}{2}\right)w^2\theta_0\right) + \frac{\rho\pi b^4}{8}w^2\theta_0 \right\}\mathrm{e}^{\mathrm{j}wt}. \tag{11.28}$$

Taking the equations (11.27) and (11.28) into equation (11.13), the matrix form equation can be obtained:

$$\begin{bmatrix} K_h - w^2 m + \pi\rho W_0 cC(k)\mathrm{j}w - w^2\rho\pi b^2 & -w^2 S + \rho\pi b^2 W_0 \mathrm{j}w + \rho\pi b^2\left(x_\mathrm{f} - \frac{c}{2}\right)w^2 + \pi\rho W_0 cC(k)\left(U + \left(\frac{3}{4}c - x_\mathrm{f}\right)\mathrm{j}w\right) \\ -w^2 S - \pi\rho W_0 ec^2 C(k)\mathrm{j}w + \left(x_\mathrm{f} - \frac{c}{2}\right)\rho\pi b^2 w^2 & K_\theta - w^2 I_\theta + \left(\frac{3}{4}c - x_\mathrm{f}\right)\rho\pi b^2 W_0\mathrm{j}w - \pi\rho W_0 ec^2 C(k)\left(W_0 + \left(\frac{3}{4}c - x_\mathrm{f}\right)\mathrm{j}w\right) - \left(x_\mathrm{f} - \frac{c}{2}\right)^2\rho\pi b^2 w^2 - \frac{\rho\pi b^4}{8}w^2 \end{bmatrix} \begin{Bmatrix} h_0 \\ \theta_0 \end{Bmatrix} = 0. \tag{11.29}$$

Equation (11.29) is the aerodynamic and structural coupling model with the vibrating determinant

$$D = \begin{vmatrix} K_h - w^2 m + \pi\rho W_0 cC(k)\mathrm{j}w - w^2\rho\pi b^2 & -w^2 S + \rho\pi b^2 W_0 \mathrm{j}w + \rho\pi b^2\left(x_\mathrm{f} - \frac{c}{2}\right)w^2 + \pi\rho W_0 cC(k)\left(W_0 + \left(\frac{3}{4}c - x_\mathrm{f}\right)\mathrm{j}w\right) \\ -w^2 S - \pi\rho W_0 ec^2 C(k)\mathrm{j}w + \left(x_\mathrm{f} - \frac{c}{2}\right)\rho\pi b^2 w^2 & K_\theta - w^2 I_\theta + \left(\frac{3}{4}c - x_\mathrm{f}\right)\rho\pi b^2 W_0\mathrm{j}w - \pi\rho W_0 ec^2 C(k)\left(W_0 + \left(\frac{3}{4}c - x_\mathrm{f}\right)\mathrm{j}w\right) - \left(x_\mathrm{f} - \frac{c}{2}\right)^2\rho\pi b^2 w^2 - \frac{\rho\pi b^4}{8}w^2 \end{vmatrix}. \tag{11.30}$$

11.3.4 Aeroelastic analysis of a wind turbine airfoil section

11.3.4.1 The stability analysis of quasi-steady aerodynamic-structural coupling

Combining $Z = Z_0\mathrm{e}^{\lambda t}$ and equation (11.25), the problem turns into solving for the eigenvalue value:

$$(\lambda E - Q)Z_0\mathrm{e}^{\lambda t} = 0. \tag{11.31}$$

The main parameters are determined according to the sixteenth airfoil section data of National Renewable Energy Laboratory (NREL) 5 MW wind turbine blade [220]. The mass per unit length is 68.772 kg with chord of 2.086 m. As the positions of elastic center and gravity center have a large influence on the flutter speed, the chordwise position of elastic center (x_f) is set initially as 0.35c and that of gravity center x_c is 0.40c [219]. For blades of a multi-megawatt level wind turbine, the aeroelastic instability generally occurs in the coupling of the second-order flap and first-order torsion coupling. Normally, the second-order flapwise natural frequency is over 2.0 Hz and the first-order torsion natural frequency is about 6.0 Hz [74]. The rotational radius relative to the mass center is 0.25c [221]. The main parameters are listed in Tab. 11.2.

Tab. 11.2: Parameters for an airfoil section of a multi-megawatt level wind turbine.

Mass per unit length m (kg)	68.772
Air densityρ (kg/m^3)	1.225
Chord c (m)	2.086
Chordwise position of elastic center x_f (m)	0.35c
Chordwise position of gravity center x_c (m)	0.40c
Bending natural frequency f_h (Hz)	2.0
Torsional natural frequency f_θ (Hz)	6.0
Radius relative to the mass center r_{cg} (m)	0.25c
Distance between elastic and aerodynamic center e	$e = x_f/c - \frac{1}{4}$

The eigenvalues can be calculated through taking the parameters of Tab. 11.2 into equation (11.31) and setting the initial relative velocity. The natural frequency is $\omega_n = |\lambda|$ and the damping ratio is $\xi = \mathrm{Re}(\lambda)/\omega_n$.

As can be observed from Fig. 11.15 and 11.16, although the varying trend of natural frequencies of bending and torsion is different (one is a decreasing, the other one is slightly increasing), they are getting closer with increasing relative velocity W_0. The damping ratio of bending is declining and that of torsion decreases slightly to a negative value at first and then increased above zero. When the damping ratio reaches zero, the unstable phenomenon (flutter) occurs. The relative inflow speed when damping equals zero is 85 m/s. It is called the flutter speed, with a flutter frequency at that time of 4.084 Hz.

In order to analyze the displacement and torsion of an airfoil section before and after the flutter, the fourth-order Runge–Kutta method is utilized to solve the differential equation (11.23). The variation of displacement and torsional angle at relative velocity of 82 m/s, 84.85 m/s and 84.85 m/s are shown in Fig. 11.17–11.19. When the relative velocity is 82 m/s, the amplitude is decreasing gradually, which means that the system is stable. When the relative velocity is 84.85 m/s, the displacement and torsion are in simple harmonic vibrations meaning that the system is in a critical state. When the relative velocity is 86 m/s, the values are divergent and the system is in dynamic

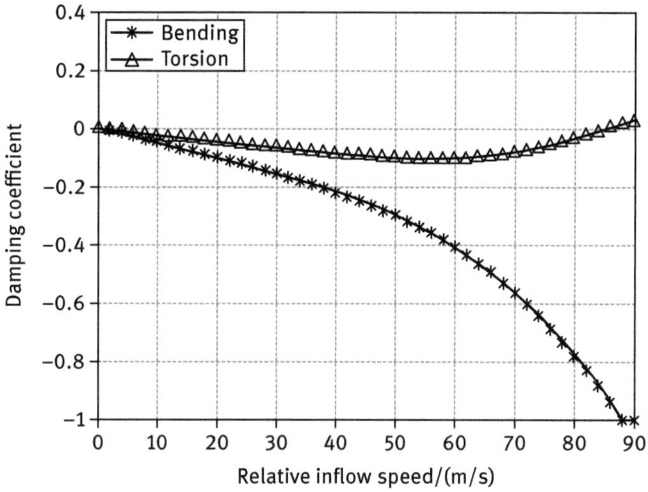

Fig. 11.15: Relationship between W_0 and ξ.

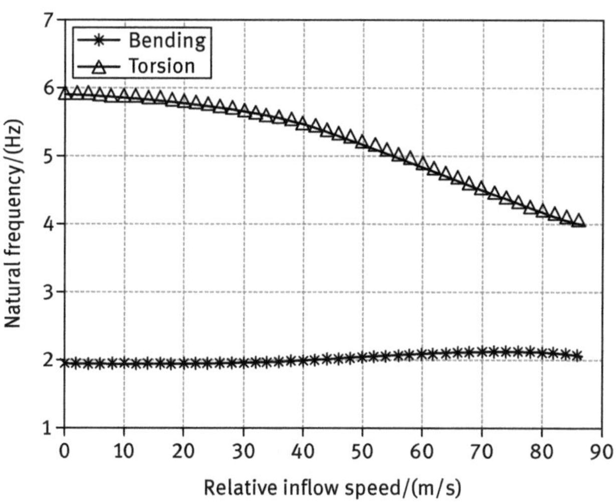

Fig. 11.16: Relationship between W_0 and w_n.

instability. What is more, the relative velocity at harmonic vibration (84.85 m/s) is consistent with the above obtained flutter critical value (85 m/s).

The Routh–Hurwitz theory is used to determine the stability of an airfoil section and the influence of chordwise position of elastic and gravity center on the flutter speed. The eigenvalues matrix $|\lambda E - Q| = 0$ of equation (11.25) can be transformed into a polynomial of λ:

$$a_4\lambda^4 + a_3\lambda^3 + a_2\lambda^2 + a_1\lambda + a_0. \tag{11.32}$$

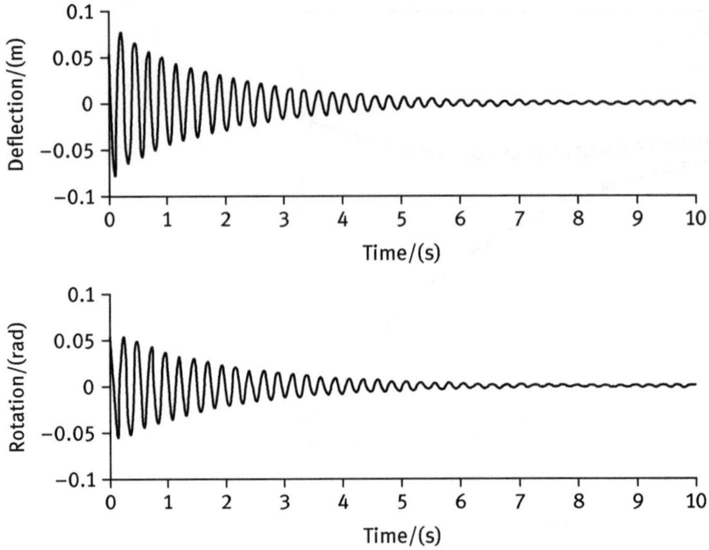

Fig. 11.17: Displacement and torsion when $W_0 = 82$ m/s.

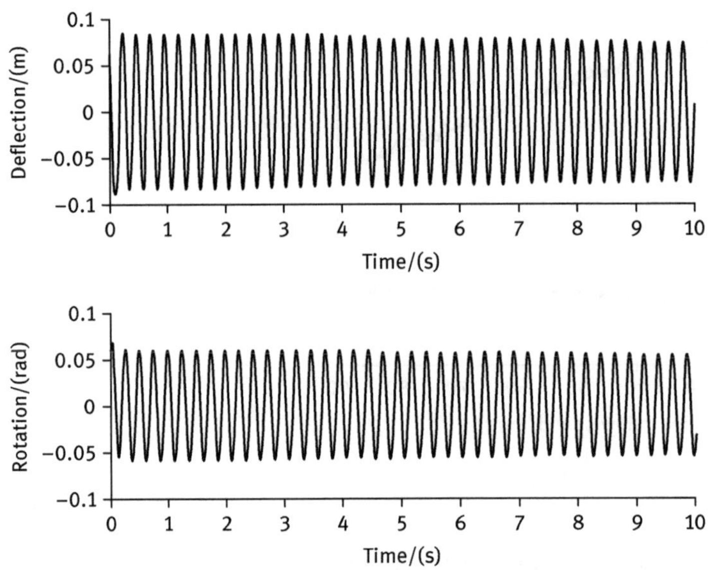

Fig. 11.18: Displacement and torsion when $W_0 = 84.85$ m/s

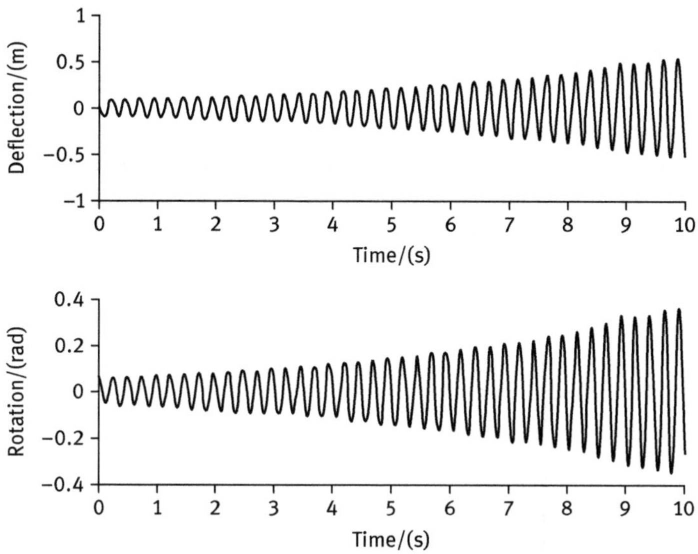

Fig. 11.19: Displacement and torsion when $W_0 = 86$ m/s.

The Routh–Hurwitz theory judges whether the system is stable based on the relationship between the coefficients of the above equation.

(1) If any of the polynomial coefficients a_i is negative or zero and there is at least one positive number among the other coefficients, the system is unstable.

(2) If at least one of the elements in the first column of matrix H changes its sign, the system is unstable. In which,

$$H = \begin{pmatrix} a_4 & a_2 & a_0 \\ a_3 & a_1 & 0 \\ b_1 & b_2 & 0 \\ c_1 & 0 & 0 \\ d_1 & 0 & 0 \end{pmatrix},$$

$$b_1 = (a_3 a_2 - a_4 a_1)/a_3,$$

$$b_2 = (a_3 a_0 - a_4 \times 0)/a_3 = a_0,$$

$$c_1 = (b_1 a_1 - b_2 a_3)/b_1,$$

$$d_1 = (c_1 b_2 - b_1 \times 0)/c_1 = b_2 = a_0.$$

The divergence speed and flutter speed of the airfoil section can be directly obtained from the matrix H.

(3) When $a < 0$, the conditions for static divergence are obtained $K_\theta < \rho W_0^2 e c^2 \pi$.

When $c_1 = 0$ or $a_3a_2a_1 - a_4a_1^2 - a_0a_3^2 = 0$, by solving this equation, the relative velocity W_0 is obtained. Two of the results for equation (11.32) are 0 and the other two are the flutter speed.

As can be seen from Fig. 11.20, the chordwise position of the elastic center has a large influence on the flutter speed and divergence speed. With the elastic axis moving gradually towards the airfoil trailing edge, the divergence speed decreases gradually and the flutter speed increases. When the elastic center is on the right of the gravity

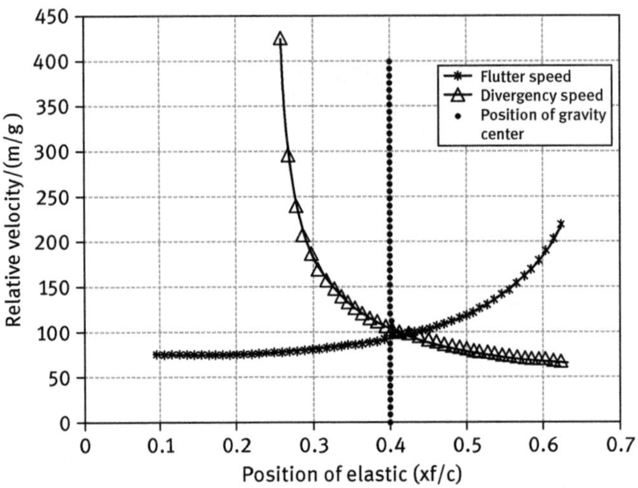

Fig. 11.20: The influence of chordwise position of elastic center on flutter speed.

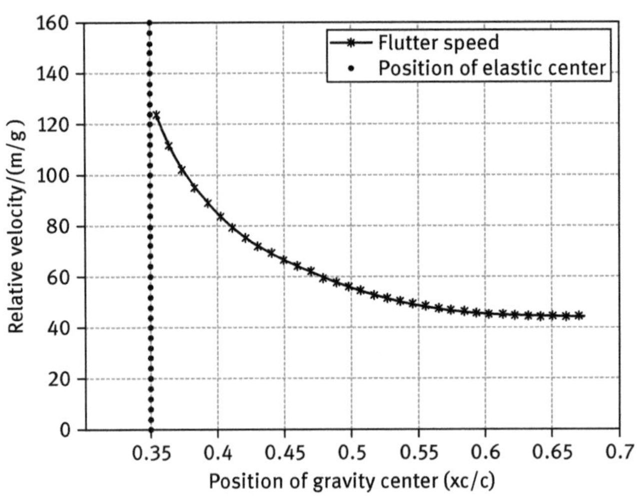

Fig. 11.21: The influence of chordwise position of gravity center on flutter speed.

center, the flutter phenomenon will not occur. The chordwise position of the gravity center also has a large influence on the flutter speed and divergence speed. With the center of gravity moves toward the trailing edge, the flutter speed decreases gradually. So increasing the flutter speed of the blade can be achieved through moving the gravity center towards the leading edge. Combining the Fig. 11.20 and 11.21, with the elastic center located at $0.35c$ and the gravity center located at $0.40c$, the flutter speed can be calculated respectively as 83.56 m/s and 84.36 m/s. They are basically consistent with the analysis discussed above. So improving the flutter velocity of the blade for megawatt composite wind turbines or avoiding the occurrence of flutter can be done through reasonable design of the internal structure.

11.3.4.2 The stability analysis of unsteady aerodynamic and structural coupling

When the equation (11.30) has a solution, the flutter determinant D equals zero, which means that $\mathrm{Re}(D) = 0$ and $\mathrm{Im}(D) = 0$. So the system of nonlinear equations about w and W_0 are obtained. Adopting Newton iteration method with an appropriate initial value, the flutter speed can be calculated. The variation of flutter speed and divergence speed with respect to the variation of the chordwise position of the elastic center is shown in Fig. 11.22. If the position of the gravity center is fixed, with the chordwise position of the elastic center varying from 0.1 to 0.7, the flutter speed increases gradually and divergence speed decreases. When the chordwise position of the elastic center is larger than that of the gravity center ($0.4c$), the flutter speed is larger than the divergence speed, meaning flutter no longer occurs. The variation of flutter speed and divergence speed with respect to the variation of the chordwise position of the

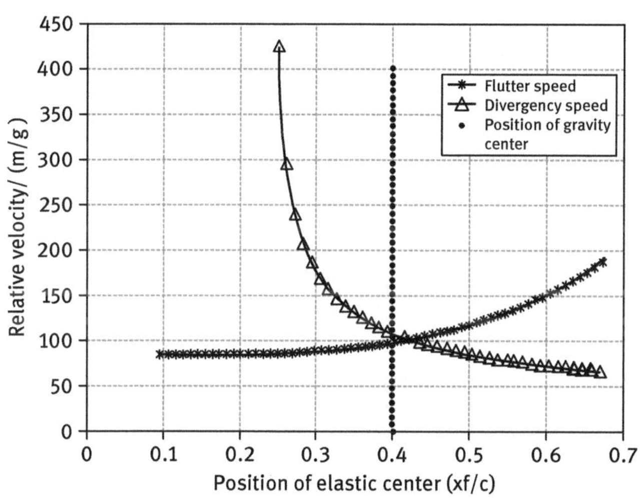

Fig. 11.22: The influence of chordwise position of the elastic center on flutter speed (unsteady).

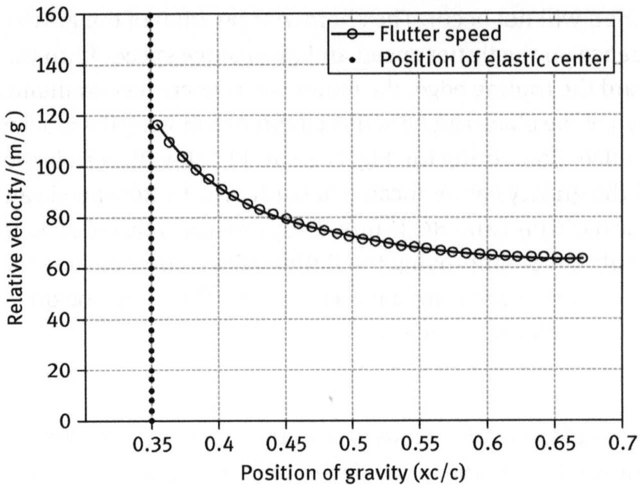

Fig. 11.23: The influence of chordwise position of the gravity center on flutter speed (unsteady).

gravity center is shown in Fig. 11.23. If the position of the elastic center is fixed, with the chordwise position of the gravity center varying from 0.35 to 0.7, the flutter speed decreases. The trend of flutter speed with respect to gravity center position is opposite to the circumstance of elastic center position, which is similar to the quasi-steady aerodynamic and structural coupling model.

The differences between the numerical results of the quasi-steady and unsteady models were compared. In flutter calculations, the commonly used methods are $V-g$ method and $p-k$ method [222]. As the values of the $p-k$ method can be compared with the results of blade flutter tests (although there is no relation to the blade flutter test), the $p-k$ method is chosen to calculate the flutter velocity of a blade for a wind turbine of multi-megawatt level. If we consider the blade to move arbitrarily, the hypothesis of aerodynamic force is more complex. What is more, it is the critical flutter state, with a simple harmonic motion, that was often the concern and study focus of researchers. Therefore, when calculating the unsteady aerodynamic force using Theodorsen theory, assuming the left part in equations (11.13) to be

$$\begin{cases} h = h_0 e^{pt}, \\ \theta = \theta_0 e^{pt}, \end{cases} \quad \text{i.e.} \quad \frac{d}{dt} = p. \tag{11.33}$$

Combining equations (11.33), (11.13), (11.19) and (11.20), we obtain that

$$\left(p^2 M_s + K_s - \tfrac{1}{2}\rho W_0^2 Q(jk)\right) q = 0, \tag{11.34}$$

where

$$q = \begin{bmatrix} h & \theta \end{bmatrix}^T,$$

$$Q = \begin{bmatrix} -4\pi C(k)\mathrm{j}k & -2\pi c C(k) - 2\pi b \mathrm{j}k \\ & - 4\pi C(k)\left(\frac{3}{4}c - x_{\mathrm{f}}\right)\mathrm{j}k + 2\pi b^2 k^2 \\ 4\pi e c C(k)\mathrm{j}k & 2\pi e c^2 C(k) - 2\left(\frac{3}{4}c - x_{\mathrm{f}}\right)\pi b \mathrm{j}k \\ - 2\pi\left(x_{\mathrm{f}} - \frac{c}{2}\right)k^2 & + 2\pi e c C(k)\left(\frac{3}{4}c - x_{\mathrm{f}}\right)\mathrm{j}k \\ & + 2\pi\left(x_{\mathrm{f}} - \frac{c}{2}\right)^2 k^2 + \pi\frac{b^2}{4}k^2 \end{bmatrix}.$$

The equation (11.34) is calculated iteratively using the $p-k$ method whose process is shown in Fig. 11.24.

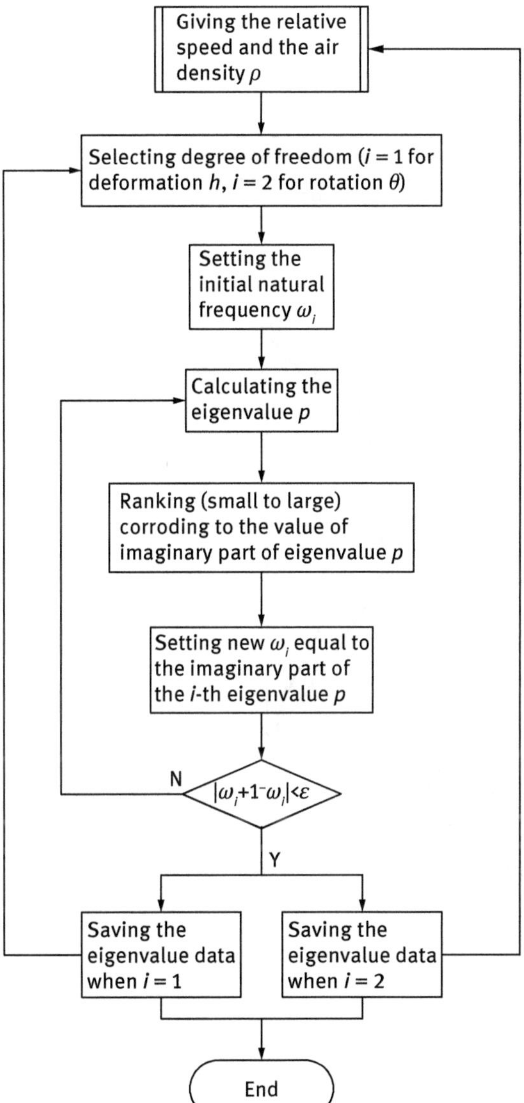

Fig. 11.24: Flowchart of the $p-k$ method.

A comparison of this method with the aeroelastic module in NASTRAN software was made as shown in Fig. 11.25 and 11.26. When the relative velocity is about 94 m/s, flutter occurs. The calculated flutter speeds of these two methods are similar, with relative error of only 2.174 %. The flutter frequency calculated by this method is 3.648 Hz and that by NASTRAN is 4.222 Hz. Although there is a certain gap between the two results, the trend is basically the same. What is more, the results were also compared with the quasi-steady coupling model and listed in Tab. 11.3. It can be seen that the flutter speed of the quasi-steady model is smaller than that of the unsteady model. The main reason is that the unsteady model considers the influence of the trailing edge free vortex whose induction can increase the flutter speed. Therefore, the results from the quasi-steady model tends to be conservative.

Tab. 11.3: Comparison of different models.

	Quasi-steady model	Unsteady model	NASTRAN
Flutter speed (m/s)	84	94	92
Flutter speed (Hz)	4.084	3.648	4.222

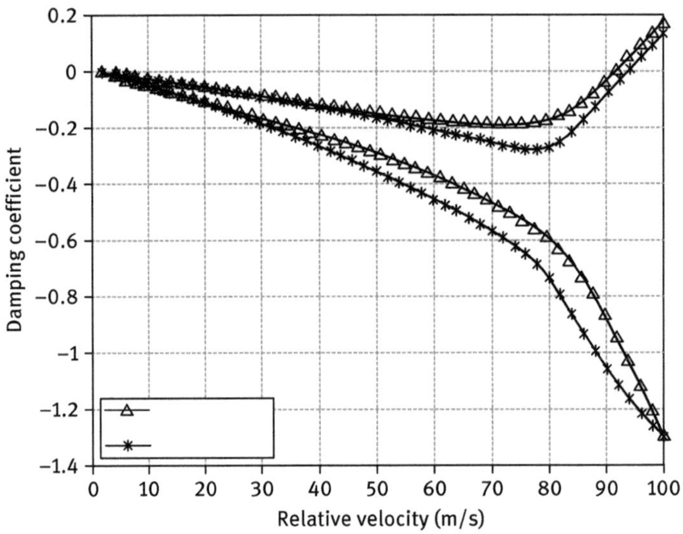

Fig. 11.25: The relationship between W_0 and ξ.

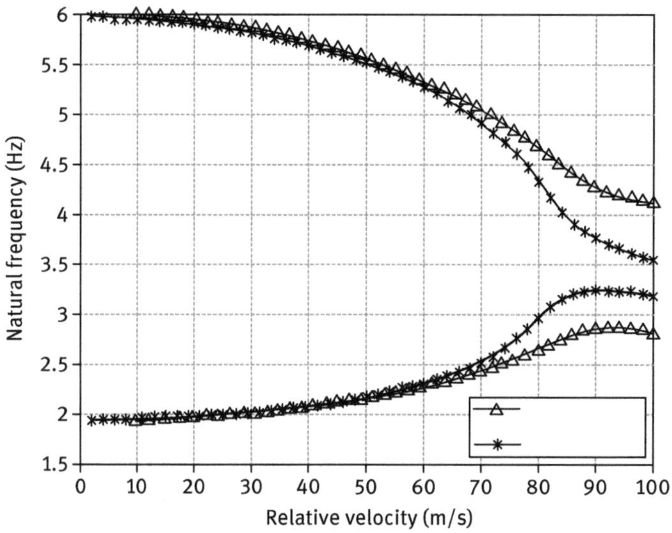

Fig. 11.26: The relationship between W_0 and ω_n.

11.4 The dynamic stall and aeroelastic analysis of the wind turbine blade

The improved Beddoes–Leishman dynamic stall model was utilized to study nonlinear aeroelastic performance of a wind turbine airfoil section with three degrees of freedom. The influences of linear and nonlinear aerodynamic force models on blade vibration response were compared. The effects of airfoil camber on aeroelastic limit-cycle-oscillation were deeply studied. The effects of structural damping on the response of a nonlinear aeroelastic system are discussed.

11.4.1 The structural kinematics model

The model of an airfoil section with three degrees of freedom is shown in Fig. 11.27. The degrees of freedom are the vertical displacement of the rigidity center y, the horizontal displacement of airfoil section x, and rotation angle around rigidity center α. The stiffness of the airfoil section is simulated with two tension springs (K_x and K_y) and a torsion spring (K_α). x_f is the chordwise position of the rigidity center and x_c is the chordwise position of the mass center. The hypothesis is adopted that the airfoil sections are rigid without deformation and the elastic, aerodynamic, mass centers of the blade sections are located in the same line.

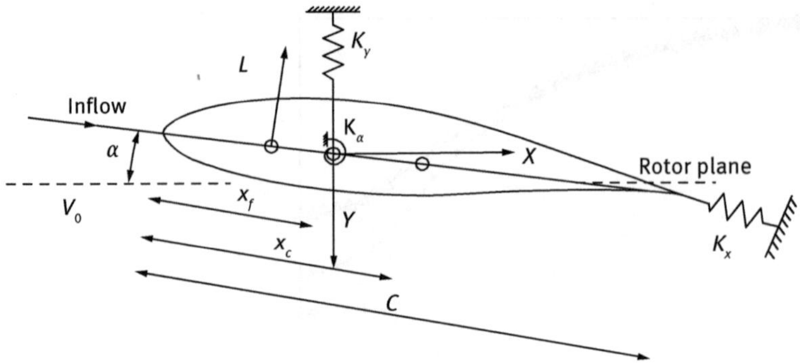

Fig. 11.27: Model of airfoil section with three degrees of freedom.

Under the action of gravity, drag, elastic force and aerodynamic force, the differential equation of motion of the structure with three degrees of freedom is as follows [223]:

$$m\ddot{x} - S_\alpha \sin(\alpha)\ddot{\alpha} + c_x\dot{x} + K_x x = D_{\text{aed}}, \tag{11.35}$$

$$m\ddot{y} - S_\alpha \cos(\alpha)\ddot{\alpha} + c_y\dot{x} + K_y x = L_{\text{aed}}, \tag{11.36}$$

$$-S_\alpha \sin(\alpha)\ddot{x} - S_{\text{d}}\alpha \cos(\alpha)\ddot{y} + I_\alpha\ddot{\alpha} + c_\alpha\dot{\alpha} + K_\alpha\alpha = M_{\text{aed}}, \tag{11.37}$$

where $m = \int_0^c \bar{\rho}\,\mathrm{d}x$ is the blade mass per unit length; $\bar{\rho}$ is the chordwise density of the airfoil section; $S = m(x_{\text{c}} - x_{\text{f}})$ is the inertia about the Y_{f} axis (the origin of Y_{f} is the rigidity center, the direction of Y_{f} is perpendicular to chord) per unit length; $I_\alpha = \int_0^c \bar{\rho}(x - x_{\text{f}})^2\,\mathrm{d}x$ is mass moment of inertia about the Y_{f} axis per unit length; r_{cg} is the rotational radius with respect to the gravity center; $c_{\text{dof}} = 2\varepsilon_{\text{dof}} w_{\text{dof}} m$ is the structural damping coefficient of each degree of freedom (ε_{dof} is the damping ratio of each degree of freedom, w_{dof} is the frequency of each degree of freedom); D_{aed}, L_{aed}, M_{aed} are lift, drag and moment of an airfoil respectively.

11.4.2 The aerodynamic model

The improved Beddoes–Leishman dynamic stall model, a semi-empirical model, was utilized. The aerodynamic model was simplified by Hansen et al. [216] and was used to perform the stability analysis for a multi-megawatt wind turbine during the linear elastic range. However, they did not take the nonlinear effects of the aerodynamic force into account. The nonlinear aerodynamic model is considered in this section to analyze the stability characteristics of an airfoil section through three degrees of freedom.

11.4.2.1 Lift force

The lift force on the airfoil can be divided into a cyclic component and a noncyclic component. The noncyclic component is related to the instantaneous motion of the airfoil, but not affected by the disturbed flow. The cyclic component is mainly caused by the vortex shed from the airfoil's trailing edge. In the present model, the effective angle of attack is introduced to take account of this cyclic component. Expressions for noncyclic lift and cyclic lift are shown in equation (11.39) and (11.40).

$$L_{\text{aed}} = L_{\text{nc}} + L_{\text{circ}}, \tag{11.38}$$

$$L_{\text{nc}} = \rho b^2 \pi (V_0 - \dot{x})\dot{\alpha} - \rho b^2 \pi \ddot{y} - \rho b^3 \pi \xi_{\text{ea}} \ddot{\alpha}, \tag{11.39}$$

$$L_{\text{circ}} = (\partial C_1/\partial \alpha)\rho h (V_0 - \dot{x})^2 (\alpha_{\text{eff}} - \alpha_{\text{lift},0}), \tag{11.40}$$

where ρ is air density; b is half of the chord; V_0 is inflow velocity which is equal to inflow velocity subtracted from vibration velocity along the x direction $V_0 - \dot{x}$; ξ_{ea} is the ratio of chordwise position of the elastic center to b which is x_f/b; $\partial C_1/\partial \alpha$ is the slope of the lift coefficient with respect to angle of attack, $\alpha_{\text{lift},0}$ is the angle of attack when the lift coefficient equals zero. α_{eff} is the effective angle of attack which is expressed as follows:

$$\alpha_{\text{eff}} = \alpha_{3/4}\left(1 - \sum_{i=1}^{3} A_i\right) + \sum_{i=1}^{3} z_i, \tag{11.41}$$

$$\dot{z}_i = -\left(\frac{V_0 - \dot{x}}{b} b_i - \frac{\ddot{x}}{V_0 - \dot{x}}\right) z_i + \frac{V_0 - \dot{x}}{b} b_i A_i \alpha_{3/4}, \tag{11.42}$$

$$\alpha_{3/4} = \alpha - \frac{\dot{y}}{V_0 - \dot{x}} + \frac{0.5 - \xi_{\text{ea}}}{V_0 - \dot{x}} \dot{\alpha}. \tag{11.43}$$

The coefficients A_i and B_i are the response parameters of the airfoil, which can be obtained through experiment. The coefficients of the Risø B1-18 airfoil are listed in Tab. 11.4. Z_i is the additional variable responsible for the time delay due to flow disturbances and its first-order differential equation is shown in equation (11.42). $\alpha_{3/4}$ is the "quasi-steady" angle of attack obtained at the three-fourths position on the chord which is shown in equation (11.43).

Tab. 11.4: Coefficients from experimental data of Risø B1-18.

Risø B1-18 airfoil	A_1	A_2	A_3	b_1	b_2	b_3
	0.0821	0.1429	0.3939	0.0199	0.7817	0.1453

11.4.2.2 Drag force

The drag force on the airfoil is composed of static drag and induced drag. Static resistance can be obtained from the static resistance curve. The induced drag is mainly caused by the x-direction component of the cyclic lift:

$$D_{\text{aed}} = D_{\text{static}} + \Delta D_{\text{induced}}, \tag{11.44}$$

$$D_{\text{static}} = \rho b(V_0 - \dot{x})C_{\text{d},(\alpha_{\text{eff}})}, \tag{11.45}$$

$$\Delta D_{\text{induced}} = L_{\text{circ}}(\alpha - \alpha_{\text{eff}}), \tag{11.46}$$

where $C_{\text{d},(\alpha_{\text{eff}})}$ is the curve of the drag coefficient at the effective angle of attack.

11.4.2.3 The moment

The aerodynamic moment is also composed of noncyclic component and a cyclic component. The noncyclic component acts on the elastic axis and the cyclic component is generated by the cyclic lift and drag force which act on the aerodynamic center of the airfoil. Moreover, considering the effect of camber, the static moment $C_{\text{m},(\alpha_{\text{eff}})}$ under the effective angle is also included in equation (11.49).

$$M_{\text{aed}} = M_{\text{nc}} + M_{\text{cric}}, \tag{11.47}$$

$$\begin{aligned} M_{\text{nc}} = & -\rho\pi b^3(V_0 - \dot{x})(0.5 - \xi_{\text{ea}})\dot{\alpha} - \rho\pi b^3\xi_{\text{ea}}\alpha\ddot{x} - \rho\pi b^3\xi_{\text{ea}}\ddot{y} \\ & - \rho\pi b^4(1/8 + \xi_{\text{ea}})\ddot{\alpha}, \end{aligned} \tag{11.48}$$

$$M_{\text{circ}} = (\xi_{\text{ea}} + 0.5)\cos(\alpha)L_{\text{circ}} + (\xi_{\text{ea}} + 0.5)\sin(\alpha)D_{\text{aed}} + \rho b^2(V_0 - \dot{x})C_{\text{m},(\alpha_{\text{eff}})}. \tag{11.49}$$

11.4.3 The aeroelastic coupling system

The aeroelastic response of the airfoil can be simulated through coupling the differential equations of flow and structure motion. Combining the equations (11.35)–(11.38), (11.44) and (11.47), the aeroelastic coupling system is obtained as shown in equation (11.50):

$$\begin{cases} m\ddot{x} - S_{\text{a}}\sin\alpha\ddot{\alpha} = & -C_x\dot{x} - K_x x + D_{\text{static}} + \Delta D_{\text{induced}}(m + \rho\pi b^2)\ddot{y} \\ & - (S_{\text{a}}\cos\alpha - \rho\pi b^3\xi_{\text{ea}})\ddot{\alpha} \\ = & -C_y y - K_y y + \rho\pi b^2(V_0 - \dot{x})\dot{\alpha} \\ & + L_{\text{circ}}(S_{\text{a}}\sin\alpha - \rho\pi b^3\xi_{\text{ea}}\alpha)\ddot{x} \\ & + (S_{\text{a}}\cos\alpha - \rho\pi b^3\xi_{\text{ea}})\ddot{y} \\ & - \left[I_{\text{ea}} + \rho\pi b^4\left(\frac{1}{8} + \xi_{\text{ea}}\right)\right]\ddot{\alpha} \\ = & [C_\alpha + \rho\pi b^3(V_0 - \dot{x})(0.5 - \xi_{\text{ea}})]\dot{\alpha} + K_\alpha\alpha - M_{\text{circ}}. \end{cases} \tag{11.50}$$

In order to solve second-order differential equations, usually they are transformed into first-order differential equations, which can be written in the form of a matrix:

$$M_{\text{nl}}\dot{X} = A_{\text{nl}}X + f_{\text{nl}}, \tag{11.51}$$

where x is the variable and can be written as follows

$$x = \{x, y, \alpha, \dot{x}, \dot{y}, \dot{\alpha}, z_1, z_2, z_3\}^{\mathrm{T}}. \tag{11.52}$$

As matrix M_{nl}, matrix A_{nl} and force vector f_{nl} are also functions of the system variables, the system is nonlinear. The traditional eigenvalue method is no longer applicable for solving the aeroelastic equations of the nonlinear system. The numerical integration method in the time domain is usually adopted. The fourth-order Runge–Kutta numerical integral method is utilized here to solve the flapwise, edgewise and torsional vibration responses. The initial value of the variable x is zero. The order of equation (11.50) is reduced, which simplifies the calculation. And it is not compulsory to transform equation (11.50) into equation (11.51) and can be solved as follows

$$\begin{cases} A_{11}\ddot{x} - A_{12}\ddot{\alpha} = A_{13}, \\ A_{21}\ddot{y} - A_{22}\ddot{\alpha} = A_{23}, \\ A_{31}\ddot{x} + A_{32}\ddot{y} - A_{33}\ddot{\alpha} = A_{34}. \end{cases} \tag{11.53}$$

Equation (11.53) is simplified into

$$\begin{cases} \ddot{x} = \dfrac{A_{13}(A_{22}A_{32} - A_{21}A_{33}) + A_{12}(A_{21}A_{34} - A_{23}A_{32})}{A_{11}(A_{32}A_{22} - A_{21}A_{33}) + A_{12}A_{21}A_{31}}, \\ \ddot{y} = \dfrac{A_{22}(A_{11}A_{34} - A_{13}A_{31}) + A_{23}(A_{12}A_{31} - A_{11}A_{33})}{A_{21}(A_{12}A_{31} - A_{11}A_{33}) + A_{11}A_{22}A_{32}}, \\ \ddot{\alpha} = \dfrac{A_{11}(A_{21}A_{34} - A_{23}A_{32}) - A_{13}A_{21}A_{31}}{A_{11}(A_{22}A_{32} - A_{21}A_{33}) + A_{12}A_{21}A_{31}}, \end{cases} \tag{11.54}$$

where

$$\begin{aligned} A_{11} &= m; \\ A_{12} &= S_{\mathrm{a}} \sin\alpha; \\ A_{13} &= -C_x\dot{x} - K_x x + D_{\mathrm{static}} + \Delta D_{\mathrm{induced}}; \\ A_{21} &= m + \rho\pi b^2; \\ A_{22} &= S_{\mathrm{a}} \cos\alpha - \rho\pi b^3 \xi_{\mathrm{ea}}; \\ A_{23} &= -C_y y - K_y y + \rho\pi b^2 (V_0 - \dot{x})\dot{\alpha} + L_{\mathrm{circ}}; \\ A_{31} &= S_{\mathrm{a}} \sin\alpha - \rho\pi b^3 \xi_{\mathrm{ea}}\alpha; \\ A_{32} &= S_{\mathrm{a}} \cos\alpha - \rho\pi b^3 \xi_{\mathrm{ea}}; \\ A_{33} &= I_{\mathrm{ea}} + \rho\pi b^4 \left(\tfrac{1}{8} + \xi_{\mathrm{ea}}\right); \\ A_{34} &= [C_\alpha + \rho\pi b^3 (V_0 - \dot{x})(0.5 - \xi_{\mathrm{ea}})]\dot{\alpha} + K_\alpha \alpha - M_{\mathrm{circ}}. \end{aligned}$$

Combining equations (11.54) and (11.42) and adopting the fourth-order Runge–Kutta numerical integral method, the flapwise, edgewise and torsional vibration responses can be obtained.

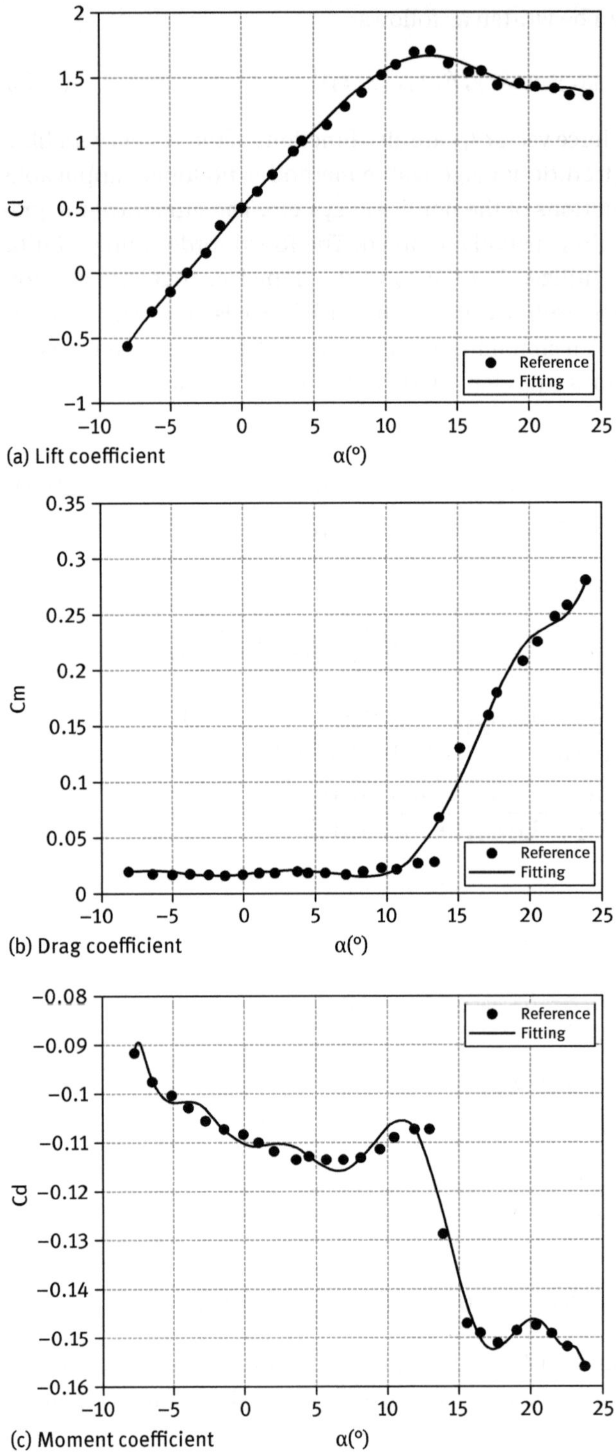

(a) Lift coefficient

(b) Drag coefficient

(c) Moment coefficient

Fig. 11.28: The aerodynamic characteristics of Risø B1-18.

The Risø B1-18 airfoil is selected and its related parameters are listed in Tab. 11.5 [106]. Considering the effect of airfoil camber, the aerodynamic coefficients are listed in Tab. 11.6. What is more, the static lift coefficient, drag coefficient and moment coefficient under the effective angle of attack also need to be calculated. The lift coefficient, drag coefficient, and torque coefficient are taken from reference [224]. The aerodynamic characteristic curve was fitted using the 8th-order polynomial as shown in Fig. 11.28. The standard deviations are respectively 0.016256, 0.007067 and 0.002209.

Tab. 11.5: The parameters of Risø B1-18.

Mass per unit length m (kg)	40
Chord c (m)	1
Chordwise position of elastic center x_f (m)	0.3
Chordwise position of gravity center x_c (m)	0.35
Flapwise natural frequency f_x (Hz)	1
Edgewise natural frequency f_y(Hz)	2
Torsional natural frequency f_α (Hz)	10
Radius with respect to mass center r_{cg} (m)	0.2

Tab. 11.6: The aerodynamic parameters of Risø B1-18.

Angle of attack with zero lift $\alpha_{lift,0}$ (°)	−3.512
Drag coefficient when angle of attack is 0° Cd_0	0.0074
Moment coefficient when angle of attack is 0° Cm_0	−0.112

11.4.4 The numerical results

In order to compare the results from the nonlinear aerodynamic numerical calculations and the linear calculations, the equation (11.51) needs linearization treatment using a first-order Taylor series expression. The detailed simplification process can be found in reference [223]. Without considering the viscous damping, the numerical integral method in time domain is used to solve the control equation. The calculated flutter speed is 137.2 m/s and that described in the literature [223] is 142.3 m/s. The relative error of only 3.584 % shows the feasibility of this method. Without considering the airfoil camber and using the same method to solve equation (11.51), the calculation results were compared with those from the linear system. The variation of displacement with respect to time along each degree of freedom is shown in Fig. 11.29. It can be seen that the differences between the nonlinear and the linear systems are large. The vibration follows simple harmonic motion at critical flutter speed in the linear system. However, the nonlinear system turns into a stable state at relative speed of 117.5 m/s and did not reach the critical flutter speed of 137.2 m/s calculated from the linear system.

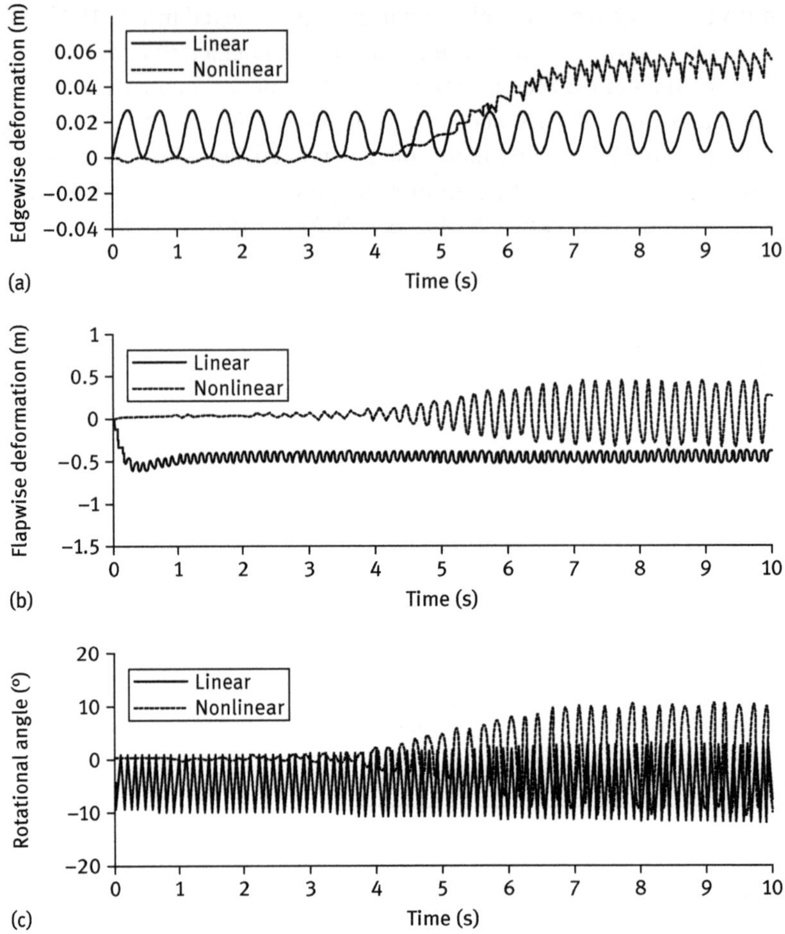

Fig. 11.29: Vibrations from the linear and nonlinear systems.

The effect of camber on the aeroelastic stability will be studied next. The static aerodynamic force coefficients are shown in Tab. 11.6. The coefficients not listed in Tab. 11.6 remain unchanged. The calculation results were compared with those from calculations without considering camber effects, as shown in Fig. 11.30. The flapwise, edgewise and torsional limit-cycle-oscillation considering camber effects of the airfoil come into a stable state earlier (with relative inflow speed of 118 m/s) than those without considering camber effects (with relative inflow speed of 117.5 m/s). The flapwise displacement is about 1.5 m when the vibration reaches stable state compared with 0 m from computations without considering camber. The lift coefficient of airfoils with camber is higher than that of a symmetric airfoil in the main angle of attack zone. When the angle of attack is zero, the lift coefficient of an airfoil with camber is greater than zero and that of a symmetric airfoil is zero. Therefore, the flapwise stable state is

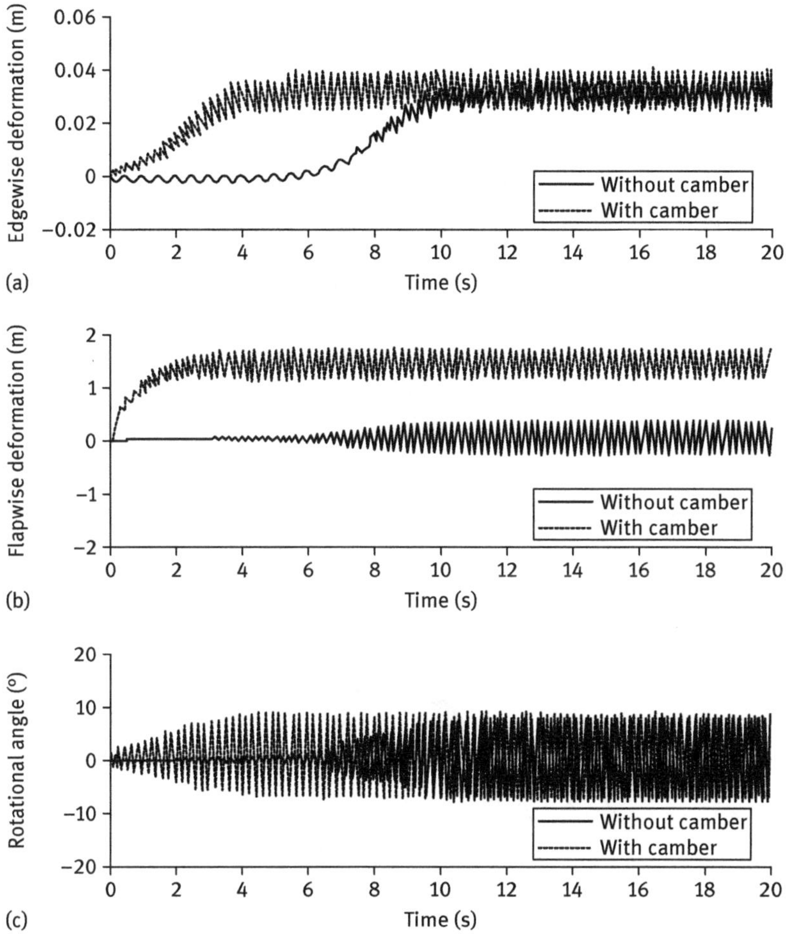

Fig. 11.30: Vibrations of an airfoil with and without camber.

achieved when the blade is away from the initial position. The variations of flapwise, edgewise and torsional speed with respect to their respective displacement of an airfoil with camber are given Fig. 11.31–11.33. The limit-cycle-oscillation will appear when the vibration is in stable state due to the nonlinear aerodynamic force, which is shown as the tracks with high density in these figures.

The effect of structural damping on the aeroelastic stability will be studied next. Assuming the same damping ratio of 0.08 % along the directions of the three degrees of freedom, the structural damping coefficients of three directions (c_x, c_y and c_α) can be obtained, which are 0.8, 0.4 and 4 respectively. The calculation results were compared with those from calculations without considering damping effects, as shown in Fig. 11.34. The flapwise and torsional oscillations considering damping come into stable state finally, which implies that the existence of structural damping can effectively

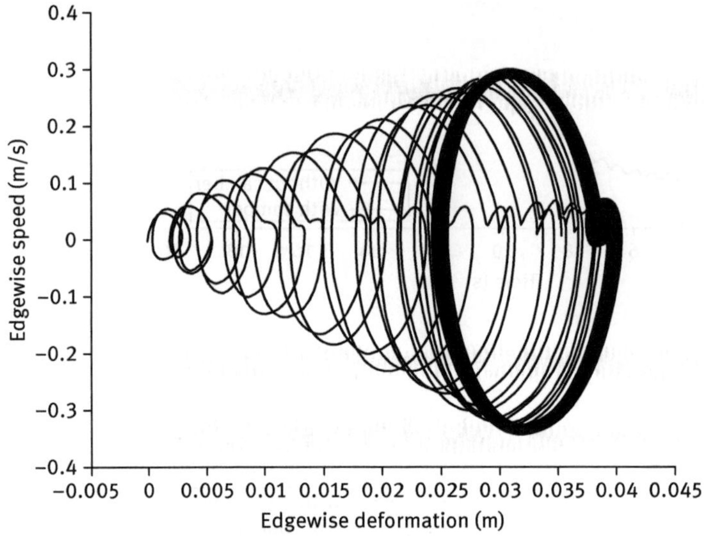

Fig. 11.31: Edgewise speed displacement relationships at relative speed of 118 m/s.

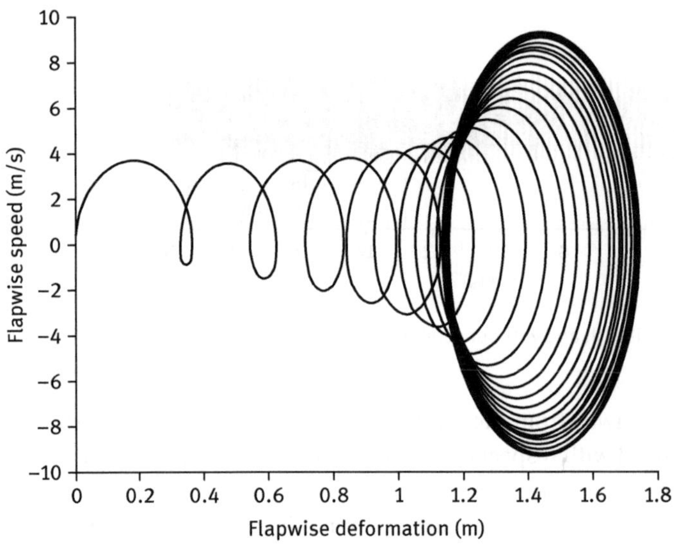

Fig. 11.32: Flapwise speed displacement relationships at relative speed of 118 m/s.

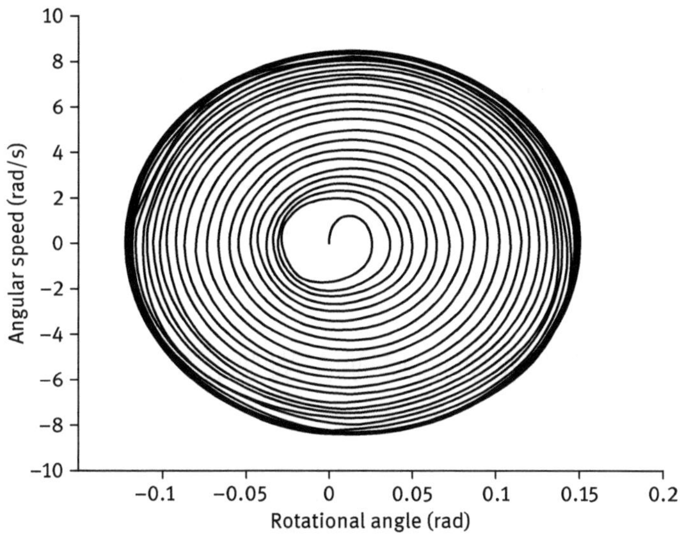

Fig. 11.33: Torsional speed angle relationships at relative speed of 118 m/s.

improve the stability of the wind turbine blade. The vibrations along the directions of the three degrees of freedom with relative inflow speed of 130 m/s are shown in Fig. 11.35. It can be seen that vibrations become more complex when the relative inflow speed is large (close to the linear flutter speed). The main reason is that the airfoil gradually goes into dynamic stall state and it becomes difficult to accurately predict the lift, drag and moment coefficients, which leads to the complexities of vibrations. The variations of flapwise, edgewise and torsional speed with respect to their respective displacement damping are given Fig. 11.36–11.38. It can be seen from these figures that chaotic motions appear.

11.5 Chapter conclusions

Based on aeroelastic theory, an aeroelastic feedback model for wind turbine blade airfoils was established. The critical condition of torsional divergence and the iterative model calculating the redistribution of aeroelastic load for a 2D airfoil section was obtained using the coupling relationship between lift coefficient, aerodynamic center, dynamic pressure and the additional torsional angle. The relevant aerodynamic parameters were obtained using XFOIL software in the angle of attack range from 0 to 6 degrees. The effects of torsion stiffness and distance from stiffness center to aerodynamic center on the divergence speed, additional elastic angle and lift force were analyzed. The influences of different airfoils on additional elastic angle and lift force were analyzed intensely. It was found that airfoil shape has some influence on the aeroelastic response system.

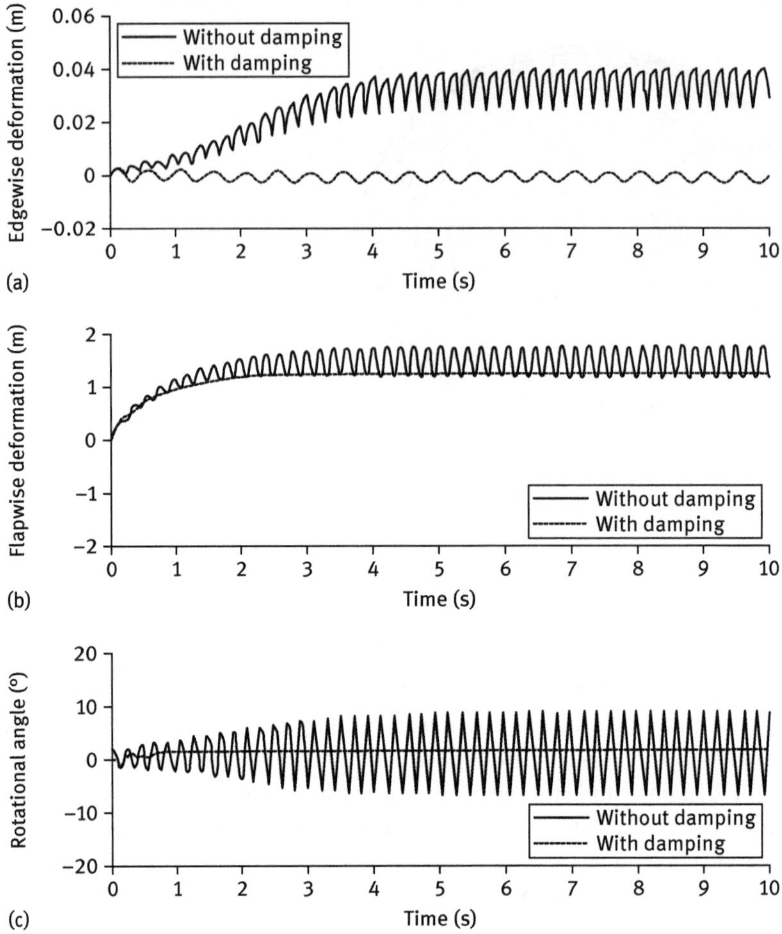

Fig. 11.34: Vibrations of an airfoil with and without damping at relative speed of 118 m/s.

Secondly, the quasi-steady and unsteady aerodynamic-structural coupling models of a 2D airfoil section were established. The aeroelastic stability analysis of a certain section on the NREL 5 MW wind turbine blade was conducted. Comparisons were made between results from the two coupling models. The results from the unsteady aerodynamic-structural coupling model were compared with those from the aeroelastic module of NASTRAN and the accuracy of the unsteady model was validated. With regard to the structural characteristics, it was found that the chordwise position of the elastic center and the gravity center will have a great influence on the flutter speed of wind turbine blades.

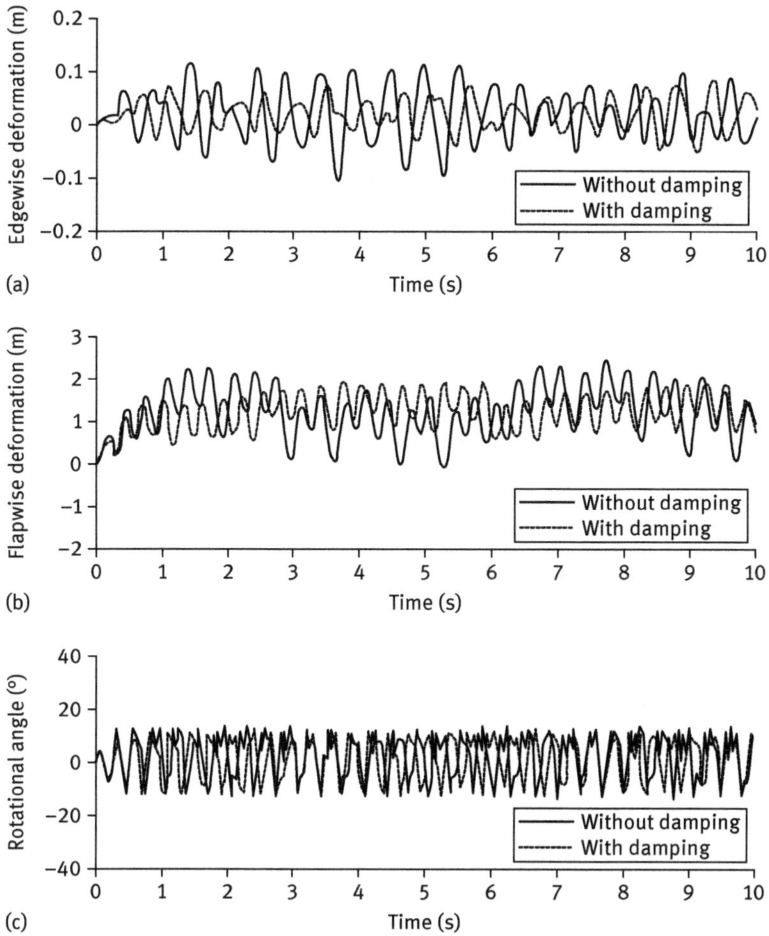

Fig. 11.35: Vibrations of an airfoil with and without damping at relative speed of 130 m/s.

Finally, the nonlinear aerodynamic-structural coupling model of a 2D airfoil section with three degrees of freedom was established using the improved Beddoes–Leishman dynamic stall model. The different vibration responses between linear and nonlinear models were compared. And large differences between nonlinear and linear systems can be found. The vibration follows simple harmonic motion at critical flutter speed in the linear system. However, the nonlinear system turns into stable state at a relative speed of 117.5 m/s before reaching the critical flutter speed of 137.2 m/s calculated from the linear system. The effects of airfoil camber on aeroelastic limit-cycle-oscillation were deeply studied. When the camber effects of an airfoil are considered, the flapwise, edgewise and torsional limit-cycle-oscillation come into stable state earlier than when camber effects are not considered. The flapwise displacement when the vibration reaches stable state is larger than that without considering camber. The lift coefficient

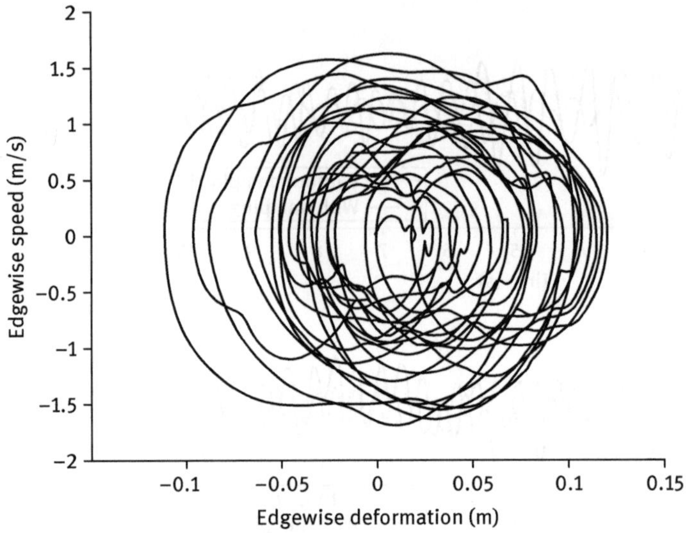

Fig. 11.36: Edgewise speed displacement relationships at relative speed of 118 m/s.

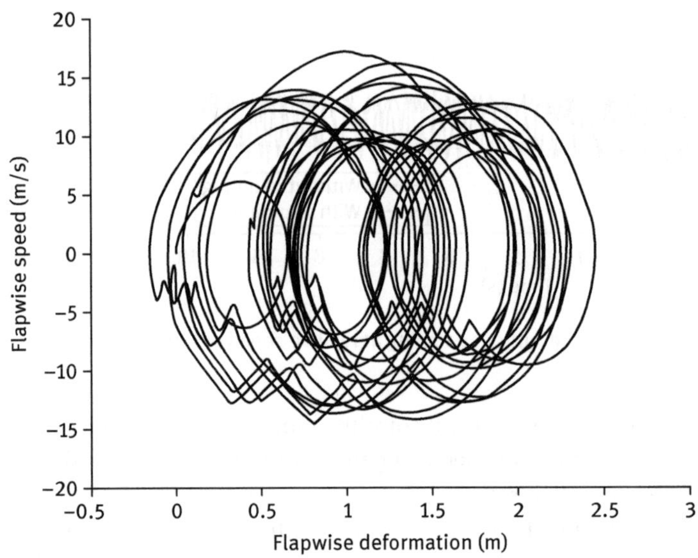

Fig. 11.37: Flapwise speed displacement relationships at relative speed of 118 m/s.

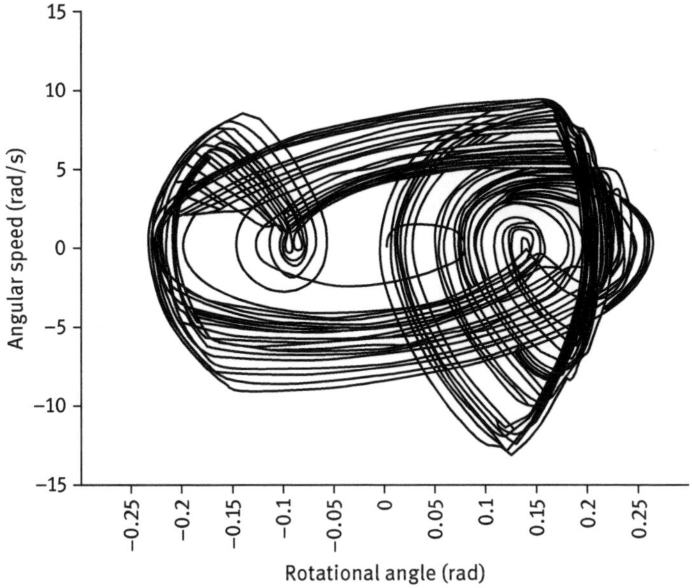

Fig. 11.38: Torsional speed angle relationships at relative speed of 118 m/s.

of airfoils with camber is higher than that of symmetric airfoils in the main angle of attack zone. The effects of structural damping on the response of a nonlinear aeroelastic system were discussed. When the relative inflow velocity is 118 m/s, the existence of structural damping can effectively improve the stability of the wind turbine blade. When the relative inflow speed is large (close to the linear flutter speed), vibrations become more complex. This is mainly because the airfoil gradually goes into dynamic stall state and it becomes difficult to accurately predict the lift, drag and moment coefficients. So the vibration responses of an elastic coupling system become more complicated with chaotic motions appearing.

References

[1] China Meteorological Administration. China wind energy resource assessment report (2009); 2010.

[2] Li JF, Shi PF and Gao H. China wind power outlook 2012. Peking: China Environmental Science Press; 2012.

[3] Tangler JL, Somers DM. Status of the special purpose airfoil families SERI/CP-217-3315. Proceedings of WINDPOWER '87, San Francisco, USA, 1987, 229–335.

[4] Van Rooij RPJOM, Timmer WA. Roughness sensitivity considerations for thick rotor blade airfoils. Journal of Solar Energy Engineering. 2003;125:468–478.

[5] Timmer WA, van Rooij PRJOM. Summary of the Delft University wind turbine dedicated airfoils. Journal of Solar Energy Engineering. 2003;125:488–496.

[6] Fuglsong P, Bak C. Design and verification of the Risø-B1 airfoil family for wind turbines. ASME. 2004;126:1002–1010.

[7] Fuglsong P, Bak C. Wind tunnel tests of Risø-B-18 and Risø-B1-24, Risø-R-1375(EN). Risø National Laboratory, Denmark; January 2003.

[8] Sobieczky H. Parametric airfoils and wings. Notes on Numerical Fluid Mechanics. 1998;68:71–88.

[9] Hájek J. Parameterization of airfoils and its application in aerodynamic optimization. In: Proceedings of the 16th Annual Conference of Doctoral Students – WDS 2007. WDS'07 Proceedings of Contributed Papers part 1, June 5–8, 2007, Charles University, Prague. Czech Republic: Matfyzpress; 2007: 233–240.

[10] Bjork A. Coordinates and calculations for the FFA-W1-xxx, FFA-W2-xxx and FFA-w3-xxx series of airfoils for horizontal axis wind turbines. FFA TN, Stockholm, Sweden; 1990.

[11] Soemarwoto BI, Labrujere THE. Airfoil design and optimization methods: Recent progress at NLR. International Journal for Numerical Methods in Fluids. 1999;30:217–228.

[12] Habali SM, Saleh IA. Design and testing of small mixed airfoil wind turbine blades. Renewable Energy. 1995;6(2):161–169.

[13] Larsen JW, Nielsen SRK, Krenk S. Dynamic stall model for wind turbine airfoils. Journal of Fluids and Structures. 2007;23:959–982.

[14] Baker JP, Mayda EA, van Dam CP. Experimental analysis of thick blunt trailing-edge wind turbine airfoils. Journal of Solar Energy Engineering. 2006;128(11):422–431.

[15] Bermudez L, Velazquez A, Matesanz A. Viscous–inviscid method for the simulation of turbulent unsteady wind turbine airfoil flow. Journal of Wind Engineering and Industrial Aerodynamics. 2002;90:643–661.

[16] Zhang WZ, He DX and Zhang ZS. The Design and Experiment Study for a High Airfoil at Low Reynold Numbers. Acta Aerodynamica Sinica. 1998;16(3):363–367.

[17] Qiao ZD, Song WP and Gao YW. Design and experiment of the NPU-WA airfoil family for wind turbine. Acta Aerodynamica Sinica. 2012; 30(2):260–265.

[18] Li J, Gao ZH, Huang JT, et al. Airfoil optimization based on distributed particle swarm algorithm. Acta Aerodynamica Sinica. 2011;29(4):464–469.

[19] Dai R, Wang HG, Yang AL, et al. Airfoil optimization based in aerodynamic inverse problem for wind turbine. Journal of Engineering Thermophysics. 2011;32(9):1501–1504.

[20] Sun MJ, Zhan H. Synthesis airfoil optimization by particle swarm optimization based on global information. Acta Aeronautica ET Astronautica Sinica. 2010;31(11):2166–2173.

[21] Gao ZH, Zhu J, Zhan H. Aerodynamic optimization design of airfoil based on correlation analysis. Acta Aeronautica et Astronautica Sinica. 2009;30(3):430–434.

DOI 10.1515/9783110344387-013

[22] Du L, Ning FF. Aerodynamic inverse design method for low mach number airfoils. Acta Aeronautica et Astronautica Sinica. 2011;32(7):1180–1188.

[23] Fuglsang P, Madsen HA. Optimization method for wind turbine rotors. Journal of Wind Engineering and Industrial Aerodynamics. 1999;(80):191–206.

[24] Varola A, Ilkilic C, Varol Y. Increasing the efficiency of wind turbines. Journal of Wind Engineering and Industrial Aerodynamics. 2001;(89):809–815.

[25] Kamoun B, Helali A, Afungchui D. Optimum project for horizontal axis wind turbines 'OPHWT'. Renewable Energy. 2005;(30):2019–2043.

[26] Rajakumar S, Ravindran D. Iterative approach for optimizing coefficient of power, coefficient of lift and drag of wind turbine rotor. Renewable Energy. 2012;(38):83–93.

[27] Liu X, Wang L, Pang X. Optimized linearization of chord and twist angle profiles for fixed-pitch fixed-speed wind turbine blades. Renewable Energy. 2013;(57):111–119.

[28] Maki K, Sbragio R, Vlahopoulos N. System design of a wind turbine using a multi-level optimization approach. Renewable Energy. 2012;(43):101–110.

[29] Liu X, Chen Y and Ye ZQ. Research on the HAWT rotor blades optimization. Journal of Shantou University (Natural Science). 2006;21(1):44–49.

[30] Han ZH, Wu TJ. Optimal design of wind turbine blades based on genetic algorithm. Journal of Power Engineering. 2008;28(6):955–958.

[31] Wang L, Wang TG and Luo Y. Improved NSGA-II in multi-objective optimization studies of wind turbine blades. Applied Mathematics and Mechanics. 2011;32(6):693–701.

[32] Fuglsang P, Dahl KS. Design of the new Risø-Al airfoil family for wind turbines. 1999 European Wind Energy Conference; Nice, France. Vol. 1–5, March 1999: 134–137.

[33] Fuglsang P, Bak C. Development of the Risø wind turbine airfoils. Wind Energy. 2004;(7):145–162.

[34] Jureczko M, Pawlak M, Mezyk A. Optimization of wind turbine blades. Journal of Materials Processing Technology. 2005;(167):463–471.

[35] Kong C, Bang J. Structural investigation of composite wind turbine blade considering various load cases and fatigue life. Energy. 2005;(30):2101–2114.

[36] Shokrieh MM. Simulation of fatigue in a full composite wind turbine blade. Composite Structures. 2006;(74):332–342.

[37] Maheri A, Noroozi S, Toomer CA. WTAB, a computer program for predicting the performance of horizontal axis wind turbines with adaptive blades. Renewable Energy. 2006;(31):1673–1685.

[38] Maheri A, Noroozi S, Toomer CA. Efficient meshing of a wind turbine blade using force adaptive mesh sizing functions. Renewable Energy. 2007;(32):95–104.

[39] Jensen FM, Falzon BG. Structural testing and numerical simulation of a 34m composite wind turbine blade. Composite Structures. 2006;(76):52–61.

[40] Paluch B, Faye A. Combining a finite element programme and a genetic algorithm to optimize structures with variable thickness. Composite Structures. 2008;(83):284–294.

[41] Johansen LS, Lund E. Failure optimization of geometrically linear/nonlinear laminated composite structures using two-step hierarchical model adaptivity. Comput. Methods Appl. Mech. Engrg. 2009;(198): 2421–2438.

[42] Marin JC, Barroso A. Study of fatigue in wind turbine blades. Engineering Failure Analysis. 2009;(16):656–668.

[43] Ping F, Farzaneh M. A CFD approach for modeling the rime-ice accretion process on a horizontal-axis wind turbine. Journal of Wind Engineering and Industrial Aerodynamics. 2009;(10):1–8.

[44] Lund E. Bucking topology optimization of laminated multi-material composite shell structures. Composite Structures. 2009;(91):158–167.

[45] Overgaard LCT, Lund E, Thomsen OT. Structural collapse of a wind turbine blade. Part A: Static test and equivalent single layered models. Composites: Part A. 2010;(41):257–270.

[46] Maalawia KY, Negmb HM. Optimal frequency design of wind turbine blades. Journal of Wind Engineering and Industrial Aerodynamics. 2002;(90):961–986.
[47] Li DY, Ye ZQ, Bao NS, et al. Vibration mODAL aNALYSIS OF the rotating rotor of horizontal axis wind turbine. Acta Energiae Solaris Sinica. 2004;25(1):72–77.
[48] Chen Y. Design of large grp wind turbine blades. Fiber Reinforced Plastics/Composites. 1998;(4):17–20.
[49] Liao CC, Zhao XL, Xu JZ. Blade layers optimization of wind turbines using FAST and improved PSO algorithm. Renewable Energy. 2011;(2):1–7.
[50] Fernandes da Silva G, Marin JC, Barroso A. Evaluation of shear flow in composite wind turbine blades. Composite Structures. 2011;93:1983–1841.
[51] Nicholls-Lee RF, Turnock SR, Boyd SW. Application of bend-twist coupled blades for horizontal axis tidal turbines. Renewable Energy. 2013;(50):541–550.
[52] Mulizwan MJ, Karimirad M, Moan T. Dynamic response and power performance of a combined Spar-type floating wind turbine and coaxial floating wave energy converter. Renewable Energy. 2013;(50):47–57.
[53] Kim H, Lee S, Son E, Lee S, et al. Aerodynamic noise analysis of large horizontal axis wind turbines considering fluid-structure interaction. Renewable Energy. 2012;(42):46–53.
[54] Lee S, Kim H, Son E, Lee S. Effects of design parameters on aerodynamic performance of a counter-rotating wind turbine. Renewable Energy. 2012;(42):140–144.
[55] Sebastian T, Lackner MA. Development of a free vortex wake method code for offshore floating wind turbines. Renewable Energy. 2012;(46):269–275.
[56] Guezuraga B, Zauner R, Polz W. Life cycle assessment of two different 2 MW class wind turbines. Renewable Energy. 2012;(37):37–44.
[57] Hameed Z, Vatn J, Heggset J. Challenges in the reliability and maintainability data collection for offshore wind turbines. Renewable Energy. 2011;(36):2154–2165.
[58] Lozano-Minguez E, Kolios AJ, Brennan EP. Multi-criteria assessment of offshore wind turbine support structures. Renewable Energy. 2011;(36):2831–2837.
[59] Tadamasa A, Zangenen M. Numerical prediction of wind turbine noise. Renewable Energy. 2011;(36):1902–1912.
[60] Kusiak A, Li W. The prediction and diagnosis of wind turbine faults. Renewable Energy. 2011;(36):16–23.
[61] Lanzafame R, Messina M. Horizontal axis wind turbine working at maximum power coefficient continuously. Renewable Energy. 2010;(35):301–306.
[62] Gonzalez JS, Gonzalez Rodriguez AG, Mora JC, et al. Optimization of wind farm turbines using an evolutive algorithm. Renewable Energy. 2010;(35):1671–1681.
[63] Kusiak A, Zheng H, Song Z. Power optimization of wind turbines with data mining and evolutionary computation. Renewable Energy. 2010;(35):695–702.
[64] Mishnaevsky L Jr, Brondsted P, Nijssen R, et al. Materials of large wind turbine blades: recent results in testing and modeling. Wind Energy. 2012;(15):83–97.
[65] Zavvos A, Bang D, McDonald AS, Polinder H, Mueller M. Structural analysis and optimisation of transverse flux permanent magnet machines for 5 and 10 MW direct drive wind turbines. Wind Energy. 2012;(15):19–43.
[66] Sieros G, Chaviaropoulos P, Sorensen JD, et al. Upscaling wind turbines: theoretical and practical aspects and their impact on the cost of energy. Wind Energy. 2012;(15):3–17.
[67] Maheri A, Isikveren AT. Performance prediction of wind turbines utilizing passive smart blades: approaches and evaluation. Wind Energy. 2010;(13):255–265.
[68] Gray CS, Watson SJ. Physics of failure approach to wind turbine condition based on maintenance. Wind Energy. 2010;(13):395–405.

[69] Chen H, Yu W, Capellaro M. A critical assessment of computer tools for calculating composite wind turbine blade properties. Wind Energy. 2010;(13):497–516.

[70] Yeh CP, Hwang WS. Numerical simulation of the casting process of a wind turbine rotor hub. Wind Energy. 2010;(13):727–736.

[71] Chaviaropoulos PK. Flap/lead-lag aeroelastic stability of wind turbine blade sections. Wind Energy. 1999;(2):99–112

[72] Chaviaropoulos PK, Soerensen NN. Viscous and aeroelastic effects on wind turbine. Wind Energy. 2003;(6):387–403.

[73] Hansen MH. Aeroelastic stability analysis of wind turbine using an eigenvalue approach. Wind Energy. 2004;(7):133–143.

[74] Hansen MOL, Sørensen JN, Voutsinas S, et al. State of the art in wind turbine aerodynamics and aeroelasticity. Progress in Aerospace Sciences. 2006;(42):285–330.

[75] Hoogedoorn E, Jacobs BG, Beyene A. Aero-elastic behavior of a flexible blade for wind turbine application: A 2D computational study. Energy. 2010;(35):778–785.

[76] Baxevanou CA, Chaviaropoulos PK, Voutsinas SG, et al. Evaluation study of a Navier–Stokes CFD aeroelastic model of wind turbine airfoils in classical flutter. Journal of Wind Engineering and Industrial Aerodynamics. 2008;(96):1425–1443.

[77] Chattot JJ. Helicoidal vortex model for wind turbine aeroelastic simulation. Computers and Structures. 2007;(85):1072–1079.

[78] Maheri A, Noroozi S, Vinney J. Combined analytical/FEA-based coupled aero structure simulation of a wind turbine with bend–twist adaptive blades. Renewable Energy. 2007;(32):916–930.

[79] Sarkar S, Bijl H. Nonlinear aeroelastic behavior of an oscillating airfoil during stall-induced vibration. Journal of Fluids and Structures. 2008;(24):757–777.

[80] Simao Ferreira CJ, van Zuijlen A, Bijl H, et al. Simulating dynamic stall in a two-dimensional vertical-axis wind turbine: verification and validation with particle image velocimetry data. Wind Energy. 2010;(13):1–17.

[81] Bergami L, Gaunaa M. Stability investigation of an airfoil section with active flap control. Wind Energy. 2010;(13):151–166.

[82] Bertagnolio F, Rasmussen F, Sørensen NN, et al. A stochastic model for the simulation of wind turbine blades in static stalls. Wind Energy. 2010;(13):323–338.

[83] lvanell S, Mikkelsen R, Sørensen JN, et al. Stability analysis of the tip vortices of a wind turbine. Wind Energy. 2010;(13):705–715.

[84] Andersen PB, Gaunaa M, Bak C, et al. A dynamic stall model for airfoils with deformable trailing edges. Wind Energy. 2009;12:734–751.

[85] Lee JW, Lee JS, Han JH, et al. Aeroelastic analysis of wind turbine blades based on modified strip theory. Journal of Wind Engineering and Industrial Aerodynamics. 2012;(110):62–69.

[86] Kirch A, Clobes M, Peil U. Aeroelastic divergence and flutter: Critical comments on the regulations of EN 1991-1-4. Journal of Wind Engineering and Industrial Aerodynamics. 2011;(99):1221–1226.

[87] Matsumoto M, Matsumiya H, Fujiwara S, et al. New consideration on flutter properties based on step-by-step analysis. Journal of Wind Engineering and Industrial Aerodynamics. 2010;(98):429–437.

[88] Cao RJ, Liu X, Ma HM. Aero-elastic stability model of wind turbine blade based on pressure representation method and its application. Acta Energiae Solaris Sinica. 2003;(2):227–231.

[89] Jin Y, Yuan X. Aeroelastic analysis on an airfoil's flutter and flutter control technique of blowing. Acta Energiae Solaris Sinica. 2002;23(4):403–407.

[90] Ren YS, Ling XH. Flap/lead-lag nonlinear aeroelastic stability of a wind turbine blade system during dynamic stall. Journal of Vibration and Shock. 2010;29(1):121–124, 241.

[91] Ren YS, Liu TR and Yang SL. Aeroelastic stability analysis of composite wind turbine blade dynamic stall. Journal of Mechanical Engineering. 2011;47(12):113–125.
[92] He DX. Wind engineering and industrial aerodynamics. Peking: National Defence of Industry Press; 2006.
[93] Xu HF. Fundamentals of aerodynamics. Peking: National Defence of Industry Press; 1979.
[94] Qian YJ. Aerodynamics. Peking: Beijing University of Aeronautics and Astronautics Press; 2005.
[95] Yi Z, Wu X and Zhou LZ. Low speed aerodynamics. Peking: Press of Metallurgy Industry; 2005.
[96] Derla M. Two-dimensional transonic aerodynamic design and analysis using the Euler equations. Massachusetts Institute of Technology; 1985.
[97] Burton T, Sharpe D, Jenkins N, et al. Wind energy handbook. New York: John Wiley & Sons Ltd.; 2001.
[98] Lu ZL. Aerodynamics. Peking: Beijing University of Aeronautics and Astronautics Press; 2009.
[99] Drela M. XFOIL, an analysis and design system for low Reynolds number airfoils. In: Low Reynolds number aerodynamics. Lecture Notes in Engineering. Springer-Verlag; 1989.
[100] Somers DM, Design and experimental results for the S809 airfoil. Technical Report NREL/SR-440-6918. National Renewable Energy Laboratory, Colorado, USA; January 1997.
[101] Bak C, Fuglsang P. Modification of the NACA 63-415 leading edge for better aerodynamic performance. Journal of Solar Energy Engineering. 2002;(124):327–334.
[102] Theodorsen T. Theory of wing sections of arbitrary shape. Report NACA-TR-411. National Advisory Committee for Aeronautics, Washington, D. C., USA; 1931:229–240.
[103] Dong YK. Series (First edition). Shanghai: Shanghai Science and Technology Press; 1982.
[104] Abbott IH, Von Doenhoff AE. Theory of wing sections. New York: Dover Publications, Inc.; 1959.
[105] Shen WZ, Sørensen JN. Quasi-3D Navier-Stokes model for a rotating airfoil. Journal of Computational Physics. 1999;(150):518–548.
[106] Bertagnolio F, Sørensen N, Johansen J, et al. Wind turbine airfoil catalogue. Roskilde, Denmark: Risø National Laboratory; August 2001.
[107] Kulfan BM, Bussoletti JE. Fundamental parametric geometry representations for aircraft component shapes. AIAA-2006-6948, 11th AIAA/ISSMO Multidisciplinary Analysis and Optimization Conference: The Modeling and Simulation Frontier for Multidisciplinary Design Optimization. 6–8 September 2006.
[108] Kulfan BM. Universal parametric geometry representation method – CST. 45th AIAA Aerospace Sciences Meeting and Exhibit; 8–11 January 2007.
[109] Mongomerie BOG, Brand AJ, Bosschers J, et al. Three-dimensional effects in stall. ECN, Netherlands: ECN-Renewable Energy; 1996.
[110] Timmer WA, Schaffarczyk AP. The effect of roughness at high Reynolds numbers on the performance of aerofoil DU 97-W-300Mod. Wind Energy. 2004;136(7):295–307.
[111] Bosschers J, Montgomery B, Brand A. Influence of blade rotation on the sectional aerodynamics of rotational blades. 22nd European Rotorcraft Forum 1996; Bristol, England.
[112] van Rooij RPJOM. Modification of the boundary layer calculation in RFOIL for improved airfoil stall prediction. Report IW-96087R. TU-Delft, the Netherlands.
[113] Eppler R, Somers DM. Low speed airfoil design and analysis. Advanced Technology Airfoil Research. Volume I. NASA CP-2045, part 1; 1979:73–99.
[114] Eppler R, Somers DM. A computer program for the design and analysis of low-speed airfoils. NASA TM-80210; 1980.
[115] Tangler JL, Somers DM. Status of the special-purpose airfoil families. SERI/TP-217-3246; Dec. 1987.
[116] Tangler JL, Somers DM. NREL airfoil families for HAWT's. Proc. WINDPOWER'95, Washington D.C.; 1995: 117–123.

[117] Filippone A. Airfoil inverse design and optimization by means of viscous-inviscid techniques. Journal of Wind Engineering and Industrial Aerodynamics. 1995;(56):123–136.

[118] Thokala P, Martins J. Variable-complexity optimization applied to airfoil design. Engineering Optimization. 2007;39(3):271–286.

[119] Chen P, Du MY, Liu JP. Development status and key aerodynamic problems of wind turbine dedicated airfoils. Power System and Clean Energy. 2009; 25(2):36–40.

[120] Christian B. Research in aeroelasticity EFP-2006. Report RISø-R-1611(EN). Risø National Laboratory: Roskilde, Denmark; 2007.

[121] Bertagnolio F, Sørensen NN, Rasmussen F. New insight into the flow around a wind turbine airfoil section. Journal of Solar Energy Engineering. 2003;(127):214–222.

[122] Fuglsang P, Bak C. Wind tunnel tests of the Risø-A1-18, Risø-A1-21 and Risø-A1-24 airfoils. Report Risø-R-1112(EN), Risø National Laboratory, Roskilde, Denmark, 1999.

[123] Bagley JD. The behavior of adaptive systems which employ genetic and correlation algorithms. Doctoral dissertation, University of Michigan; 1967.

[124] Holland JH. Adaptation in natural and artificial systems: An introductory analysis with applications to biology, control, and artificial intelligence. 1st edition, Ann Arbor, MI: The University of Michigan Press; 1975. 2nd edition, Cambridge, MA: MIT Press; 1992.

[125] De Long KA. An analysis of the behavior of a class of genetic adaptive systems. PhD dissertation, University of Michigan. NO. 76-9381; 1975.

[126] De Long KA. Learning with genetic algorithms: An overview machine leaning. IEEE Transactions on Neural Networks. 1996;(3):122–138.

[127] Smith SE. A learning system based on genetic adaptive algorithm. Doctoral dissertation, University of Pittsburgh; 1980.

[128] Goldberg DE. Genetic algorithms in search, optimization and machine learning. Addison-Wesley Publishing Company; 1989.

[129] Davis L (editor). Handbook of genetic algorithms. New York: Van Nostrand Reinhold; 1991.

[130] Grefenstette JJ. Optimization of control parameters for genetic algorithms. IEEE Transactions on Systems, Man, and Cybernetics. 1986;16(1):122–128.

[131] Bauer RJ. Genetic algorithms and investment strategies. New York: John Wiley & Sons, Inc.; 1991.

[132] Srinivas M, Patnaik LM. Adaptive probabilities of crossover and mutation in genetic algorithms. IEEE Transactions on Systems, Man, and Cybernetics. 1994;24(4):656–667.

[133] Chambers L. The practical handbook of genetic algorithms applications. New York: CRC Press; 2001.

[134] Chaffarczyk AP, Winkler H, Freudenreich K, et al. Reynolds number effects on thick aerodynamic profiles for wind turbines. European Wind Energy Conference, Madrid, Spain; 2003:16–19.

[135] Brooks TF, Pope, DS, Marcolini MA. Airfoil self-noise and prediction. NASA Reference Publication 1218. National Aeronautics and Space Administration, USA; 1989.

[136] Zhu WJ, Heilskov N, Shen WZ, Sorensen JN. Modeling of aerodynamically generated noise from wind turbines. Journal of Solar Energy Engineering. 2005;127(4):517–528.

[137] Zhu WJ, Sørensen JN, Shen WZ. An aerodynamic noise propagation model for wind turbines. Wind Engineering. 2005;29(2):129–143.

[138] Zhu WJ. Aero-acoustic computations of wind turbines. PhD dissertation. Technical University of Denmark; 2007.

[139] Brooks TF, Schlinker RH. Progress in rotor broadband noise research. Vertica. 1983;7(4):287–307.

[140] Brooks TF, Hodgson TH. Trailing edge noise prediction from measured surface pressure. Sound & Vibration. 1981;78(1):69–117.

[141] Brooks TF, Marcolini MA. Scaling of airfoil self-noise using measured flow parameters. AIAA Journal. 1985;23(2):207–213.
[142] Paterson RW, Amiet RK, Munch CL. Isolated airfoil-tip vortex interaction noise. AIAA Journal. 1974; NO. 74–194
[143] Fink MR, Bailey DA. Airframe noise reduction studies and clean-airframe noise investigation. NASA CR-159311; 1980.
[144] Tam CKW. Discrete tones of isolated airfoils. Acoust Soc. America. 1974;55(6):1173–1177.
[145] Wright SE. The acoustic spectrum of axial flow machines. Sound & Vibration. 1976;45(2):165–223.
[146] Flink MR. Fine structure of airfoil tone frequency. UTRC78-10, United Technologies Research Center; 1978.
[147] George AR, Najjar FE, Kim YN. Noise due to tip vortex formation on lifting rotors. AIAA-80-1010; 1980.
[148] Brooks TF, Marcolini MA. Airfoil tip vortex formation noise. AIAA Journal. 1986;24(2):246–252.
[149] NWTC Design Codes (NAFNoise by Pat Moriaty)[EB/OL][2006-4-12]. http://wind.nrel.gov/DesignCodes/simulators/NAFNoise/.
[150] Ye ZQ, Huang JX, Chen Y, et al. Experimental study on aerodynamic performances of new airfoils for wind turbine. Acta Energiae Solaris Sinica. 2003;24(4):548–554.
[151] Homas B. Research in aeroelasticity EFP-2007-II. Report RISø-R-1698(EN). Risø National Laboratory: Roskilde, Denmark; 2009.
[152] Michelsen JA. Basis3D – a platform for development of multiblock PDE solvers. Report AFM. 1992;92:5.
[153] Sørensen NN. General purpose flow solver applied over hills. Report Risø-R-827(EN). Risø National Laboratory: Roskilde, Denmark; 1995.
[154] Kennedy J, Eberhart RC. Particle swarm optimization. In: Proceedings of the IEEE International Conference on Neural Networks; 1995:1942–1948.
[155] Shi Y, Eberhart RC. A modified particle swarm optimizer. In: Proceedings of the IEEE International Conference on Evolutionary Computation. Piscataway: IEEE; 1998:69–73.
[156] Shi Y, Eberhart RC. Fuzzy adaptive particle swarm optimization. In: Proceedings of the Congress on Evolutionary Computation. Seoul, Korea; 2001:101–106.
[157] Chatterjee A, Siarry P. Nonlinear inertia weight variation for dynamic adaptation in particle swarm optimization. Computers and Operations Research. 2006;33(3):859–871.
[158] Xi ZX, Xie YJ, Guo Q, et al. The design characteristics of nf-3 wind tunnel. Aerodynamic Experiment and Measurement & Control. 1996;10(4):40–49.
[159] Fan JC. Handbook of wind tunnel test. Peking: Aviation Industry Press; 2002.
[160] GJB 180-1986. The specification of airplane design model for low speed wind tunnel.
[161] Zhou RX, Xi ZX, Xie YJ, et al. Investigation & discussion on the methods of experimental measurement of airfoil drag. Aerodynamic Experiment and Measurement & Control. 1995;9(4):64–68.
[162] Jiao YQ, Zhou RX, Xi ZX, et al. Experimental research on aerodynamics performance of fan airfoil. Experiments and Measurements in Fluid Mechanics. 2004;3:47–49.
[163] Xiao CS. Wind tunnel test report of TH18724-2 airfoil. Xian: Center For Aerodynamics Design Research of Northwerstern Polytechnical University; 2010.
[164] Xiao CS. Wind tunnel test report of Airfoil2121. Xian: Center For Aerodynamics Design Research of Northwerstern Polytechnical University; 2010.
[165] Marcos VH. Aerodynamic performance analysis of horizontal axis wind turbines. Renewable Energy. 1994;4(5):505–518.
[166] Maalawi KY, Badr MA. A practical approach for selecting optimum wind rotors. Renewable Energy. 2003;(28):803–822.

[167] Benini E, Toffolo A. Optimal design of horizontal-axis wind turbines using blade-element theory and evolutionary computation. Journal of Solar Energy Engineering. 2002;124(11):357–363.
[168] Prandtl L, Betz A. Vier Abhandlungen zur Hydrodynamik und Aerodynamik. Göttinger Nachr.: Göttingen; 1927:88–92.
[169] Glauert H. Airplane propellers. In: Durand WF, editor. Aerodynamic theory. New York: Dover; 1963.
[170] Wilson RE, Lissaman PBS. Applied aerodynamics of wind power machines. Oregon State University Report NSF/RA/N-74113; 1974.
[171] De Vries O. Fluid dynamic aspects of wind energy conversion. AGARD Report AG-243; 1979; 4:1–50.
[172] Shen WZ, Mikkelsen R, Sørensen JN, Bak C. Tip loss correction for wind turbines computations. Wind Energy. 2005;(8):457–475.
[173] Snel H, Schepers JG, Montgomerie B. The MEXICO project (Model Experiment in Controlled Conditions): The database and first results of data processing and interpretation. Journal of Physics: Conference Series 2007;(75):012014.
[174] Snel H. Aerodynamic pre-design and loads for the MEXICO rotor. ECN MEXICO WP3; August 10, 2001.
[175] Omri G. MEXICO blade airfoil CAD modeling. ECN DOC-MIN-03-02; July, 2003.
[176] Schepers JG, Snel H. Model experiment in controlled conditions. ECN Final Report ECN-E–07-042; February, 2007.
[177] Snel H, Houwink R, Bosschers J. Sectional prediction of life coefficients on rotating wind turbine blades in stall. ECN Report: ECN-C-93-052; 1994.
[178] Jackson KJ, Zuteck MD, van Dam CP, etc. Innovative design approaches for large wind turbine blades. Wind Energy. 2005;(8):141–171.
[179] Mathematics department of Chongqing University. Higher mathematics (Part II). Chongqing: Chongqing University Press; 2000.
[180] Griffin DA. NREL Advanced Research Turbine (ART) aerodynamic design of ART-2B rotor blades. NREL/SR-500-28473. National Renewable Energy Laboratory, Colorado, USA; August 2000.
[181] Lawlor SP. Wind turbine rotor blade. United States Patent: 5474425; 1995.
[182] Jonkman J. NRELOffshrBsline5 MW. NWTC/NREL TP-500-38060. Golden, CO: National Renewable Energy Laboratory/Cole Boulevard: National Wind Technology Center; February 15, 2007.
[183] Tony B, David S, Nick J, et al. Wind energy handbook. New York: John Wiley & Son, Ltd.; 2001.
[184] Hansen MOL. Aerodynamics of wind turbines (second edition). UK: Earthscan; 2008.
[185] Bazilevs Y, Hsu M C, et al. 3D simulation of wind turbine rotors at full scale. International Journal for Numerical Methods in Fluids. 2011;(65):b207–235.
[186] Liu W, Yin JC, Chen P, et al. Dynamic analysis and aeroelastic stability analysis of large com posite wind turbine blades. Acta Aerodynamica Sinica. 2011;29(3):391–395.
[187] Grujicic M, Arakere G, Pandurangan B, et al. Multidisciplinary design optimization for glass-fiber epoxy-matrix composite 5 MW horizontal-axis wind-turbine blades. Journal of Material Engineering and Performance. 2010;(19):1116–1127.
[188] Rashedi A, Sridhar I, Tseng KJ. Multi-objective material selection for wind turbine blade and tower: Ashby's approach. Materials & Design. 2012;(37):521–532.
[189] Zhang SS, Zhuang Z. Composite materials and viscoelastic mechanics. Peking: Mechanical Industry Press; 2005.
[190] Zhang XD, Liu W. Fundamentals of composite material mechanics. Xian: Northwestern Polytechnic University Press; 2010.
[191] Mandell JF, Samborsky DD. DOE/MSU composite material fatigue database. Sandia Technical Report: SAND97-3002; 2010.

[192] Locke J, Valencia U. Design studies for twist-coupled wind turbine blades. Sandia Technical Report: SAND2004-0522; 2004.

[193] Griffith D, Ashwill TD. The Sandia 100-meter all-glass baseline wind turbine blade: SNL100-00. Sandia Report: SAND2011-3779; 2004.

[194] Griffin DA. Wind PACT turbine design scaling studies technical area 1-composite blades for 80 to 120 meter rotor. National Renewable Energy Laboratory, Colorado, USA; 2001.

[195] Li XJ. Research on the aerodynamic performance calculation and structural design of large-scale wind turbine rotor blade [PhD thesis]. Wuhan, China: Wuhan University of technology; 2008.

[196] Fei JF, Zhang XY, Li ZQ, et al. Modeling and calculation of wind turbine blade using cad and cae. Journal of Solid Mechanics. 2008;29:37–40.

[197] Zhang LD, Ren LC, Chen RS, et al. Shape design and 3d modeling study for blades of wind turbine. Acta Energiae Solaris Sinica. 2009;29(9):1177–1180.

[198] Zhou PZ, Xiao JY, Zeng JC, et al. Structural analysis of large-scale composite wind turbine blade based on ANSYS. Journal of National University of Defense Technology. 2010;32(2):46–50.

[199] Bir GS. User's Guide to PreComp. NREL/TP-500-38929. National Renewable Energy Laboratory, Colorado, USA; January 2006.

[200] Ma TL, Ma DL and Zhang S. Application of Parameterization of Analysis Modei in Airplane Multidisciplinary Design Optimization. Acta Aeronautica ET Astronautica Sinica. 2008;29(6):1576–1580.

[201] Sun J, Song YD, Sun ZG. Integrated optimization design of structure and material of composite guide vane. Acta Mechanica Solida Sinica. 2011;32(6):595–602.

[202] Qiao W, Yao WX. Equivalent bending stiffness method for stacking sequence optimization of composite stiffened panel. Chinese Journal of Computational Mechanics. 2011;28(1):158–162.

[203] Cheng WY, Chang Y, Cui DG, et al. Composite wing optimization design based on distributed computing. Acta Materiae Compositae Sinica. 2007;24(1):167–171.

[204] Song YD, Sun J, Sun ZG, et al. Topological optimization design of microstructure of composite materials based on genetic algorithm. Acta Mechnica Solida Sinica. 2009;30(4):416–423.

[205] Kong XR, Qin YL and Luo WB. Improved Gaussian RBF response surface method-based model updating for the honeycomb sandwich panel. Acta Materiae Compositae Sinica. 2011;28(5):220–227.

[206] Qin YL, Kong XR, Luo WB. Finite element model updating of airplane wing based on Gaussian radial basis function response surface. Journal of Beijing University of Aeronautics and Astronautics. 2011;37(11):1465–1470.

[207] Qin YL, Kong XR, Luo WB. RSM-based FEM model updating for a carbon fiber honeycomb sandwich panel. Journal of Vibration and Shock. 2011;30(7):71–76, 86.

[208] Zhang ZY, Xie SS, Wang HT, et al. Contacting optimization in high pressure spool finite element parameters updating based on adaptive cloud PSO model. Journal of Aerospace Power. 2011;26(12):2798–2804.

[209] Øye S. FLEX4 Simulation of wind turbine dynamics. Proc. Int. Energy Agency, Annex XI, 28th Meeting of Experts. Lyngby, Denmark; 11–12 April, 71–77.

[210] Chen GB, Yang C, Zou CQ. Fundamentals of aeroelastic design. Peking: Beijing University of Aeronautics and Astronautics Press; 2014. p. 14–24.

[211] Newman JC. Newman PA, Taylor AC, et al. Efficient nonlinear static aeroelastic wing analysis. Computers and Fluids. 1999;28(4/5):615–628.

[212] Hepp JP. Static aeroelastic analysis of the ARW-2 wing including correlation with experiment. Durham: Duke University; 2005:15–19.

[213] Puterbaughd M, Beyene A. Parametric dependence of a morphing wind turbine blade on material elasticity. Energy. 2011;36(1):466–474.

[214] Hoogedoorn E, Jacobs GB, Beyene A. Aero-elastic behavior of a flexible blade for wind turbine application: A 2D computational study. Energy. 2010;35(2):778–785.

[215] Li DY, Fang J, Lv WG, et al. Unsteady aerodynamic characteristic analysis of horizontal axis wind turbine blade. Journal of Shenyang University of Technology. 2011;33(1):36–40, 61.

[216] Hansen MH, Gaunaa M, Madsen HA. A Beddoes-Leishnam type dynamic stall model in state-space and indicial formulations. Technical Report 1354. Risø National Laboratory: Roskilde, Denmark; 2004.

[217] Gladwell GML. A modern course in aeroelasticity. New York: Kluwer Academic Publishers; 2004:5–16.

[218] Fung YC. An introduction to the theory of aeroelasticity. New York: Dover Publications; 1993:92–94.

[219] Liu XY. Blade design and stability analysis of wind turbine. Xian: Northwestern Polytechnical University; 2004. p. 47–50.

[220] Jonkman J, Butterfield S, Musial W, et al. Definition of a 5-MW reference wind turbine for offshore system development. NREL/TP-500-38060. National Renewable Energy Laboratory, Colorado, USA; February 2009.

[221] Theodorsen T, Garrick IE. Mechanism of flutter, theoretical and experimental investigation of flutter problem. Technical Report. National Advisory Committee for Aeronautics; 1940.

[222] Zhao YH. Aeroelastics and control. Peking: Science Press; 2002. p. 121–125.

[223] Bergami L. Aeroservoelastic stability of a 2D airfoil section equipped with a trailing edge flap. Risø Report 1663. Risø National Laboratory for Sustainable Energy: Roskilde, Denmark; 2008.

[224] Fuglsang P, Bak C. Wind tunnel tests of Risø-B1-18 and Risø -B1-24. Risø Report 1375. Risø National Laboratory: Roskilde, Denmark; 2003.

Bei Fragen zur Produktsicherheit wenden Sie sich bitte an:
If you have any questions regarding product safety,
please contact:

Walter de Gruyter GmbH
Genthiner Straße 13
10785 Berlin
productsafety@degruyterbrill.com